SURGICAL ANATOMY FOR MASTERY OF OPEN OPERATIONS

A Multimedia Curriculum for Training Surgery Residents

SURGICAL ANATOMY FOR MASTERY OF OPEN OPERATIONS

A Multimedia Curriculum for Training Surgery Residents

MARK O. JENSEN, MD, FACS
Professor and Director of Surgical Anatomy
Department of Surgery
University of North Dakota School of Medicine and Health Sciences
Adjunct Professor, Electrical Engineering
North Dakota State University
VA Medical Center – Fargo
Fargo, North Dakota

Illustrations by Gary Baune

Wolters Kluwer

Philadelphia • Baltimore • New York • London
Buenos Aires • Hong Kong • Sydney • Tokyo

Acquisitions Editor: Keith Donnellan
Development Editor: Sean McGuire
Editorial Coordinator: Alexis Pozonsky
Marketing Manager: Leah Thompson
Production Project Manager: Bridgett Dougherty
Design Coordinator: Elaine Kasmer
Illustrator: Gary Baune
Manufacturing Coordinator: Beth Welsh
Prepress Vendor: SPi Global

9 8 7 6 5 4 3 2 1

Printed in China

Library of Congress Cataloging-in-Publication Data
Names: Jensen, Mark O., editor.
Title: Surgical anatomy for mastery of open operations : a multimedia curriculum for training surgery residents / [edited by] Mark O. Jensen ; illustrations by Gary Baune.
Description: First edition. | Philadelphia : Wolters Kluwer, [2019] | Includes index.
Identifiers: LCCN 2017060418 | ISBN 9781496388575 (paperback)
Subjects: | MESH: Anatomy, Regional | Surgical Procedures, Operative
Classification: LCC RD31 | NLM WO 101 | DDC 617—dc23 LC record available at https://lccn.loc.gov/2017060418

LWW.com

RRS1802

Contributing Anatomist

Thomas H. Quinn, PhD
Professor of Anatomy and Surgery
Department of Biomedical Sciences
Creighton University School of Medicine
Omaha, Nebraska

Contributing Surgeons

Cornelius Dyke, MD, FACS
Associate Professor of Surgery and Associate Dean
Department of Surgery
University of North Dakota School of Medicine and Health Sciences
Cardiothoracic Surgery, Sanford Health – Fargo
Grand Forks, North Dakota

Linda B. Lindquist, MD
Clinical Associate Professor of Surgery
Department of Surgery
University of North Dakota School of Medicine and Health Sciences
Grand Forks, North Dakota

Kurt D. Lindquist, MD, FACS
Clinical Professor of Surgery
Department of Surgery
University of North Dakota School of Medicine and Health Sciences
General and Vascular Surgery, VA Medical Center – Fargo
Grand Forks, North Dakota

Denise M. Rondeau, MD, FACOG
Clinical Assistant Professor and Academic Coordinator for Student Education
Department of Obstetrics and Gynecology
University of North Dakota School of Medicine and Health Sciences
Grand Forks, North Dakota

Robert P. Sticca, MD, FACS
Chairman and Professor
Program Director, General Surgery Residency
Department of Surgery
University of North Dakota School of Medicine and Health Sciences
Chair, Division of Surgical Oncology, Sanford Health – Fargo
Grand Forks, North Dakota

Andrew Terrell, MD, FACS
Clinical Instructor of Surgery
Department of Surgery
University of North Dakota School of Medicine and Health Sciences
Head and Neck Surgery, Sanford Health – Fargo
Grand Forks, North Dakota

Reviewing Surgeons

Robert J. Bates, MD, FACS
Department of Urology
Jamestown Regional Medical Center
Jamestown, North Dakota

Michael S. Bouton, MD, MA, FACS
Clinical Assistant Professor of Surgery
University of North Dakota School of Medicine and Health Sciences
Chair, Division of Breast Surgery, Sanford Health – Fargo
Grand Forks, North Dakota

Jason M. Erpelding, MD, FAAOS
Clinical Assistant Professor of Surgery
University of North Dakota School of Medicine and Health Sciences
Sanford Health Department of Orthopedics & Sports Medicine – Fargo
Grand Forks, North Dakota

John W. Jones Jr, MD, PhD, FACS
Clinical Assistant Professor
Department of Surgery
University of North Dakota School of Medicine and Health Sciences
Otolaryngology/Head and Neck Surgery, VA Medical Center – Fargo
Grand Forks, North Dakota

Jay M. MacGregor, MD, FACS
Instructor of Surgery, Director of Surgical Education
Department of Surgery
University of North Dakota School of Medicine and Health Sciences
Colorectal Surgery, VA Medical Center – Fargo
Grand Forks, North Dakota

Michael Traynor, MD, FACS
Clinical Professor of Surgery
Department of Surgery
University of North Dakota School of Medicine and Health Sciences
General and Endocrine Surgery, Sanford Health – Fargo
Grand Forks, North Dakota

Thomas Wambach, MD
Clinical Instructor of Surgery
Department of Surgery
University of North Dakota School of Medicine and Health Sciences
General Surgery, VA Medical Center – Fargo
Grand Forks, North Dakota

Foreword

Dr. Mark O. Jensen and his collaborators have assembled a valuable resource for surgical residents as well as for practicing surgeons. It is neatly organized into 10 chapters and arranged according to common anatomic categories. Each chapter presents a step by step description of the components of selected open operations that are nicely complemented with detailed illustrations. The inclusion of clinical highlights or "tips" is an extremely useful feature for residents and experienced surgeons alike. There are also illustrative videos providing a guide for users and reinforcing the text descriptions and illustrations.

There are some unique features of this book that deserve emphasis. As minimally invasive options and nonoperative management of abdominal solid organ injuries have developed as safe alternatives to open operations, residents are less likely to be exposed to or participate in as many open procedures as in the past. Some operations done commonly in the past have become rare. This includes operations such as common bile duct exploration, choledochoduodenostomy, splenectomy, and partial gastric resections among others. Details and nuances about proper exposure and technical details essential to performing open operations safely and efficiently are provided. There are also thorough descriptions of operations such as total mesorectal excision which has become standard as they have been demonstrated to be associated with better patient outcomes. Chapter 3 details a variety of thoracic procedures including pulmonary resections. Chapter 9 contains orchiectomy, oophorectomy, and hysterectomy. Chapter 10 provides detailed descriptions of a wealth of vascular exposures which are necessary to treat vascular injuries or to perform open operations done less often due to the increased use of endovascular techniques. There are important aspects of each of these operations that general surgeons need to understand when called on to perform or assist with them in either elective and emergent situations.

Surgical Anatomy for Mastery of Open Operations: A Multimedia Curriculum for Training Surgery Residents fills an important niche in education of surgeons in training and in practice. It is a "go to" reference for open operations, both common and uncommon. The useful highlights and inclusion of procedures often left out or underemphasized in other operative manuals are particularly worthwhile. As open operations continue to decline in frequency, we must still be able to perform open procedures safely and skillfully. This reference helps accomplish this important goal.

Mark A. Malangoni, MD
Associate Executive Director
American Board of Surgery
Philadelphia, Pennsylvania

Preface

Structure (anatomy) supports function (physiology). Incorrect structure may lead to improper function. This book is written to help train surgical residents. My goal is improved training in the OR through better structural understanding beforehand.

Tell me and I forget, show me and I remember, let me do and I understand. This phrase written by a wise person is the basis for the training curriculum in the surgical anatomy lab at the University of North Dakota where I work.

I dedicate this book to the general surgery residents who have worked in the lab with me. Thank you for your insight, suggestions, and participation. You are the anvil on which development of the book was forged.

Acknowledgments

Thanks to Dr. David Antonenko and Dr. Robert Sticca, my department chairs over the 22 years that the lab originated, grew, and was supported. Diane, thanks for typing all the dictations over the years. Thanks also to my collaborating anatomist, Dr. Tom Quinn and my collaborating surgeons who assured accuracy. Thanks to Dr. Carol Scott-Connor, my mentor in medical publishing.

I am most deeply indebted to my family. During all the evenings and weekends devoted to this work you supported and encouraged me. To our two boys who are young men pursuing their own paths, you guys are finding your own success stories. To the love of my life, my wife and life partner, Leah Fujimoto - I love you so much.

Illustrator's Perspective

"Muscle memory," a term mentioned by one surgery resident recently, is an apt description of the Surgical Anatomy Lab curriculum presented by Dr. Jensen. As opposed to memorizing muscles, the Surgical Anatomy course brings much more to the table.

In the lab, residents have the opportunity to explore, discuss, interpret, problem solve and respond to various surgical and anatomical encounters. The low-stress, hands-on environment allows them to practice technique, work as a team, learn from their mistakes and teach each other.

Over 5 years ago, Dr. Jensen and I first met to discuss the role of illustration in refining his curriculum ideas. In that pursuit, I have observed several sessions, watched residents progress through the program and note that the author's vision is time-tested and has wavered little from the original content outline.

I am not much of a mechanic and wouldn't have the nerve to be a surgeon, but I am good at holding a trouble light and find good purpose illuminating areas of concern through medical illustration where simplify and define are the objectives. Many of my most instructive life drawing studies occurred in biology lab courses. The life drawings in this lab manual and series of short videos are all from the same hand; targeted and modified, consistently stylized and drawn from observation or built upon reference materials and other publications (some rare and out of print). Note that the drawings appear realistic, but have a schematic character and some vessels or structures may vary significantly in size depending upon the subject.

In the monochromatic cadaver, twitching and pulsing colorful markers have faded. Grayscale drawings mimic the setting efficiently and effectively using contrast, texture, and contour. In some ways, the act of drawing mimics the surgical procedure itself. Both require understanding of the concept, developing an approach, choosing the right instrument and applying it. Finding the tissue plane and following it with dissection in the lab or French curve in the studio is a good example.

Finally, the efforts of both the artist and the surgeon benefit greatly by awareness of the sensitivity of touch, which is pure "muscle memory."

Gary Baune, BA
Clinical Faculty Illustrator
Department of Surgery
University of North Dakota School of Medicine &
Health Sciences
Bayworks Studios

Table of Contents

1

Head and Neck

Outline

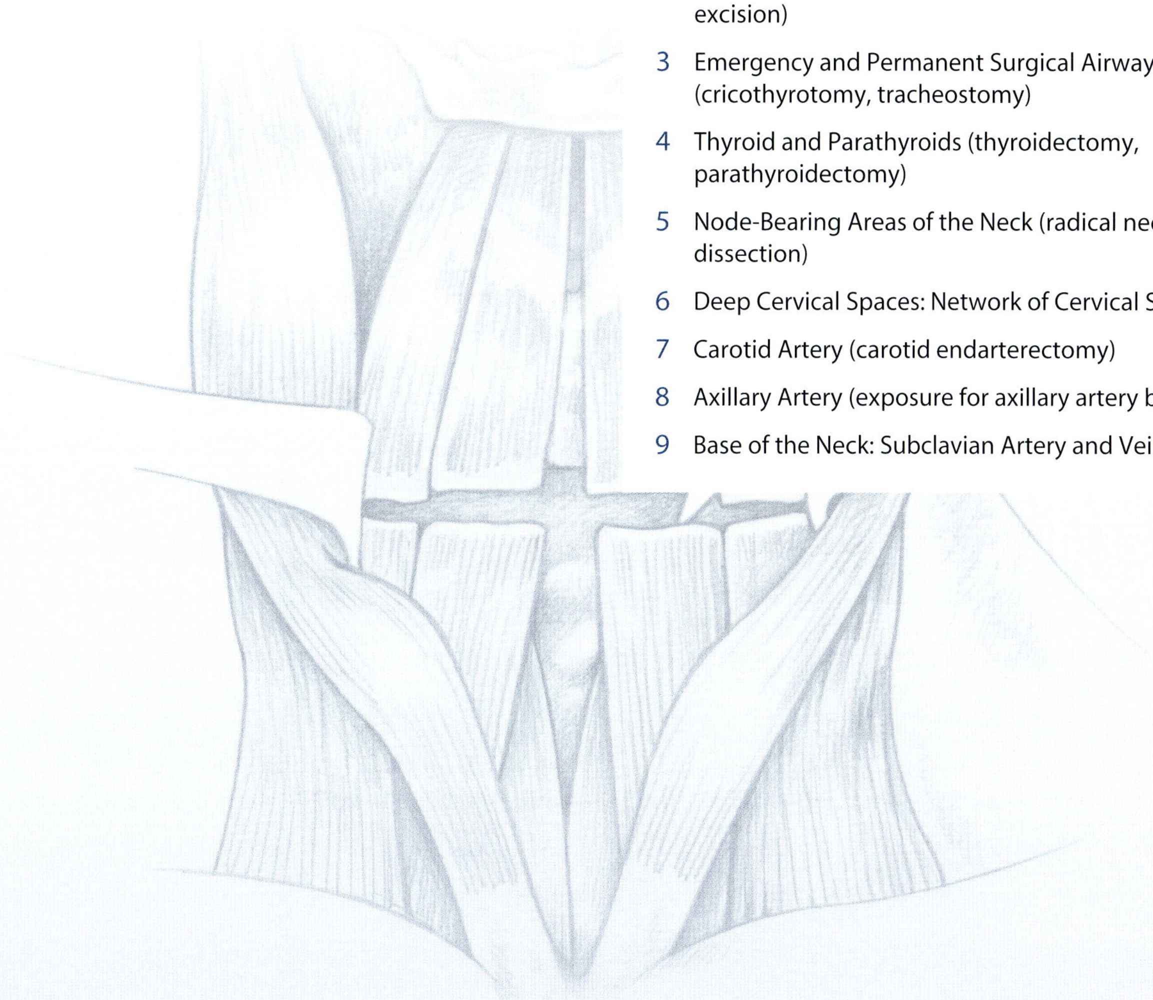

PAROTID GLAND AND FACIAL NERVE (PAROTIDECTOMY)

INCISION

Perform an S-shaped preauricular incision extending from the top of the external ear to the midneck (Fig. 1.1).

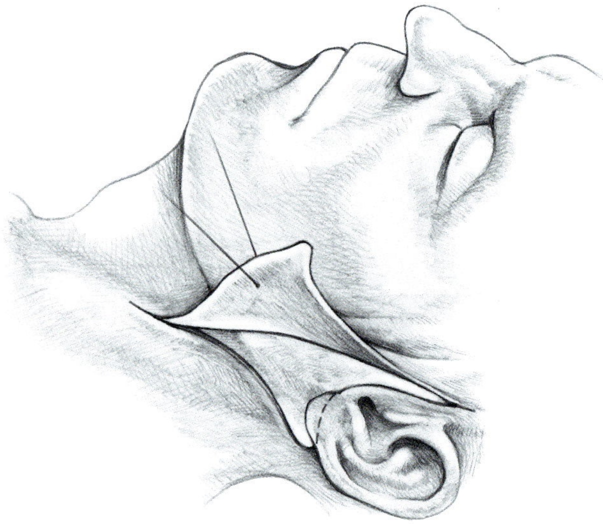

Figure 1.1 Incision for parotidectomy.

OPERATIVE PROCEDURE

Elevate the skin at the subcutaneous level forward to the midcheek. Vigorous retraction with rake or self-retaining retractors is needed. Use sharp dissection to create the skin flap (Fig. 1.2). Notice that the greater auricular nerve is superficial and may be sacrificed. There are two main methods of identifying the facial nerve, anterograde or retrograde dissection. Anterograde dissection is the preferred method and is the most often used.

- During anterograde dissection, the fascia is opened along the anterior border of the sternocleidomastoid muscle (SCM) and carried up along the preauricular area just anterior to the external cartilaginous canal. As the dissection proceeds deeper, medially the tragal pointer of the cartilaginous canal is identified. The facial nerve is typically 10-15 mm deep and inferior to this structure. Along the anterior border of the SCM, the posterior belly of the digastric muscle is identified. The facial nerve is 7-8 mm deep and superior to the posterior belly of digastric. In between the tragal pointer and the posterior belly of the digastric is located the tympanomastoid suture, which can be palpated. The main trunk of the facial nerve can subsequently be identified within 5 mm of the tympanomastoid suture.

- The second and less commonly used method is retrograde dissection of the facial nerve. Using this method, a branch of the nerve is identified and followed back to the main trunk. The most easily identified branches are the marginal mandibular nerve as it crosses the angle of mandible or the frontal nerve as it crosses the zygomatic arch in the superficial fascia. Once a branch is identified, it can be followed posteriorly back to the main trunk, and subsequent branches can then be dissected.

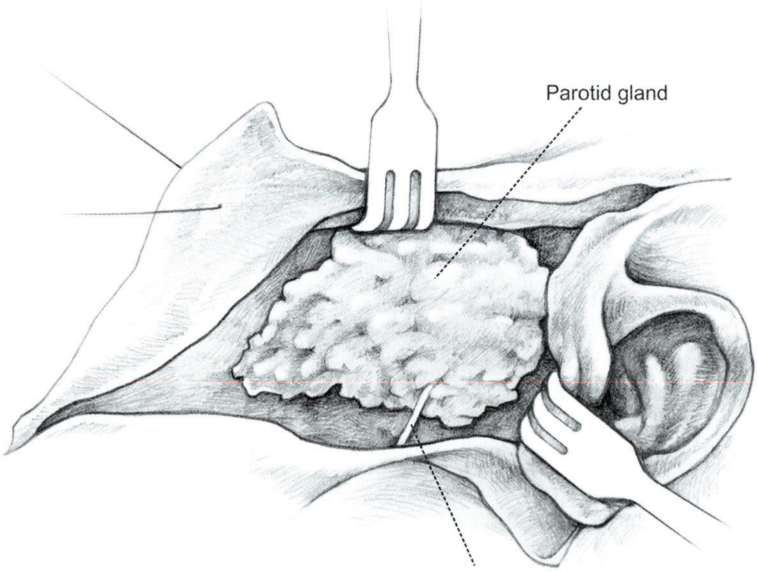

Figure 1.2 Exposure of the parotid gland.

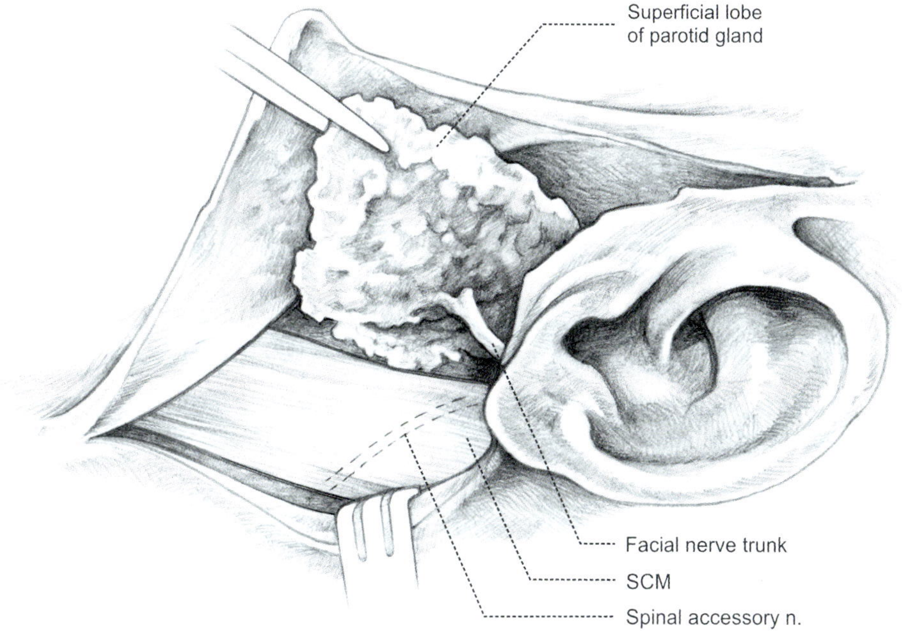

Figure 1.3 Identification of the facial nerve trunk.

Follow the facial nerve anteriorly as it enters the posterior parotid gland (Fig. 1.3). Split the parotid gland by inserting a clamp over the nerve and spreading the superficial portion out away from the deep portion of the gland (Fig. 1.4). Be sure to push the facial nerve branches flat against the deep lobe. Anteriorly, the "superficial lobe" of the parotid becomes very thin. The entire specimen is small and thin. Follow the individual branches and identify—temporal, zygomatic, buccal, marginal mandibular, and cervical. To remember these, use your hand on your ipsilateral face with your thumb as the temporal and 5th digit as your cervical branch (Fig. 1.5A).

Four categories of the branching patterns of the facial nerve relative to the origin of the buccal branch are seen in this Figure 1.6.

Next, extend the cervical portion of the neck incision inferiorly along the SCM and identify the anterior track

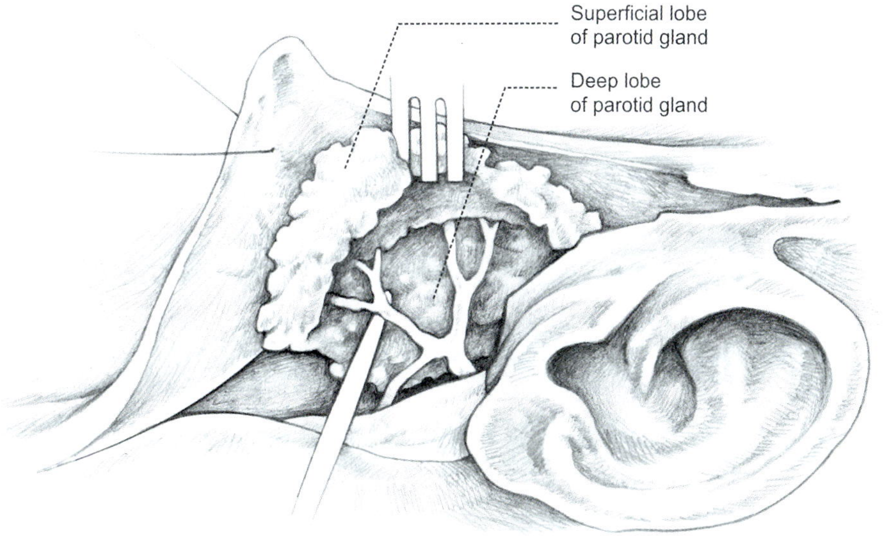

Figure 1.4 Exposure of the facial nerve branches.

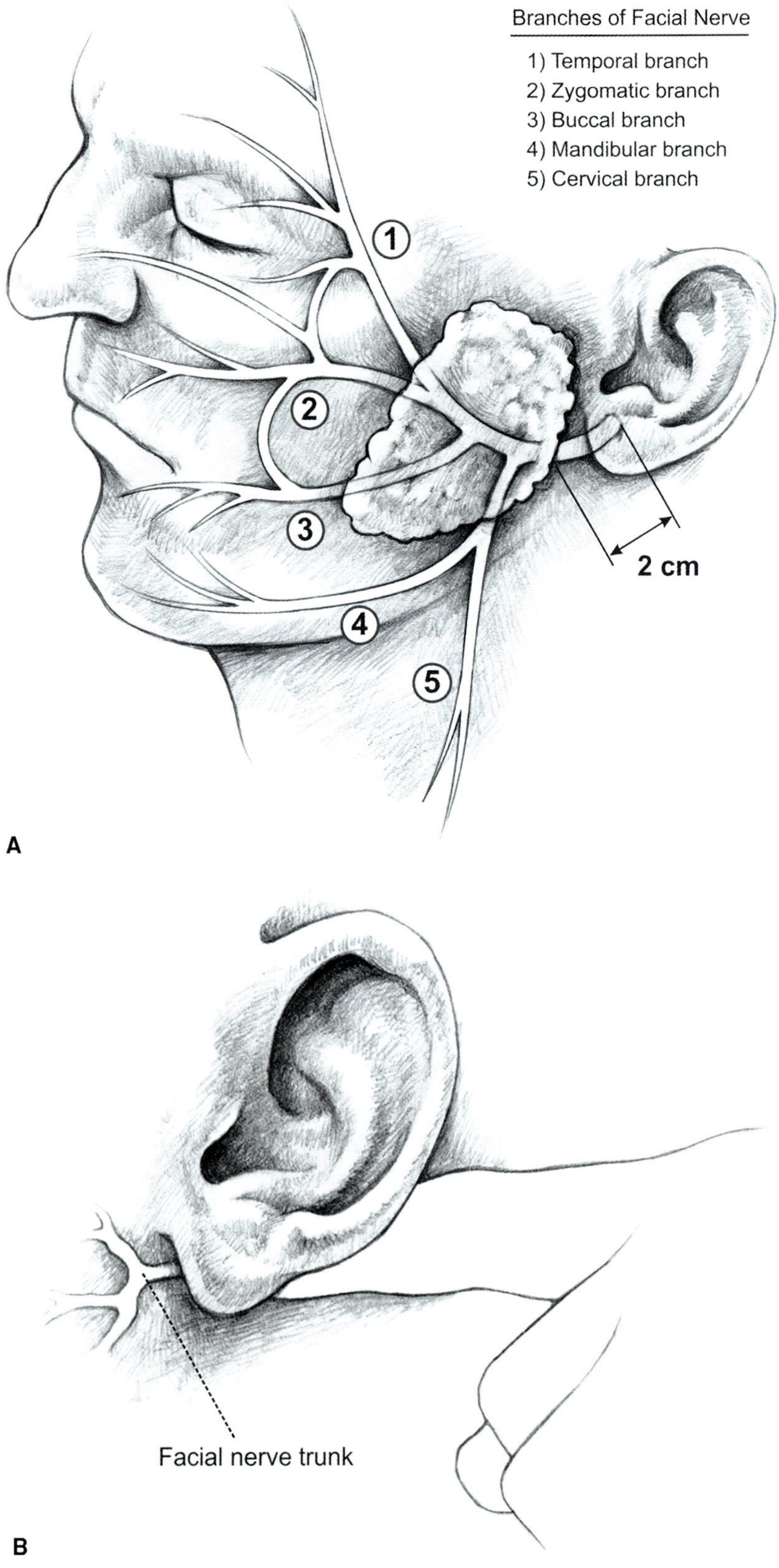

Figure 1.5 **A:** Branches of the facial nerve. **B:** Conceptual locator of the facial nerve trunk.

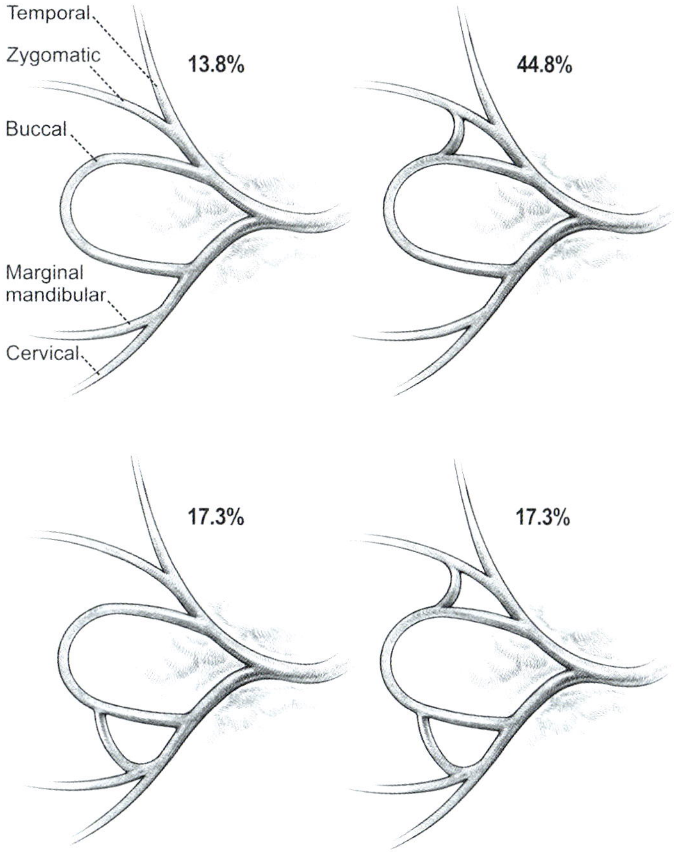

Figure 1.6 Branching patterns of the facial nerve.

of the marginal mandibular branch of the facial nerve by similar spreading dissection technique.

Dissect the facial nerve anteriorly to the anterior border of the parotid gland. Identify the parotid (Stensen) duct.

Return to the main facial nerve trunk. Identify the deep portion of the parotid gland medial to the facial nerve. Retract the nerve and excise the deep portion of the parotid gland. If you choose, you may now explore the posterior belly of the digastric muscle, the hypoglossal nerve, internal jugular vein, and spinal accessory nerve.

CLINICAL HIGHLIGHTS

- Conceptualize facial nerve location by the following process as suggested by Beahrs. Place your fingertip on the mastoid process pointing forward. A dime-sized area deep and anterior to the fingertip will have the nerve within it (Fig. 1.5B). This method uses the tympanomastoid suture as a landmark. This line is found between the posterior bony auditory canal and the mastoid portion of the temporal bone. The facial nerve is located 6 mm deep.

- The marginal mandibular branch of the facial nerve (VII) will be preserved above a transverse cervical incision made three fingerbreadths (4 cm) below the mandible. This nerve and the anterior ramus of the cervical branch supply innervation to the muscles of the corner of the mouth of the lower lip. Injury will result in persistent severe facial expressive disorder (Fig. 1.7A).

- If the anterior ramus of the cervical branch is injured, minimal drooling will resolve after 4-6 months.

- **T**o **Z**anzibar **B**y **M**otor**C**ar: facial nerve branches. **T**emporal, **Z**ygomatic, **B**uccal, **M**arginal Mandibular, **C**ervical. Or place your palm over your parotid and each digit will point to one branch (Fig. 1.5A).

- Frey syndrome is "gustatory sweating" of the ipsilateral face after division of the auriculotemporal nerve in the distribution of the sympathetic denervation.

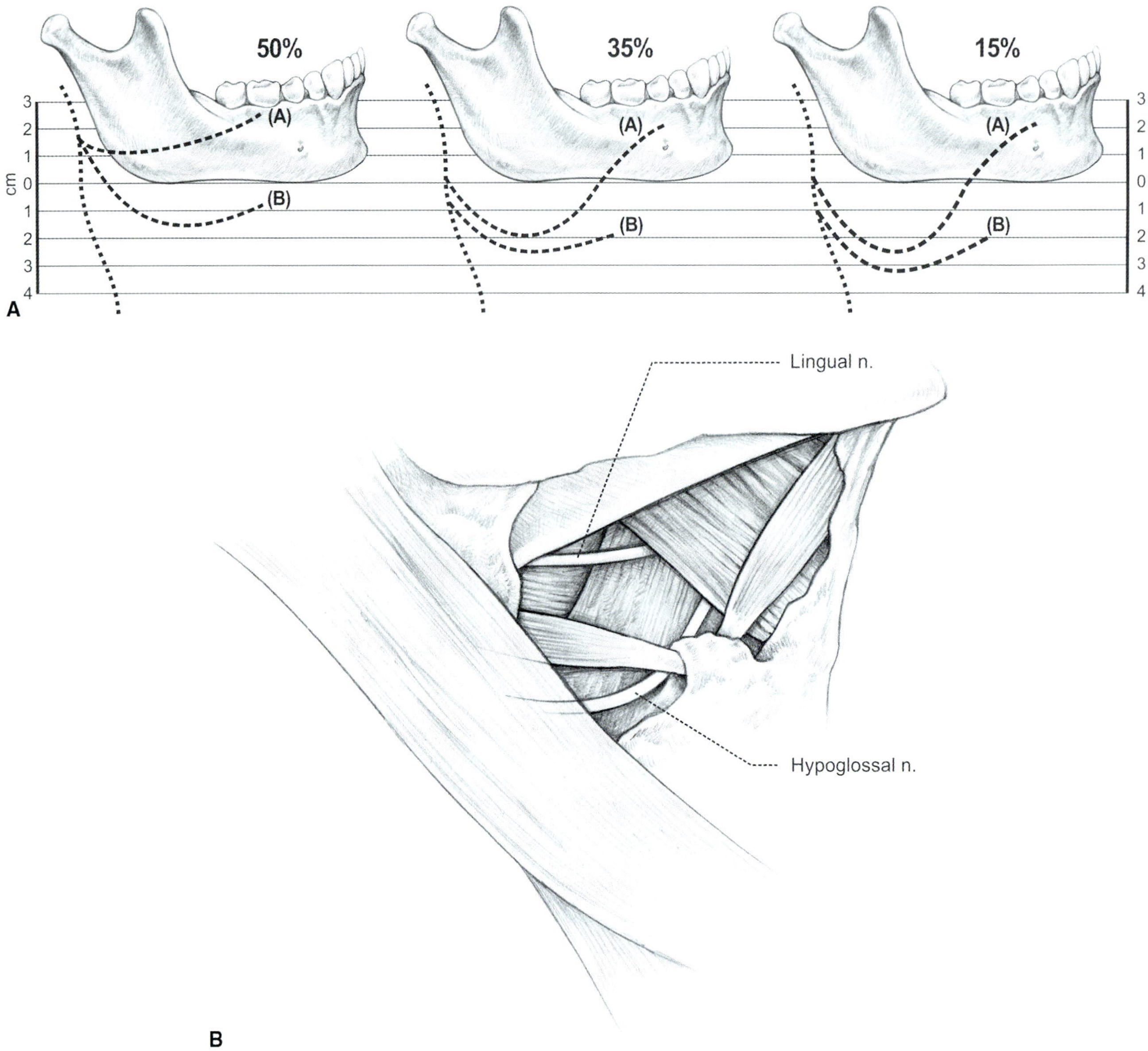

Figure 1.7 **A:** "Neural hammock" of the marginal mandibular branch (*A*) and anterior ramus of the cervical branch (*B*). **B:** Structures exposed after submandibular gland excision.

SUBMANDIBULAR GLAND (SUBMANDIBULAR GLAND EXCISION)

INCISION

For purposes of exposure and demonstration, the typical transverse incision performed in the OR is not used in the anatomy lab. Instead, continue the skin incision on the anterior border of the sternocleidomastoid to the suprasternal notch bilaterally. If you have used each side for the previous dissection, an anterior cervical flap will now be developed. Divide the skin from the tip of the chin in the midline down to the suprasternal notch. Each flap can now be elevated sharply deep to the platysma. This will expose the submandibular gland above the anterior and posterior bellies of the digastric muscle. In the operating room, exposure of the submandibular gland would be through an incision 4 cm below and parallel to the mandible extending deep to the platysma to preserve the marginal mandibular nerve. Skandalakis uses the term "neural hammock" to describe the innervation of the muscles of the lower lip. Neural hammock of the marginal mandibular branch (A) and anterior ramus of the cervical branch (B) are shown (Fig. 1.7A).

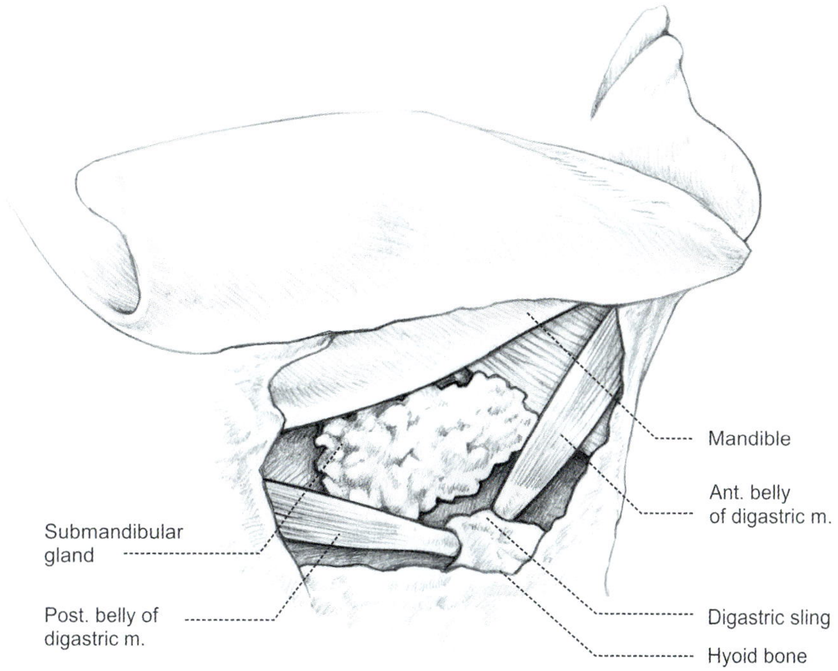

Figure 1.8 Submandibular gland located above the sling of the digastric muscle.

OPERATIVE PROCEDURE

Identify the submandibular gland and begin by separating the anterior border of the gland from the mylohyoid muscle (Fig. 1.8). Identify the underlying hypoglossal nerve and lingual veins. Continue deep dissection of the gland bluntly with the fingertip (Fig. 1.9). By anterior retraction, the facial artery is identified posteriorly. The submandibular duct (Wharton duct) and the lingual nerve are seen on the underlying hyoglossus

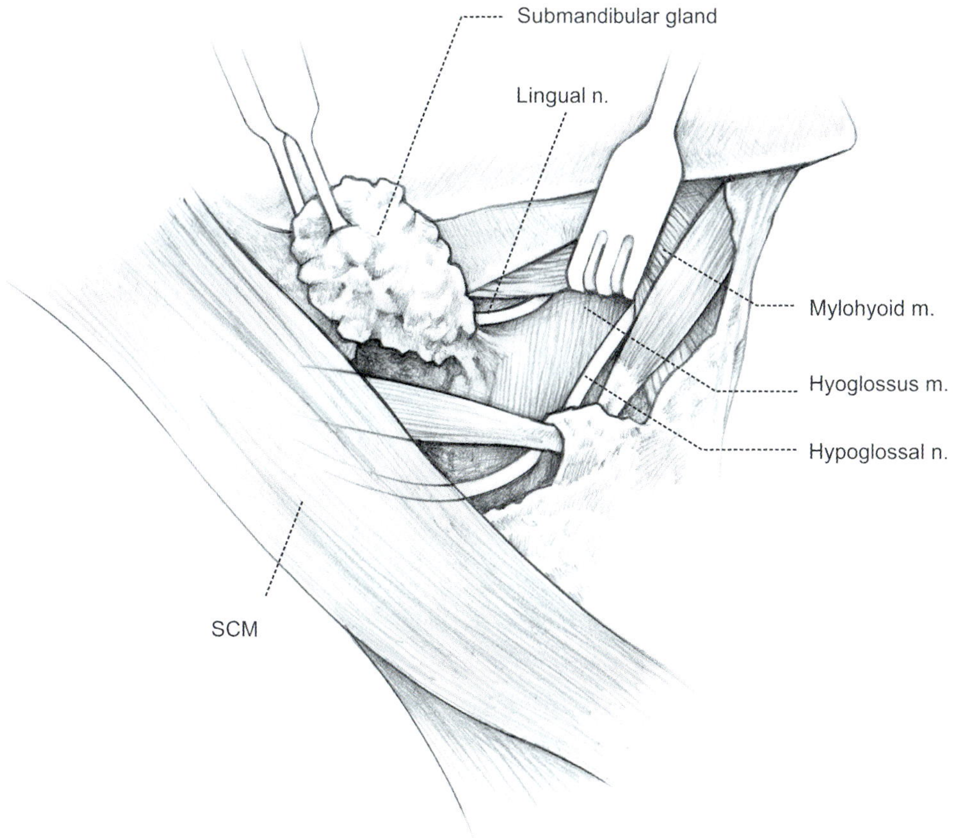

Figure 1.9 Nerves located in the bed of the submandibular gland.

muscle. Divide the duct and vascular supply of the gland. Upon completion, review the course of the hypoglossal and lingual nerves and their position relative to the digastric muscle (Fig. 1.7).

CLINICAL HIGHLIGHTS

- Note that the lingual nerve may encircle the submandibular duct during its course. Injury to the lingual nerve, a branch of the trigeminal nerve V_3, VII, and IX, will cause loss of taste of the anterior two-thirds of the tongue.
- Injury to the hypoglossal nerve will cause deviation of the tongue toward the injured side when the patient sticks the tongue out. The hypoglossal nerve (XII) is motor supply to the lateral tongue.
- The mandibular notch is a dependable landmark to find the marginal mandibular branch of the facial nerve.
- Injury to the neural hammock nerves will cause drooling involving the marginal mandibular nerve, which does not resolve, or minimal drooling involving the anterior ramus of the cervical branch resolving in 6 months.

EMERGENCY AND PERMANENT SURGICAL AIRWAYS (CRICOTHYROTOMY, TRACHEOSTOMY)

EMERGENCY CRICOTHYROTOMY

Incision

Percutaneous: A Blue Rhino or Ruesch percutaneous set will only require a small longitudinal incision at the level of the cricothyroid membrane.

Open: Open cricothyrotomy will need a 3-cm longitudinal incision above and below the level of the cricothyroid membrane. In a live patient, little bleeding will occur due to the longitudinal orientation of the neck vessels and strap muscles in the area.

Operative Procedure

Percutaneous: (Refer to the package insert for your specific percutaneous set.) Little dissection is needed. Use a Kelly, mosquito, or right-angle clamp to spread tissues in the midline. Use the introducer needle to penetrate the cricothyroid membrane and immediately angle the needle in an inferior direction in alignment with the airway (Fig. 1.10). A guide wire is inserted. Hooked

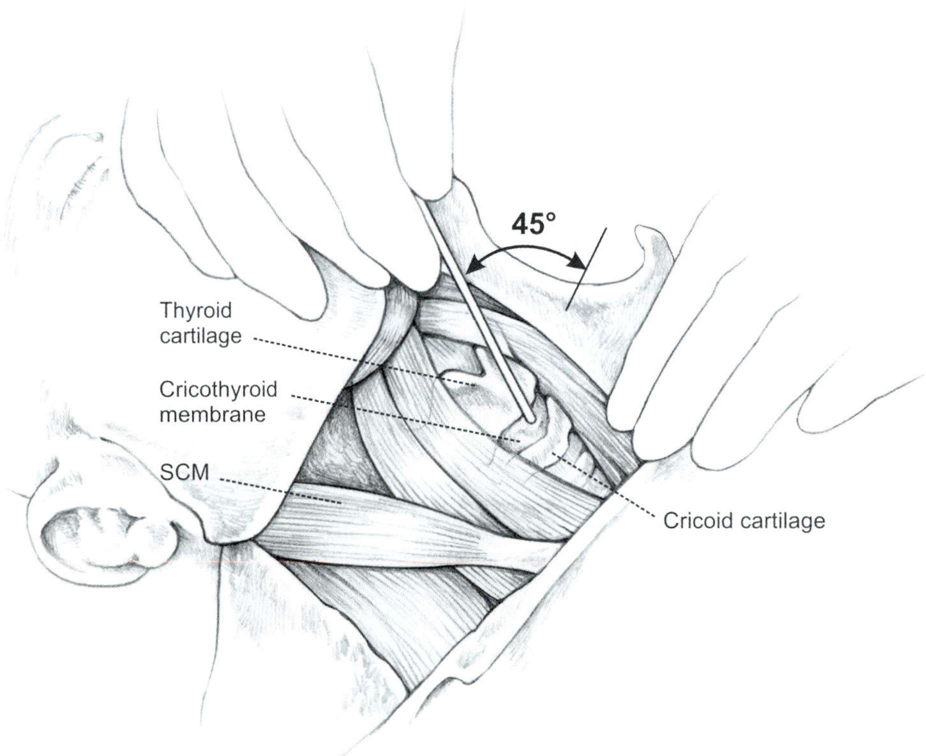

Figure 1.10 Introducer needle position for percutaneous cricothyroidotomy.

dilators are placed in size succession smallest to largest. Be sure to begin dilator insertion over the wire perpendicular to the skin. Rotate the dilator clockwise and counterclockwise to "screw" it into the airway. Do not penetrate the posterior tracheal membrane because esophageal perforation or intubation may result. Finally, place the tracheostomy, secure with twill tape around the neck, and remove the obturator.

Open: More dissection is necessary but can be done very rapidly in a clinical scenario. Use a Kelly, mosquito, or right angle clamp to spread tissues in the midline down to the cricothyroid membrane. A transverse incision in the cricothyroid membrane is made (Fig. 1.11). Place the endotracheal or tracheostomy tube into the airway under direct vision. Do not penetrate the posterior tracheal membrane because esophageal perforation or intubation may result. Secure the tube in place with a twill tape or adhesive tape.

ELECTIVE TRACHEOSTOMY

Incision

Palpate anatomic landmarks, which include the thyroid notch, the cricoid cartilage, and the sternal notch. If no incisions have been done, perform a wide transverse "collar" incision 2 finger breadths above the sternal notch, or at the inferior border of the cricoid cartilage. The incision must be wide enough on the body to fully expose the underlying anatomy.

Operative Procedure

Create small subplatysma flaps superiorly and inferiorly. Separate the strap muscles vertically (Fig. 1.12). Define the trachea with blunt dissection. Place a tracheal hook under the cricoid cartilage and elevate the trachea superiorly. Bluntly elevate the isthmus of the thyroid gland. Divide and ligate the isthmus only if necessary for tracheal exposure. Find the 3rd tracheal

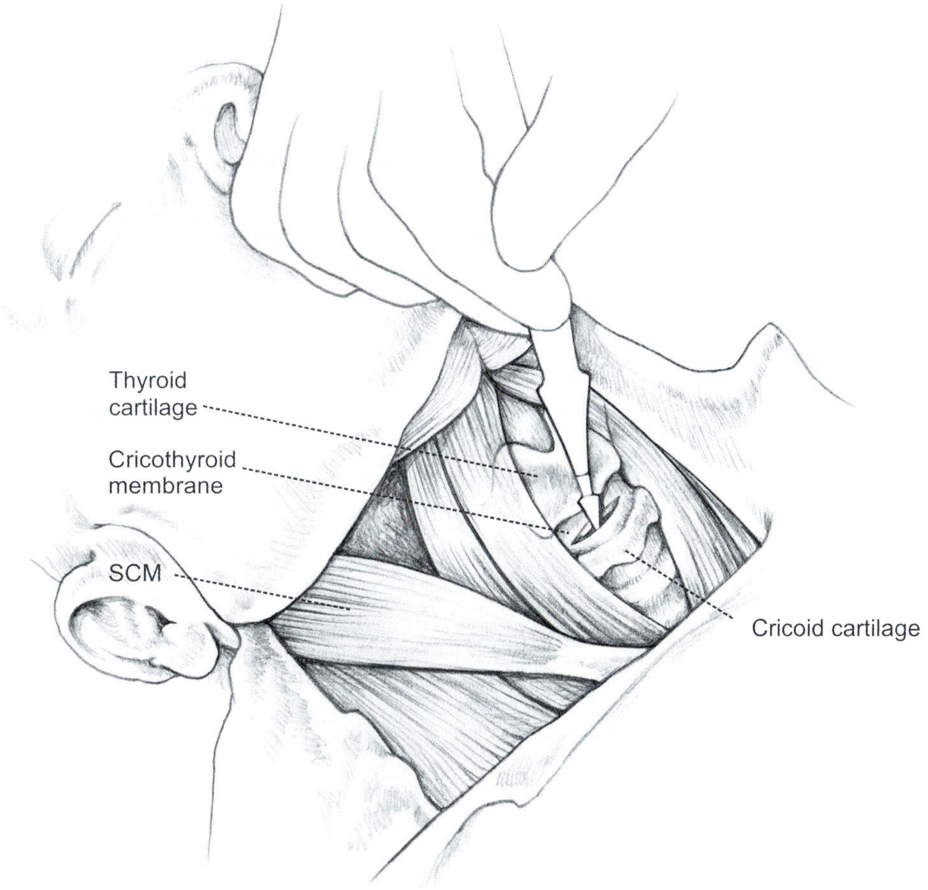

Figure 1.11 Cricothyroidotomy incision for open emergency airway.

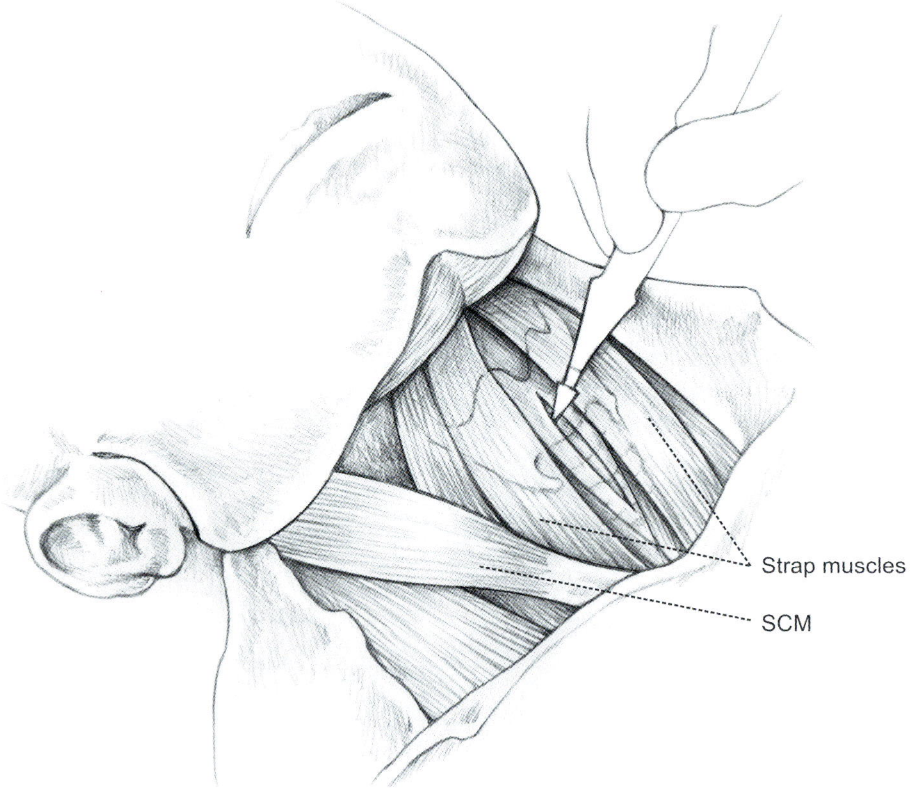

Figure 1.12 Separation of strap muscles for elective tracheostomy.

cartilage. Incise 1.5 cm of the membrane between the 2nd and 3rd cartilages transversely with a blade. Cut the 3rd cartilage longitudinally on one lateral side with a scissors. Secure the cartilage with a clamp and use your blade to cut transversely on the inferior side of the 3rd cartilage. Remove an anterior segment of the 3rd tracheal cartilage (Fig. 1.13). Place an endotracheal tube or tracheostomy tube in the airway. Do not penetrate the posterior tracheal membrane because esophageal perforation and intubation may result.

Clinical Highlights

- The worst error with any of the techniques is a misplaced tube. Most commonly the tube can be misplaced in the soft tissue of the neck between the trachea and sternum. Perforation of the posterior tracheal membrane and esophageal damage or intubation can occur. Submembranous placement of the tube can occur if it is placed between the tracheal cartilage and tracheal mucosa. Having the tube in place but inadequately secured in an emergent situation may lead to a dislodged tube.

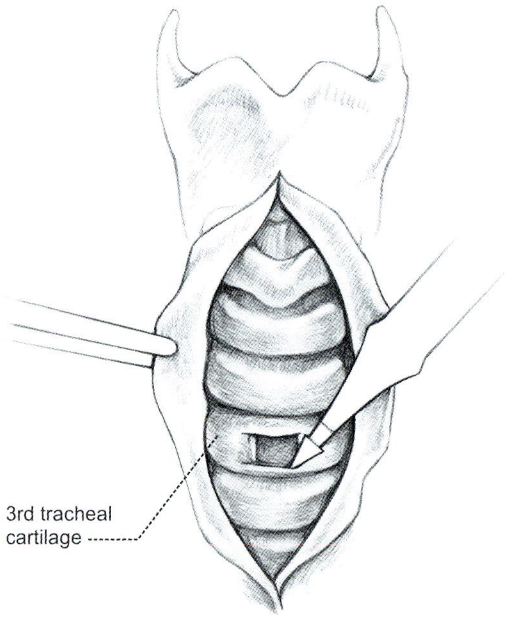

Figure 1.13 Excision of anterior segment of the third tracheal ring for tracheostomy.

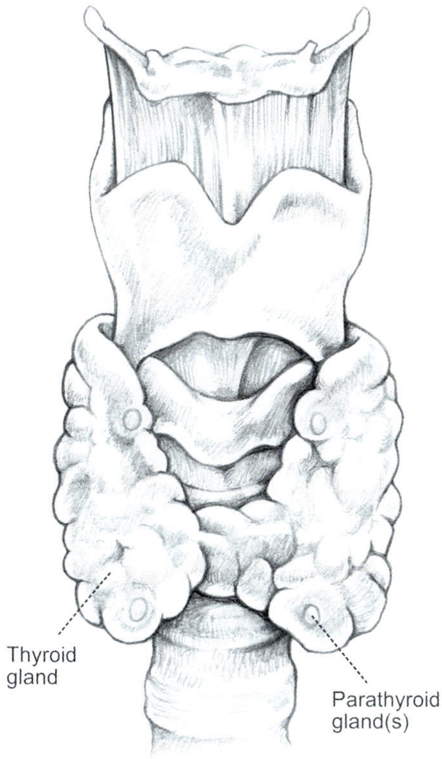

Figure 1.14 Position of thyroid, parathyroids, and trachea.

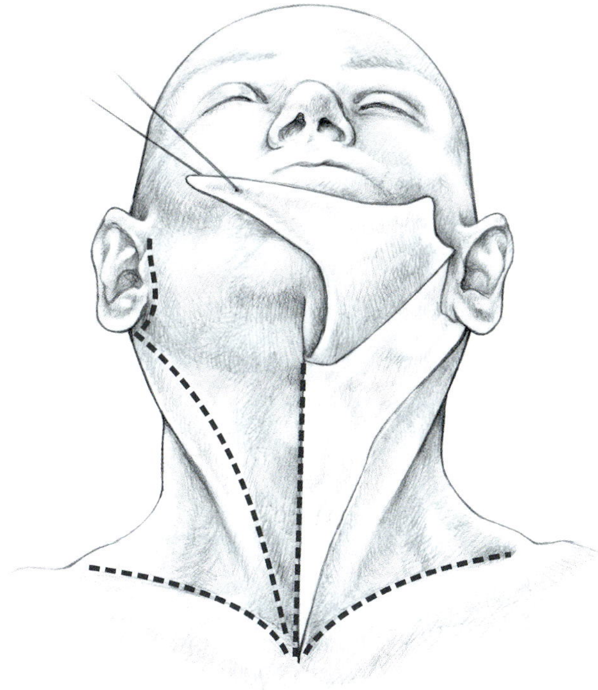

Figure 1.15 Incisions for head and neck exposure in the anatomy lab.

- Be sure to monitor O_2 saturation and/or exhaled CO_2 to ensure ventilation.
- For orientation, see Figure 1.14 to see position of thyroid and parathyroid glands.

THYROID AND PARATHYROIDS (THYROIDECTOMY, PARATHYROIDECTOMY)

INCISION

The incision must be wide enough on the body to fully expose the underlying anatomy. Use the previous incisions (Fig. 1.15), which are not the "collar" incision used in the OR. Create subplatysma flaps superiorly to the thyroid cartilage and inferiorly to the sternal notch. In the OR, you will see vertical separation of the strap muscles (see Fig. 1.12). However, for purposes of exposure in the lab, divide the strap muscles transversely across the upper third of their length (Fig. 1.16). Retraction of the strap muscles laterally gives adequate exposure in the operating room, but rarely, these muscles need to be divided when a very large tumor or goiter is encountered. Remember that the innervation to the strap muscles comes in posteriorly in the lower two-thirds of the muscle from ansa cervicalis branch of cervical branch of the facial nerve (see Fig. 1.6).

OPERATIVE PROCEDURE

Mobilize the thyroid and divide the middle thyroid veins if they still remain (Fig. 1.17). Next, identify and trace and then transect the superior thyroid artery (Fig. 1.18). Note that the external branch of the superior laryngeal nerve runs along with the superior thyroid artery and vein adjacent to the flare of the cornu of the thyroid cartilage. Note the pretracheal location of the pyramidal lobe of the thyroid and the "prelaryngeal node," which is otherwise referred to as a Delphian node.

If possible, identify the right and left thyrocervical trunks as they branch from the subclavian arteries on each side (Fig. 1.19). Trace these vessels to the thyroid gland. Rotate the thyroid gland medially during dissection on each side. During this dissection, identify the inferior thyroid vessels and note that they are highly variable and are usually asymmetric (Fig. 1.20). Running under the inferior thyroid vessels, the recurrent laryngeal nerve becomes visible.

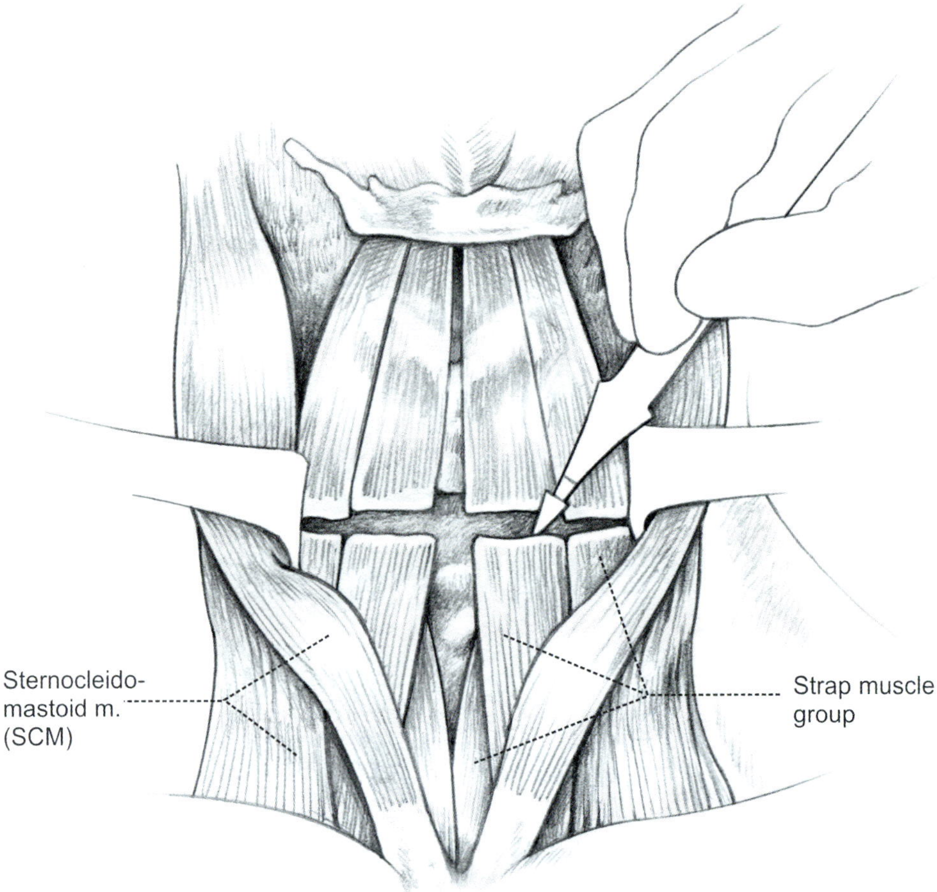

Figure 1.16 Transverse division of the strap muscles to facilitate thyroid exposure in the anatomy lab.

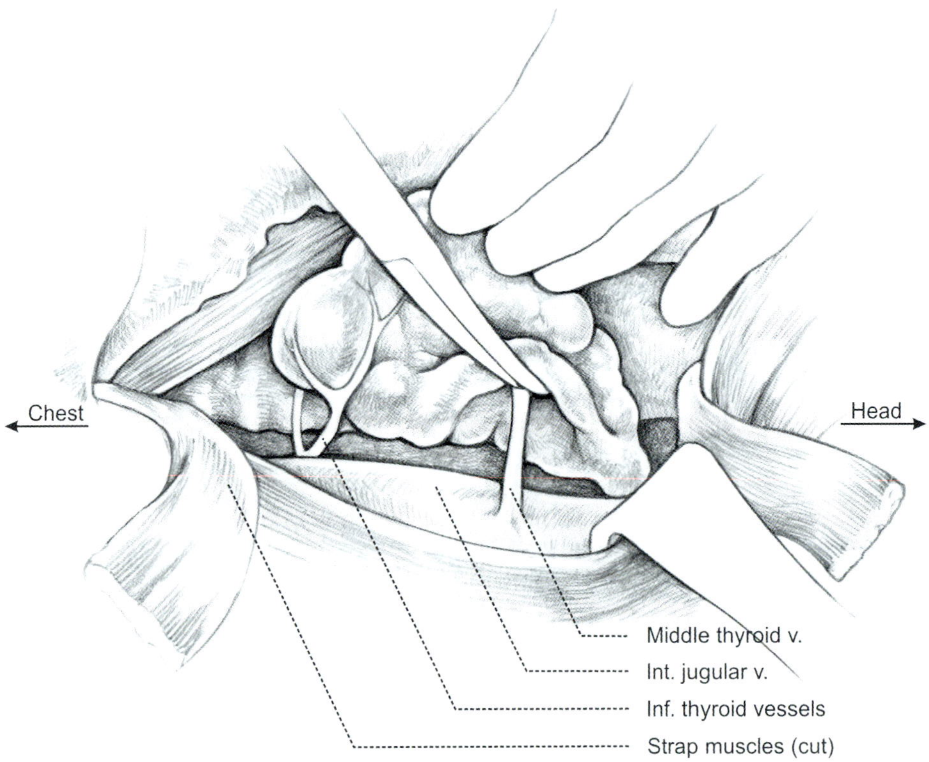

Figure 1.17 Division of the middle thyroid veins.

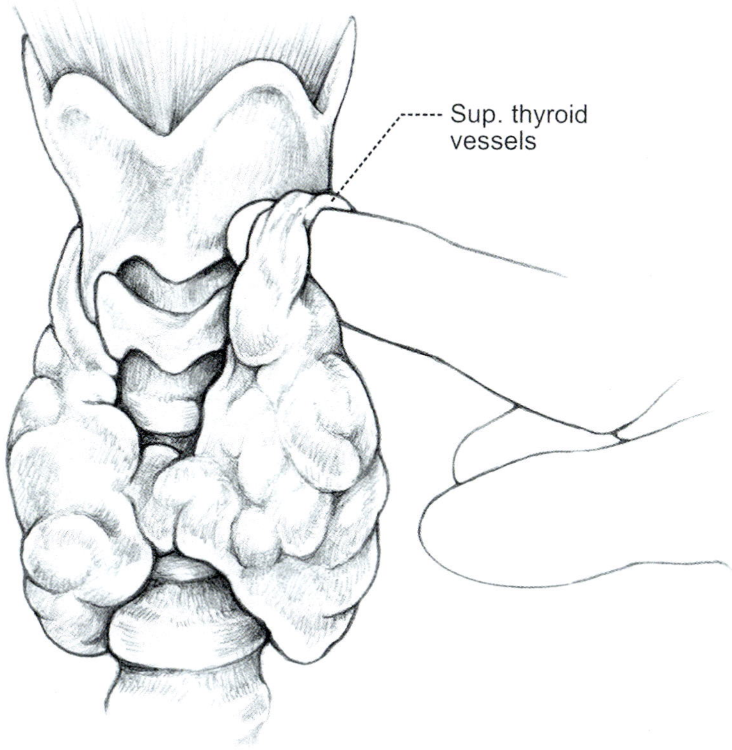

Figure 1.18 Mobilization of the superior thyroid vessels.

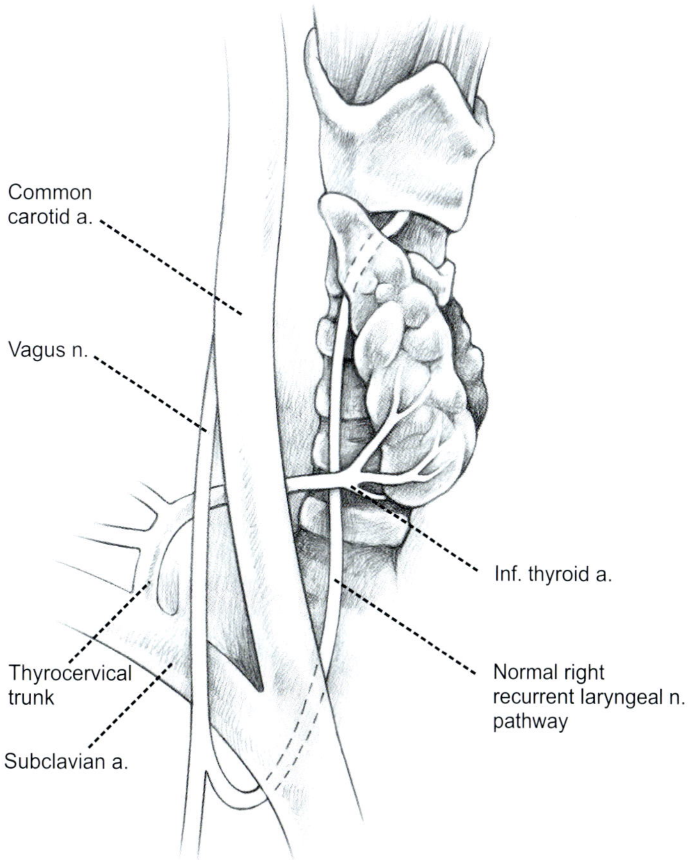

Figure 1.19 Normal right recurrent laryngeal nerve course.

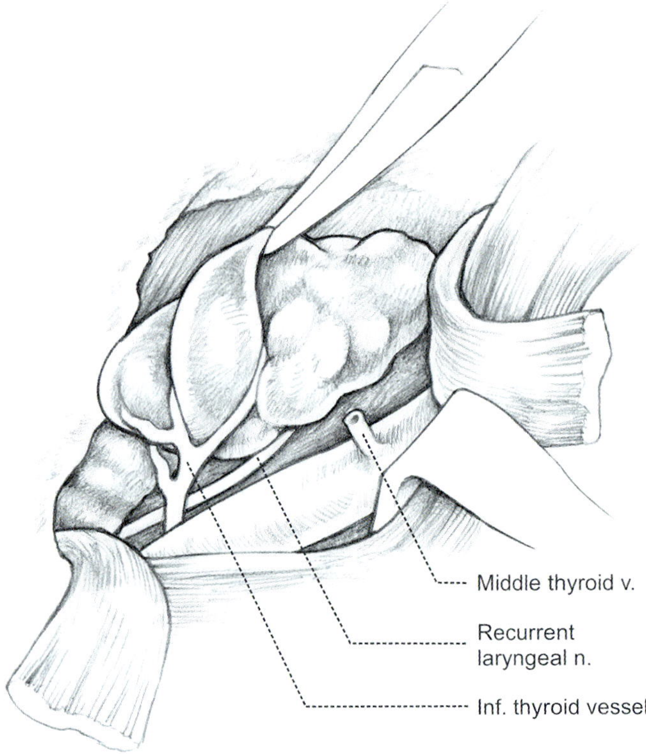

Figure 1.20 Inferior thyroid vessels' relation to the recurrent laryngeal nerve.

During this portion of the dissection, the parathyroid glands may be identified (Fig. 1.14). In the OR, remember to identify the parathyroid glands before extensive thyroid mobilization. Blood supply to the parathyroids is generally thought to originate from the inferior thyroid arteries. However, Skandalakis reports that the superior parathyroid glands may receive their blood supply from the posterior branch of the superior thyroid artery. Identify the recurrent laryngeal nerves during medial rotation of the thyroid gland as they track obliquely from the tracheoesophageal groove forward to their penetration of the inferior constrictor muscle (Figs. 1.19 to 1.21).

CLINICAL HIGHLIGHTS

Thyroid:

- Superior laryngeal nerve: internal branch—sensory, external branch—motor to cricothyroid muscle, which is necessary for singers to hit high notes.
- Right nonrecurrent laryngeal nerve: 1% of cases (Fig. 1.22).
- Left nonrecurrent laryngeal nerve: extremely rare, associated with right aortic arch.

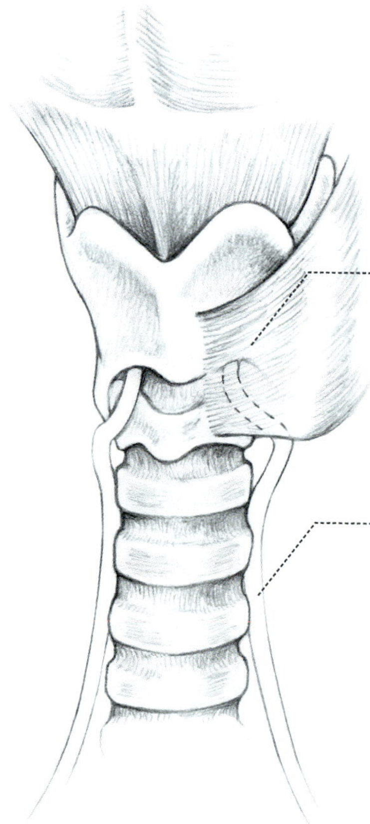

Figure 1.21 Recurrent laryngeal nerves' course to the larynx.

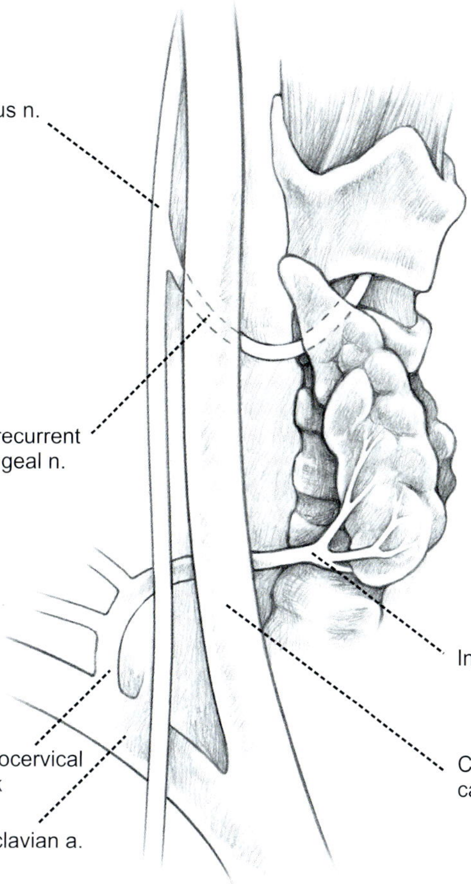

Figure 1.22 Course of nonrecurrent right laryngeal nerve.

- Subtotal thyroidectomy done for pediatric Graves disease leaves 2 g of thyroid tissue on each side.
- Level VI, anterior compartment nodal dissection may accompany malignant thyroid gland resections (Fig. 1.23).
- Remember to mobilize the superior pole early if large goiter with substernal component is present.

Parathyroid:

- To preserve parathyroids and their blood supply, gently move each away from the thyroid beginning at the parathyroid-thyroid interface.
- Third branchial pouch becomes inferior parathyroids; fourth branchial pouch becomes superior parathyroids. (To remember this, hold your arms up and make a fist on each side of your neck. Extend your thumb and index finger and flex your wrist downward. The index finger that was the third branchial pouch has become the inferior parathyroid. Your thumb was the fourth branchial pouch and has become the superior parathyroid). If on exploration, the parathyroids are not in their typical location, note that your thumb points posteriorly to the tracheoesophageal groove and your index finger points inferiorly to the anterior mediastinum and thymus. Both of these locations are accessible from the cervical incision typically used for parathyroid exploration. Rarely, the parathyroids can be located in distant sites such as the posterior mediastinum or under the aortic arch.
- Remember that only one viable parathyroid gland is needed to avoid hypoparathyroidism. If parathyroid viability is questionable, excision followed by mincing and implantation into forearm muscles or strap muscles is a reliable method to avoid hypoparathyroidism.

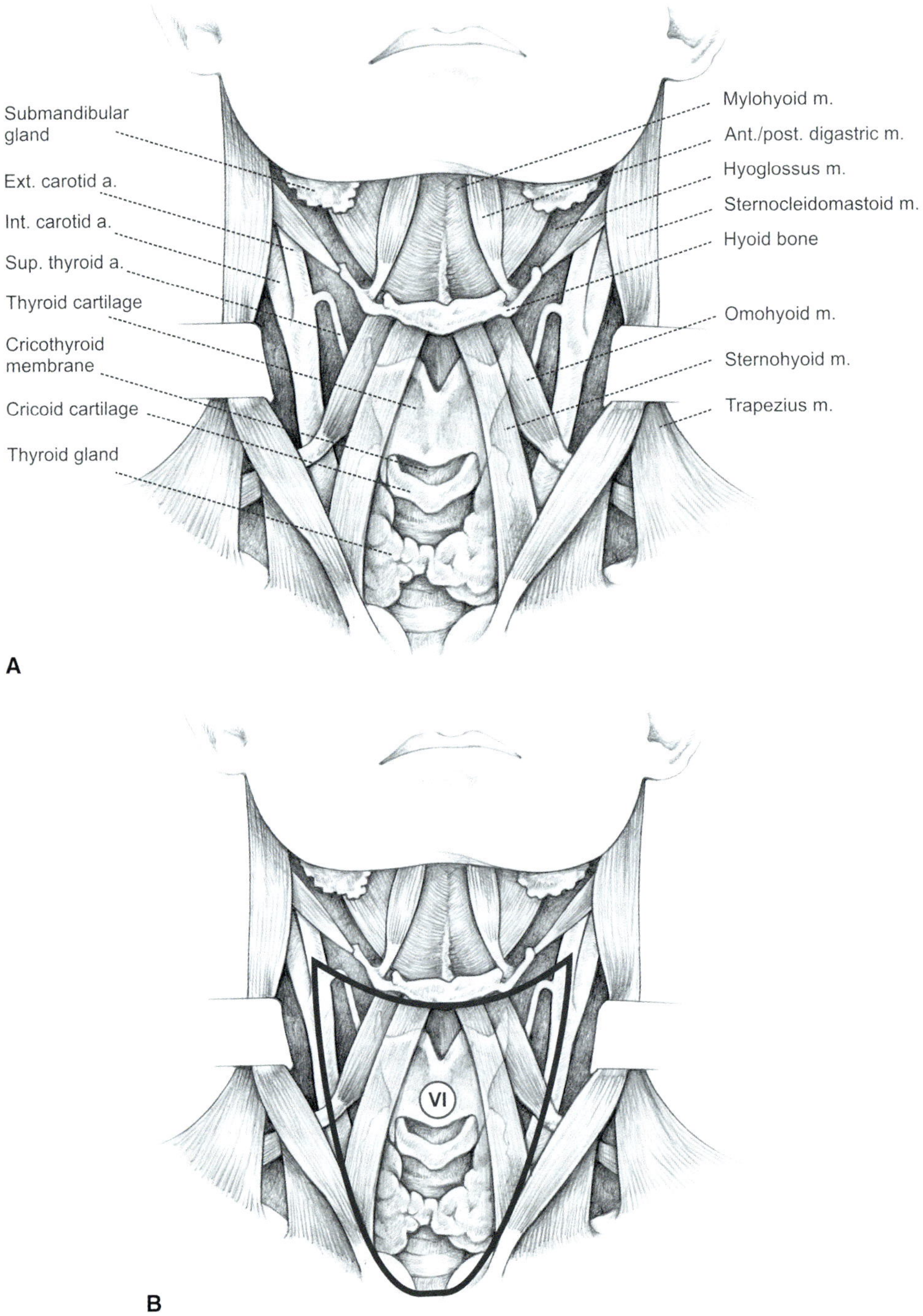

Figure 1.23 **A:** Anterior neck anatomy. **B:** Anterior neck anatomy demonstrating level VI (anterior compartment).

NODE-BEARING AREAS OF THE NECK (RADICAL NECK DISSECTION)

INCISION

Incise the skin along the clavicle laterally on each side (see Fig. 1.15). Elevate the skin flaps below the platysma. Continue this dissection to the anterior border of the trapezius muscle (see Fig. 1.15). Other incisions may be used clinically in the OR.

OPERATIVE PROCEDURE

Before starting the neck dissection, please examine the diagram of the nodal stations, termed levels, used by the American Joint Commission on Cancer (AJCC) to stage head and neck metastases (Figs. 1.23 and 1.24A).

Penetrating trauma zones of the neck are shown (Fig. 1.24B). Note the trauma zones are not the same as lymph node levels. Understand that a radical neck dissection is generally defined by the quadrilateral space containing the major lymphatic channels and lymph nodes of the neck. The classic radical neck dissection includes removal of the internal jugular vein, the spinal accessory nerve, the submandibular gland, a portion of the parotid gland, and several muscles: SCM, omohyoid, stylohyoid, and posterior belly of the digastric muscle. Many modifications have been described to meet the variations in which metastatic head and neck cancer presents itself.

You may choose to perform a classic radical neck for anatomic familiarity with neck anatomy or you may choose to perform a modern radical neck dissection for orientation prior to an OR experience.

CLASSIC

Erb's point (Fig. 1.25) is mid-SCM defined by the point where superficial branches of the cervical plexus are found at the posterior SCM. Above Erb's point nodal tissue is swept back, and above Erb's point, nodal tissue is swept forward.

At the base of the neck, divide the two bellies of the sternocleidomastoid (sternal head, clavicular head) and dissect superiorly (Fig. 1.26). Remember innervation of SCM is the spinal root of the accessory (XI) nerve and C2 and C3. Sharply incise the deep fascia from the SCM laterally along the clavicle to the anterior border of the trapezius muscle. The external jugular vein and posterior belly of the omohyoid muscle are transected (Fig. 1.26).

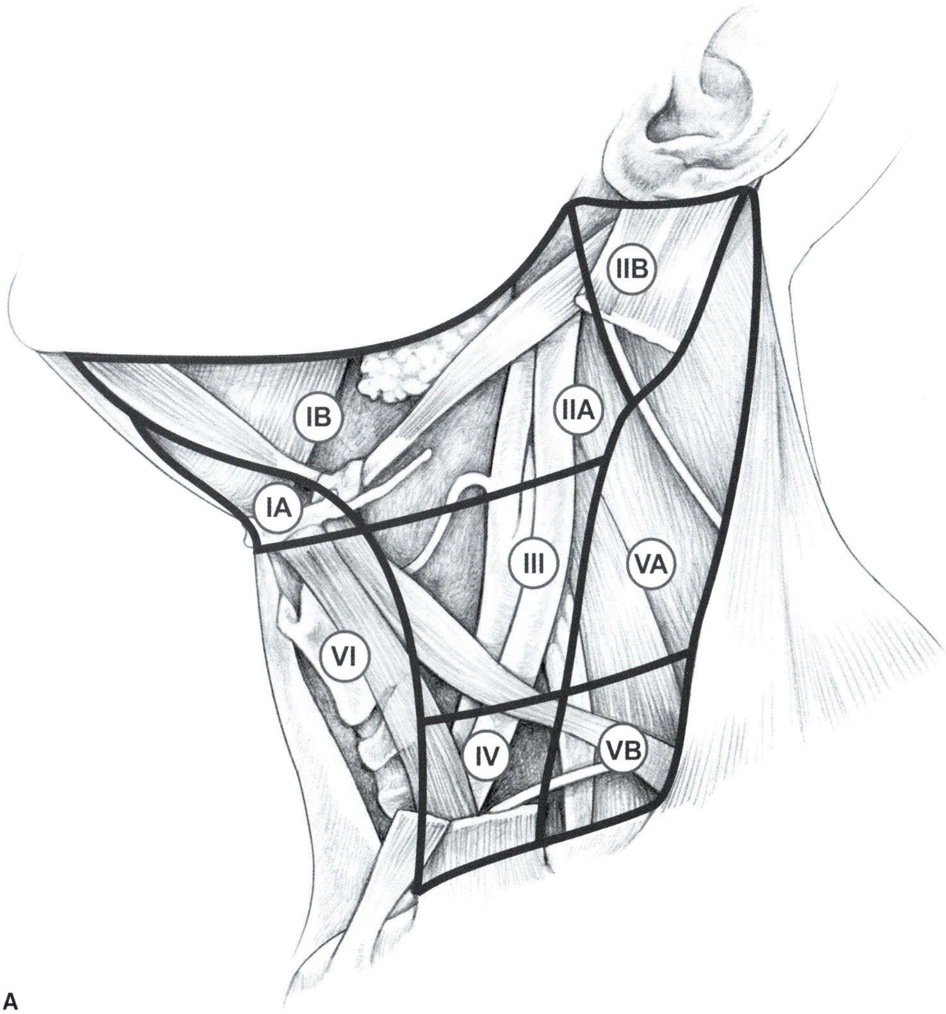

A

Figure 1.24 Lateral neck anatomy. **A:** Lateral neck anatomy demonstrating AJCC levels.

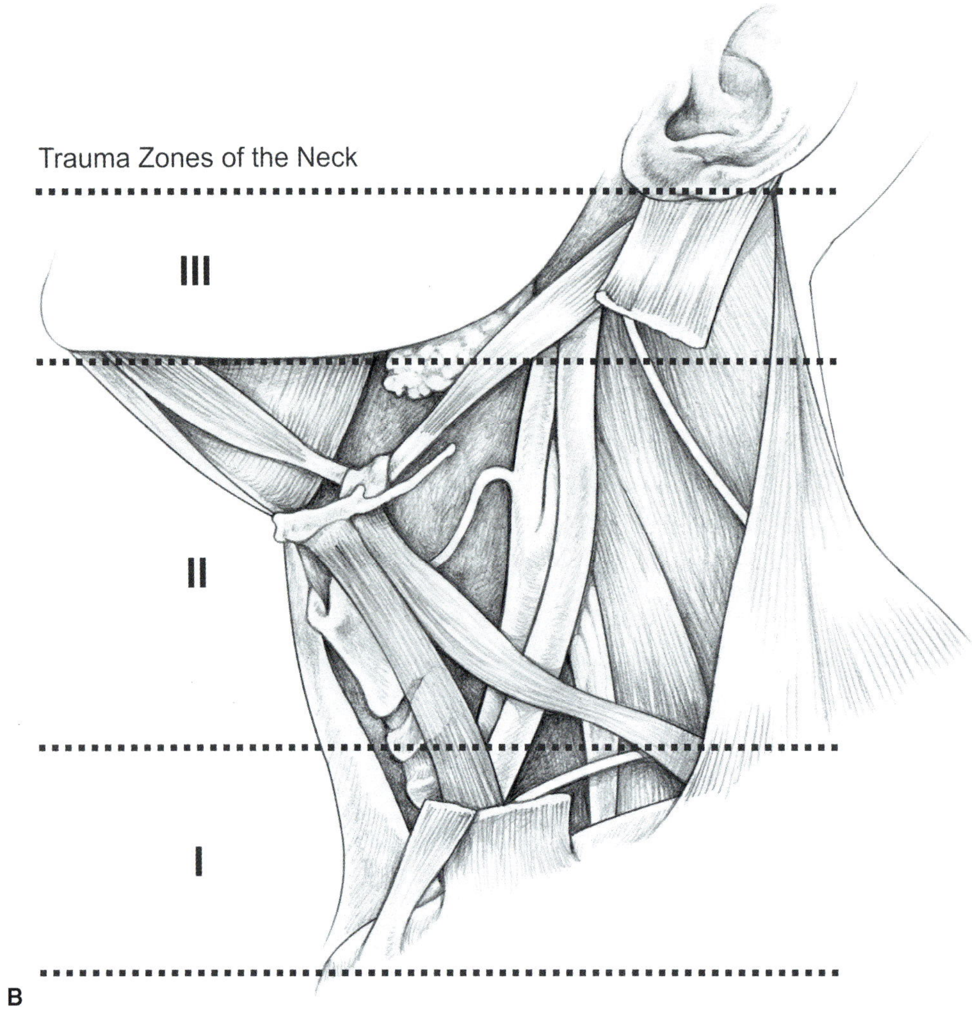

Figure 1.24 (Continued) **B:** Trauma zones of the neck.

Lift the two heads of the SCM and identify the internal jugular vein and commence block dissection of the deep structures of the neck. Leave the thoracic duct at the junction of the internal jugular and subclavian veins for later dissection. Identify the thyrocervical trunk, a branch of the subclavian artery, lateral to the jugular vein. This artery is ligated and divided. For the purposes of anatomical demonstration, divide the right thyrocervical trunk and preserve the left thyrocervical trunk. Deep to the internal jugular vein, identify the anterior scalene muscle and the phrenic nerve. The phrenic nerve traverses from lateral to medial upon the anterior scalene muscle. The next lateral structures in order will be the brachial plexus, medial scalene, posterior scalene, and levator scapulae (Fig. 1.27 and Fig. 1.31). These muscles are identified by the posterior dissection along the anterior border of the trapezius. Sacrifice or, for the purpose of anatomical demonstration, preserve the spinal accessory nerve as it traverses diagonally across these muscles. It will be necessary to sacrifice branches of the 5th and 6th cervical nerves beyond the phrenic nerve that originates from them (Fig. 1.27).

Turn your attention to the superior portion of the anterior belly of the omohyoid muscle and divide it at the hyoid insertion. Divide any attachments along the sternohyoid muscle (Fig. 1.28). Elevate the entire dissection and divide the sternocleidomastoid from the mastoid process (Fig. 1.29). Note the hypoglossal nerve as it crosses the internal and external carotid arteries and dives under the posterior belly of the digastric muscle. Note also the first branch of the external carotid artery, the superior thyroid artery as it begins its downward trajectory. Clear the submental space laterally. Follow the anterior belly of the digastric muscle back to the submandibular gland. Divide the facial vessels

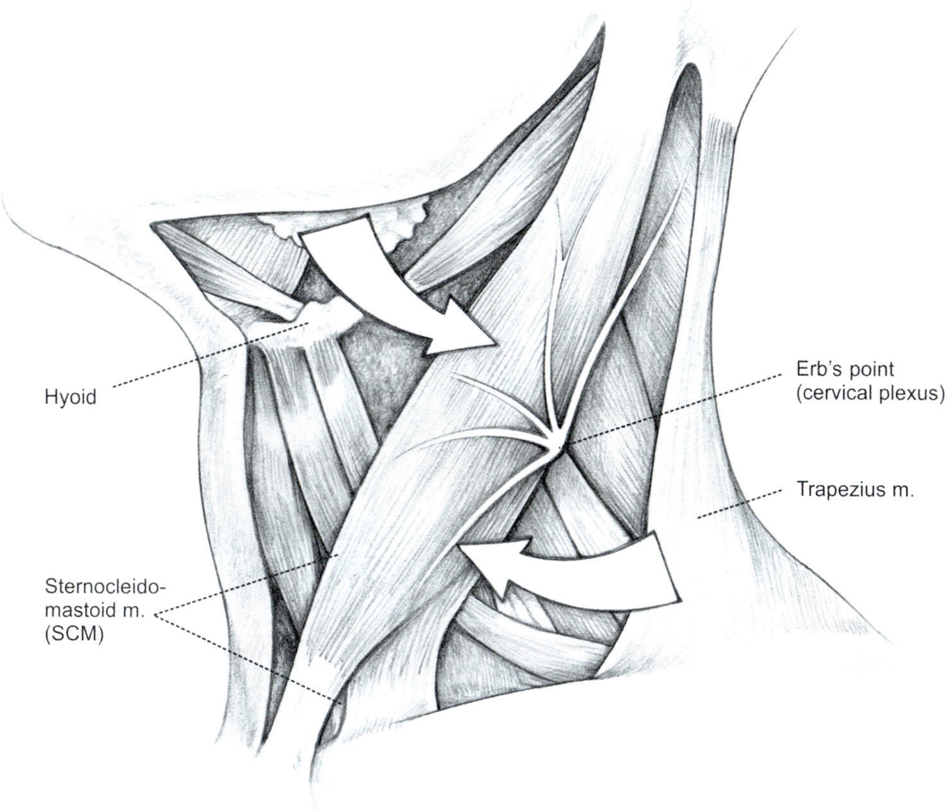

Figure 1.25 Erb's point.

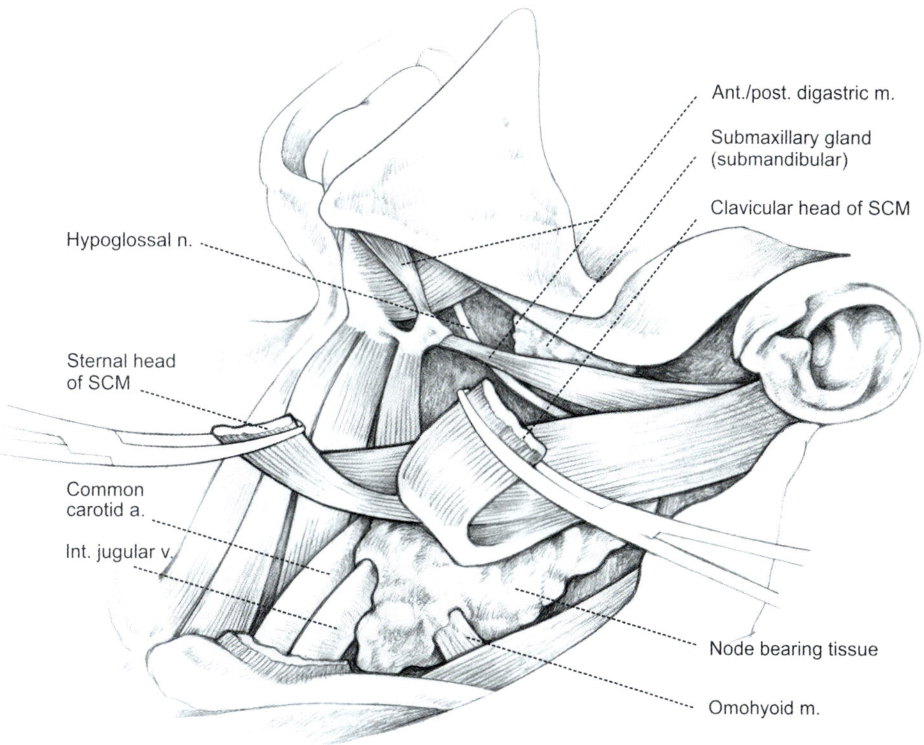

Figure 1.26 Sternocleidomastoid muscle (SCM): sternal head and clavicular head.

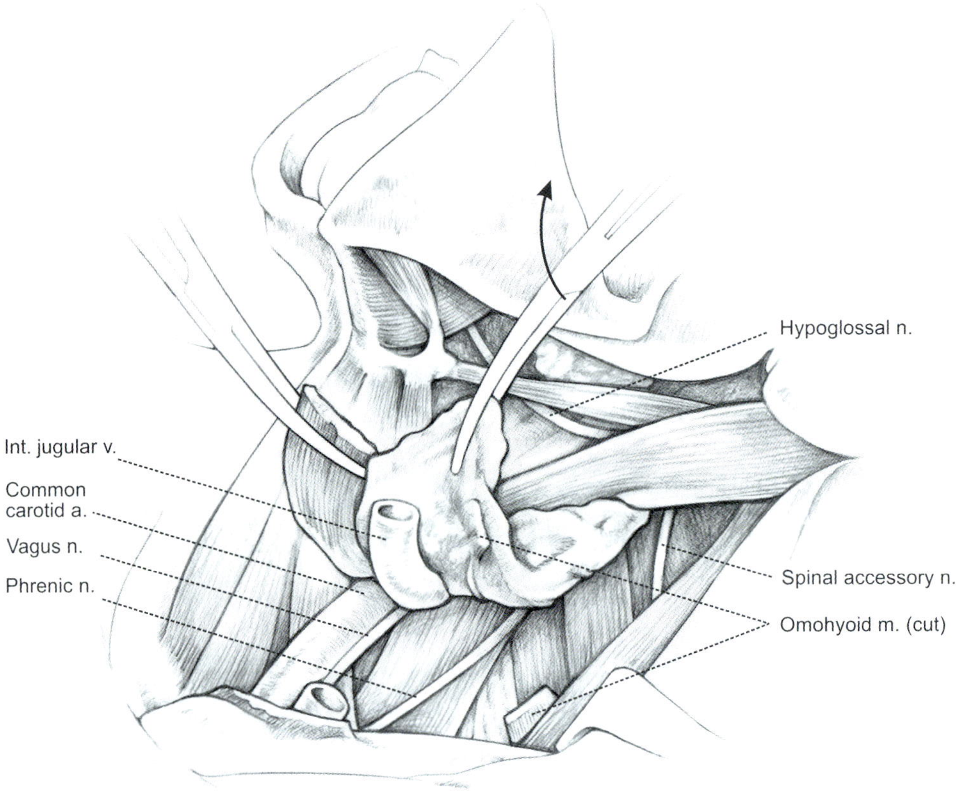

Figure 1.27 Elevation of SCM and internal jugular vein.

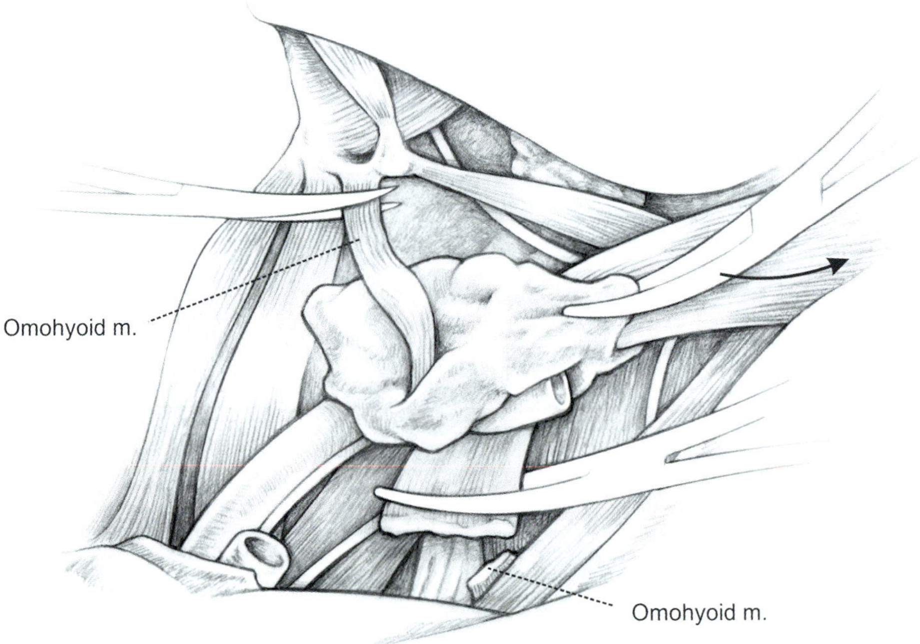

Figure 1.28 Division of anterior belly of omohyoid muscle.

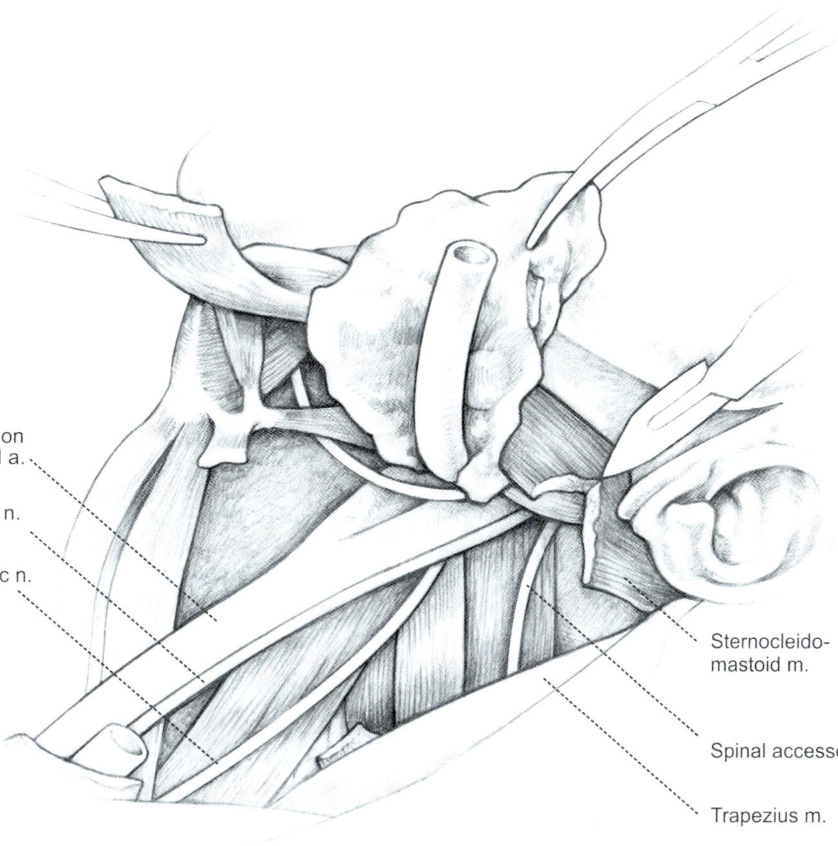

Figure 1.29 Division of SCM from mastoid process.

at the margin of the mandible. Divide the submandibular (Wharton) duct and the small branch of the lingual nerve to the gland (see Figs. 1.8 and 1.9). The stylohyoid muscle and the posterior belly of the digastric muscle are sacrificed. For purposes of anatomical demonstration, sacrifice these muscles on the right neck and preserve them on the left neck (see Fig. 1.30). Next, sweep the entire dissection specimen inferiorly to expose the spinal accessory nerve lying on the jugular vein at the mandible (Fig. 1.30). Sacrifice the jugular vein on the right side and preserve the vein on the left side. You may now remove the radical neck specimen.

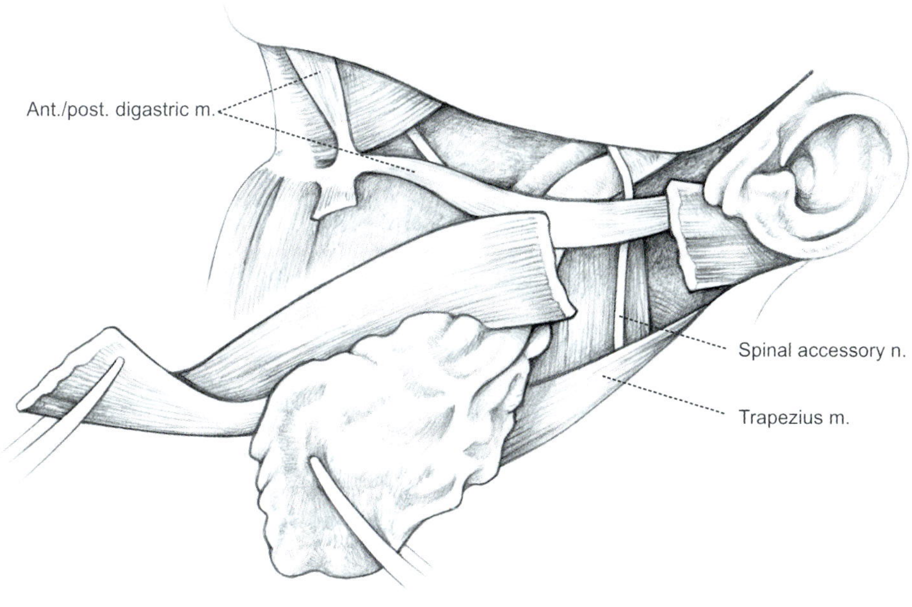

Figure 1.30 Inferior traction of specimen demonstrating the spinal accessory nerve.

MODERN

Erb's point (see Fig. 1.25) is mid-SCM defined by the point where superficial branches of the cervical plexus are found at the posterior border of the SCM. Erb's point can be used to identify the spinal accessory nerve in level V as it exits under the SCM ~1 cm above Erb's point.

As level I dissection was previously performed, start by following the posterior belly of the digastric muscle up to the mastoid and anterior border of the SCM. Identify the hypoglossal nerve as it exits deep to the digastric muscle. This will be approximately where the stylohyoid muscle crosses the digastric muscle. The hypoglossal nerve can be followed posterior to the superior aspect of the internal jugular vein (see Fig. 1.26, These illustrations demonstrate a classical approach beginning with the clavicular heads of the SCM).

Just lateral on the superior aspect of the internal jugular vein, the spinal accessory nerve can be identified (see Fig. 1.30). Dissect the superior attachment of the SCM off the mastoid tip (see Fig. 1.29). Follow this to the superior border of the trapezius muscle. Dissect out the anterior border of the trapezius muscle. Note that you can identify the spinal accessory nerve again as it enters the trapezius muscle. Follow the anterior border of the trapezius down to the clavicle. Open up the fascia following the superior border of the clavicle as seen on the completion illustration (Fig. 1.31).

Transect the inferior border of the SCM (see Fig. 1.26). Note that the superior, lateral, and inferior borders of the dissection have now been delineated. Sweep fat and lymphatic tissue off the floor of the neck identifying cervical rootlets along the way. It will be necessary to sacrifice branches of the 5th and 6th cervical nerves beyond the phrenic nerve that originates from them (see Fig. 1.27). Sweep tissue toward the internal jugular vein. Transect the lateral aspect of the omohyoid muscle as it crosses lower level V (see Fig. 1.26).

Leave the thoracic duct at the junction of the internal jugular and subclavian veins for later dissection. Next, identify the thyrocervical trunk, a branch of the subclavian artery, lateral to the jugular vein. For the purposes of anatomical demonstration, divide the thyrocervical trunk on the right side and preserve the thyrocervical trunk on the left side. Deep to the internal jugular vein, identify the anterior scalene muscle and the phrenic nerve. The phrenic nerve traverses lateral to medial upon the anterior scalene muscle. The next lateral structures in order will be the brachial plexus, medial scalene, posterior scalene, and levator scapulae (see Fig. 1.27 and Fig. 1.31). These muscles are identified

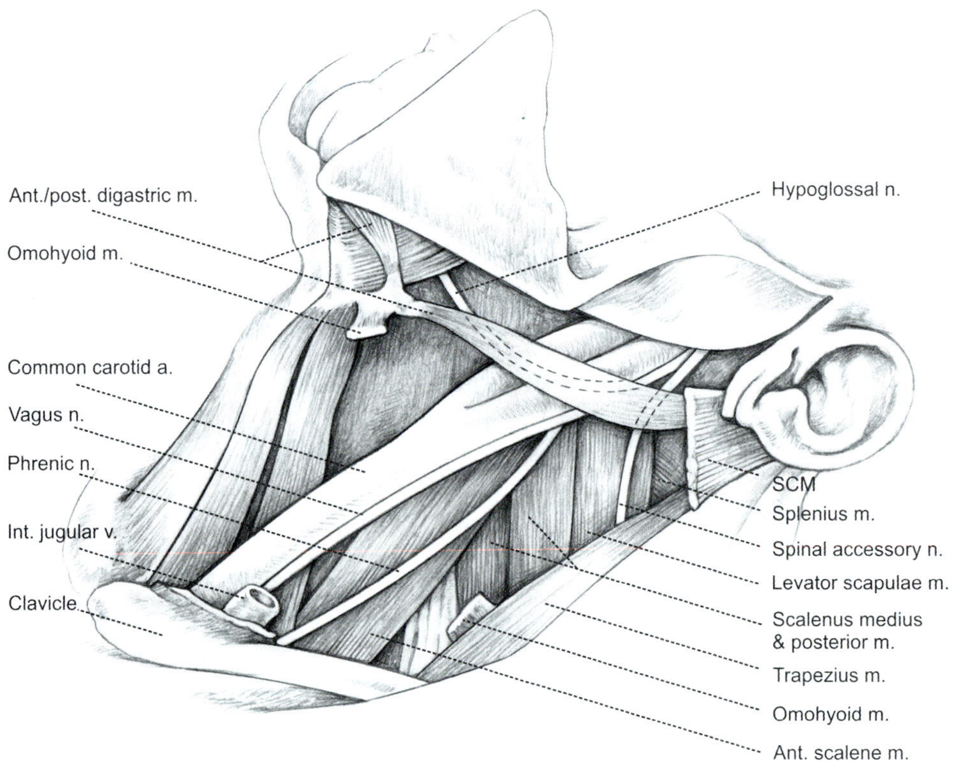

Figure 1.31 Anatomy exposed after completion of radical neck dissection.

by the posterior dissection along the anterior border of the trapezius. Preserve the accessory nerve as it traverses diagonally across these muscles.

Turn your attention to the superior portion of the anterior belly of the omohyoid muscle and divide it at the hyoid insertion (see Fig. 1.28). Note the hypoglossal nerve as it crosses the internal and external carotid arteries and dives under the posterior belly of the digastric muscle (see Fig. 1.31). Note also the first branch of the external carotid artery is the superior thyroid artery as it begins its downward trajectory (Fig. 1.32). Sweep the entire dissection specimen inferiorly (see Fig. 1.30). Preserve the jugular vein. You may now remove the radical neck specimen.

CLASSIC OR MODERN, COMPLETED

Identify the contents of the carotid sheath, the internal jugular vein, the carotid arteries, the vagus nerve, and, medially, the cervical sympathetic trunk. Identify and trace the following nerves on each side of the neck: vagus, phrenic, and sympathetic trunk located deep to the carotid artery (see Fig. 1.31).

CLINICAL HIGHLIGHTS

- Remember the spinal accessory nerve exits with the jugular vein from the jugular foramen at the base of the skull and is found in the posterior triangle of the neck. Beware of the node or mass in the posterior triangle. Open biopsy may damage the spinal accessory nerve.

- Note that radical neck dissection is rarely performed and has given way to selective neck dissection. In a selective neck dissection, only the nodal basins at risk for metastasis are removed. Also the spinal accessory nerve, internal jugular vein, and SCM are spared.

- *Brief branchial remnant review:*
  - A first cleft sinus or cyst is located anterior to the tragus of the ear and may pass above or below the facial nerve (Fig. 1.33 and Fig. 1.34—see circle anterior to the ear).
  - A second cleft cyst or sinus may be located along the lateral neck and may have a fistulous tract. That fistulous tract will pass between the internal and external carotid arteries on its way to the tonsil. Be aware of the hypoglossal nerve during this dissection (Figs. 1.33 to 1.35).
  - Rare third or fourth cleft cysts, sinuses, and fistulas may be found as low as the superior chest.

- Thyroglossal duct abnormalities typically present in the midline (Fig. 1.36). They need to be explored for fistulous tracts penetrating the hyoid. The central hyoid and fistulous tract needs to be resected to the base of the tongue (Fig. 1.37). Remember to be aware of and watch for thyroid tissue anywhere along this tract to include the base of the tongue. Some surgeons advocate ultrasound imaging of the neck to confirm a normal thyroid gland preoperatively.

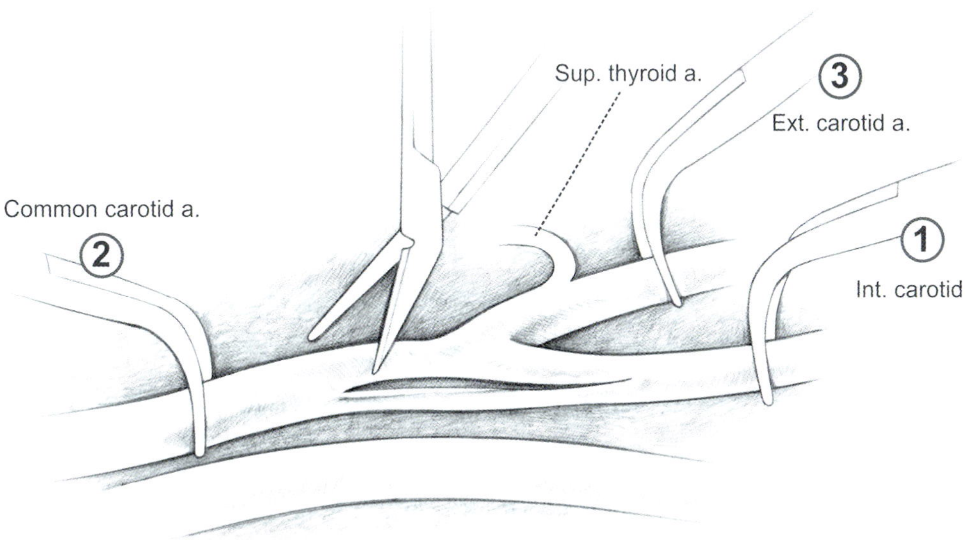

Figure 1.32 Carotid artery clamped and arteriotomy performed.

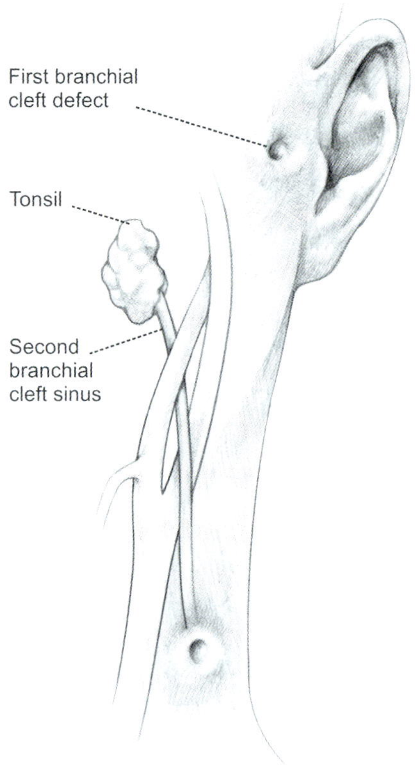

Figure 1.33 First branchial cleft defect location and second branchial cleft sinus course.

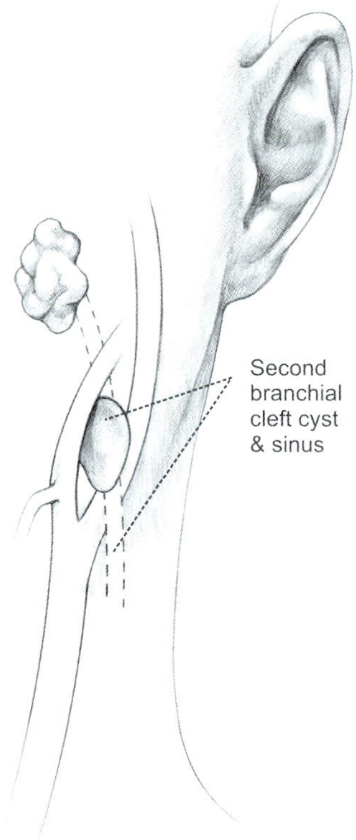

Figure 1.35 Course along which a second branchial cleft cyst may be found.

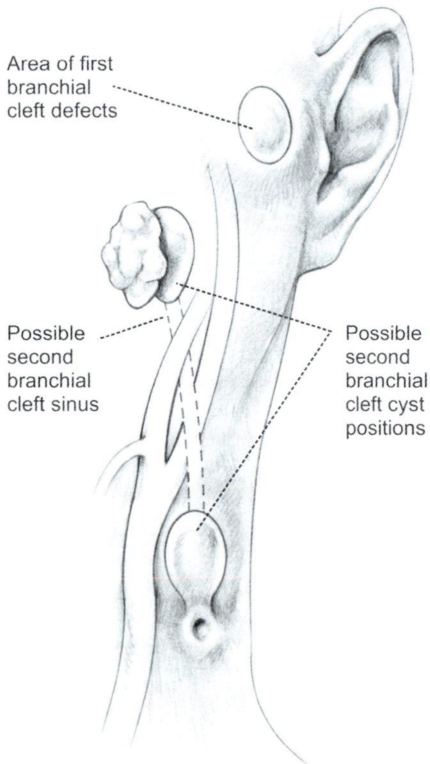

Figure 1.34 First branchial cleft defect location and second branchial cleft cyst locations.

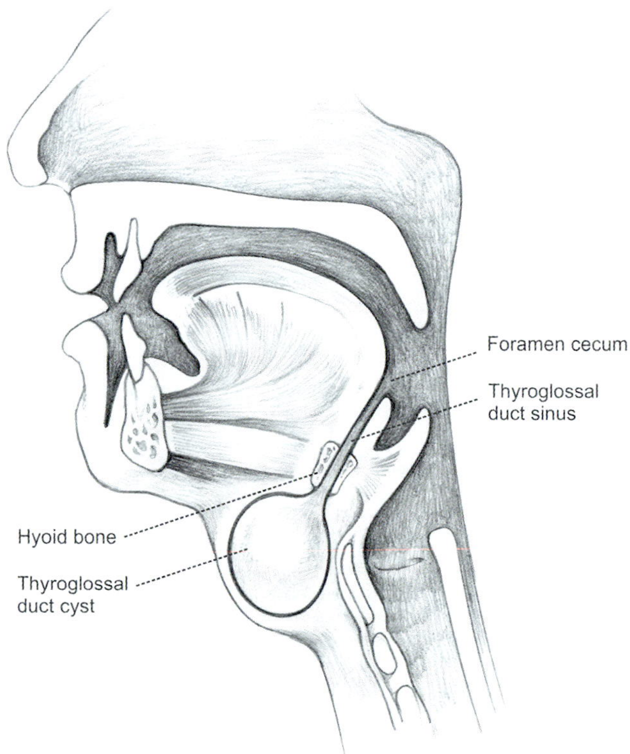

Figure 1.36 Thyroglossal duct cyst and sinus tract to base of the tongue.

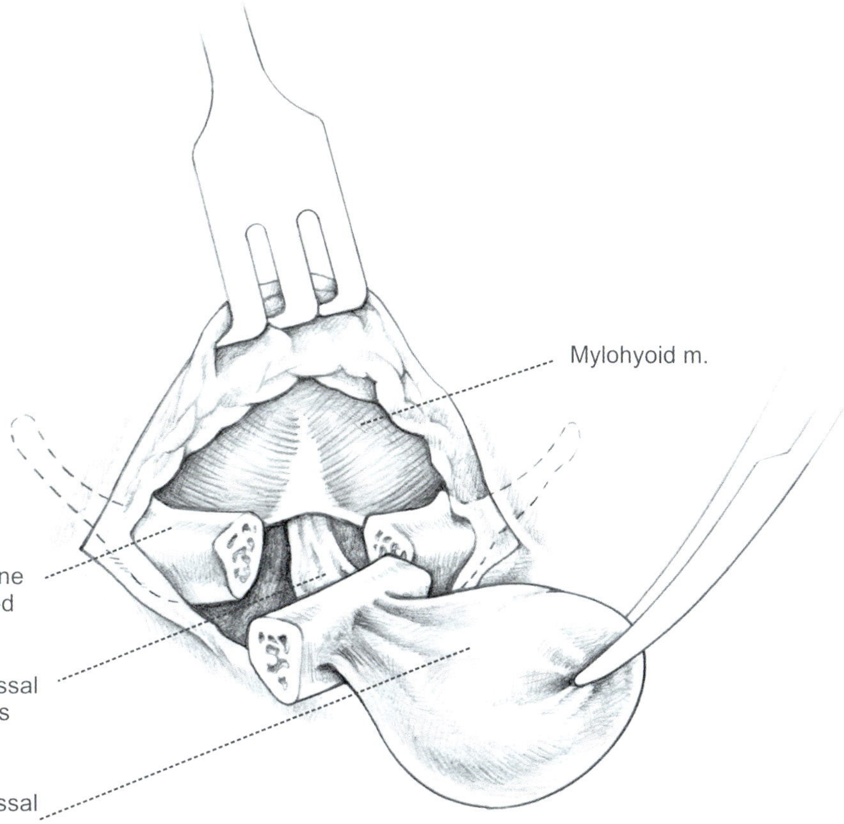

Figure 1.37 Sistrunk procedure for thyroglossal duct cyst and sinus.

DEEP CERVICAL SPACES: NETWORK OF CERVICAL SPACES

OPERATIVE PROCEDURE

The fascia of the carotid sheath has already been entered bilaterally. However, note the contiguous nature of the carotid space to the parapharyngeal space and to the prevertebral space on the accompanying diagram (Fig. 1.38). Using blunt finger dissection, enter the prevertebral space behind the esophagus. Note that inferiorly, the areolar tissue is easily dissected. An expanding abscess may invade the posterior mediastinum as easily as your finger dissected the same space.

The parapharyngeal space is the superior extension of the retroesophageal areolar tissue (Fig. 1.39). Extend your finger bluntly in the superior extension of this space and then deviate laterally into the loose adipose tissue anterior to the carotid sheath. Pharyngeal, tonsillar, and oral abscesses may invade these spaces readily. These spaces completely encircle the pharynx.

The submandibular space is dissected already. Remember that the node-bearing adipose tissue and areolar tissue removed from the submental and submandibular triangles drain inferiorly.

The sublingual space is located in the floor of the mouth superior to the mylohyoid muscle. Using a blunt clamp, penetrate the mylohyoid muscle, and using a muscle spreading technique, open the clamp. In so doing, you have practiced drainage of a sublingual abscess. See the accompanying diagram (Figs. 1.23 and 1.39).

CLINICAL HIGHLIGHTS

- The four spaces just dissected all interconnect with each other. Pathology such as infection and tumor can spread with minimal resistance through these spaces. Retropharyngeal abscess occurs in infants and adults and may compromise the airway.
- Involvement of the adjacent carotid sheath may cause major vessel thrombosis.

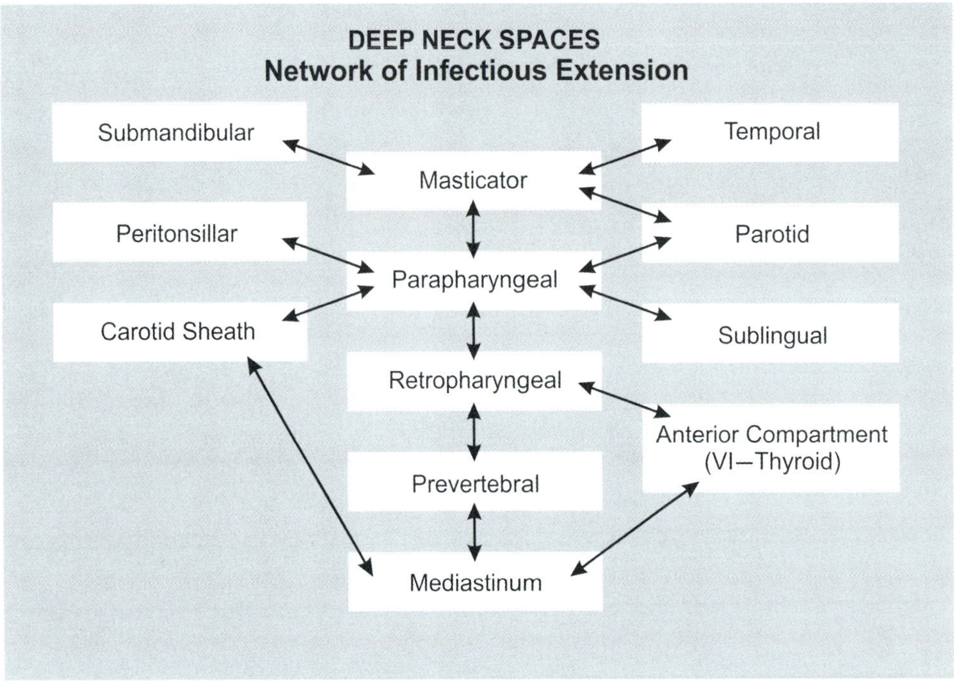

Figure 1.38 Graphic demonstrating communication paths among deep neck spaces.

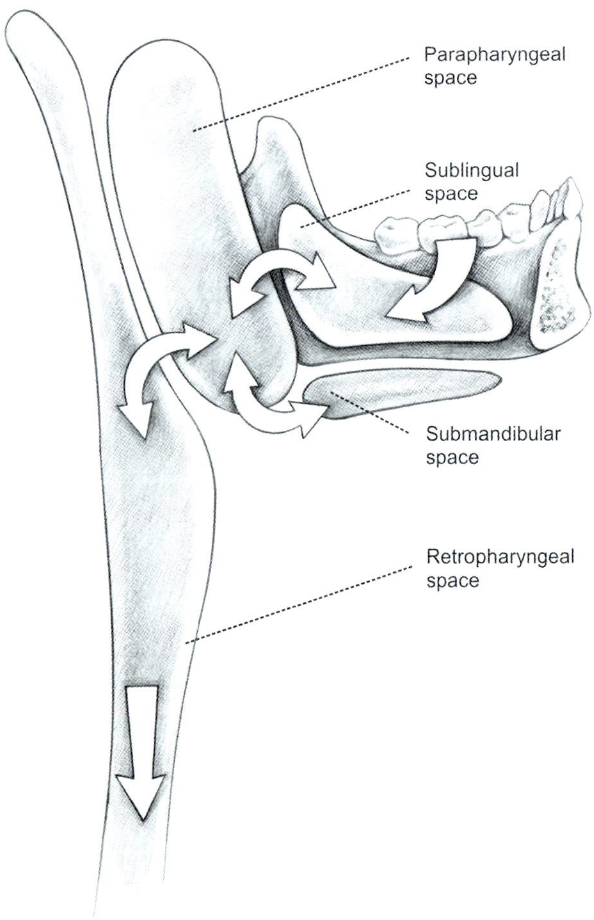

Figure 1.39 Deep neck space connections demonstrated by *arrows*.

- Ludwig's angina is a severe infection of the submandibular and sublingual spaces and is often due to dental disease. Dyspnea with airway compromise may necessitate a tracheostomy. Dysphagia is also possible.

- Superficial cervical abscess secondary to suppurative adenitis frequently occurs in children subsequent to nasal, pharyngeal, or middle ear infection. Incision and drainage of a fluctuant mass is indicated surgical therapy.

CAROTID ARTERY (CAROTID ENDARTERECTOMY)

INCISION

If you have not opened the neck, perform an incision along the entire anterior border of the SCM and expose the carotid artery. If you performed the previous procedures in this chapter, you have opened the carotid sheath bilaterally, so the carotid artery, jugular vein, vagus nerve, and hypoglossal nerve should be apparent (see Fig. 1.31). Note the relationship of the internal and external carotid arteries. Note also that the internal carotid artery has no branches. The first branch of the external carotid artery is the superior thyroid artery and descends in the neck as seen in the illustration (see Fig. 1.32). Using a self-retaining retractor, perform a carotid endarterectomy. The facial vein is an important landmark and should be sacrificed to gain exposure of the carotid bifurcation. Division of muscular branches of the external carotid artery and small branches of the internal jugular vein lying over the internal carotid artery may be sacrificed. This will allow superior mobilization of the hypoglossal nerve and exposure of the internal carotid artery distally. Transect the posterior belly of the digastric if it is still present and continue exposure of the internal carotid artery to the base of the skull (see Fig. 1.23).

The greater auricular nerve is located anterior and inferior to the ear. It should be preserved if possible (see Fig. 1.2). The superior laryngeal nerve branches from the vagus nerve and courses medial to the superior thyroid artery. The external branch of the superior laryngeal nerve is located close to the origin of the external carotid and superior thyroid arteries. Injury to this nerve results in loss of cricothyroid muscle function and causes inability to produce high tones. High tones are important to vocalists.

OPERATIVE PROCEDURE

Endarterectomy. Angled vascular clamps may be used if available. They should be placed in the order indicated (see Fig. 1.32). Utilizing a number 11 scalpel blade, make an incision proximal to the carotid bulb and incise the distal common carotid artery to find the lumen. Extend this incision superiorly into the internal carotid artery using Pott scissors. If shunt material is available, placement of the shunt may now proceed. Establish a plane of dissection with an elevator to separate the atheroma from the underlying adventitia of the distal common carotid artery to a point where the junction of the intima cephalad to the arteriotomy appears normal. This is the “feathered edge.” Continue this dissection superiorly beyond the level of the atherosclerotic plaque (Fig. 1.40). If practice repairing an arteriotomy is desired, 6-0 monofilament suture may be used in a running fashion beginning at the top and bottom of the arteriotomy to meet in the center. An autogenous patch graft using vein from the previous neck dissection is illustrated (Fig. 1.41).

CLINICAL HIGHLIGHTS

Several technical points are important to remember.

- First, if a shunt is placed, damage to the delicate internal carotid endothelium or embolization of debris may adversely affect the outcome of the operation. Realizing ischemic injury to the brain will not occur for 4 minutes without any arterial flow, arteriotomy and shunt placement should be deliberate but not rushed. Occasionally, distal plaque will mandate distal internal carotid endarterectomy first followed by shunt placement in the usual fashion.
- Technical perfection is more important than raw speed.
- During closure, the intraluminal shunt may act as a stent over which the repair is performed.
- Patching of the carotid arteriotomy is recommended routinely to reduce the incidence of restenosis. Either bovine pericardium or Dacron patch material is

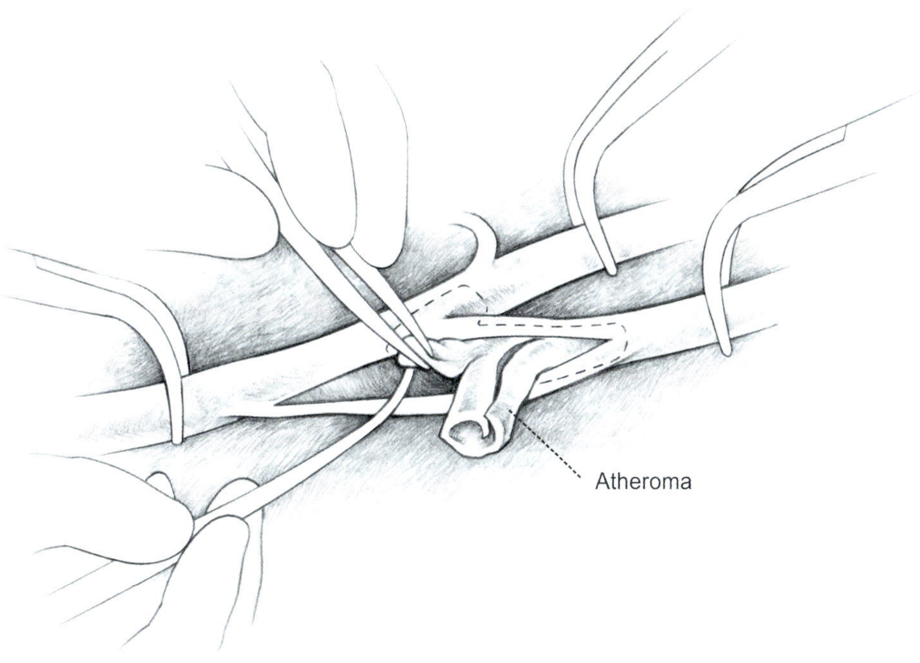

Figure 1.40 Endarterectomy demonstrating plaque separation from adventitia.

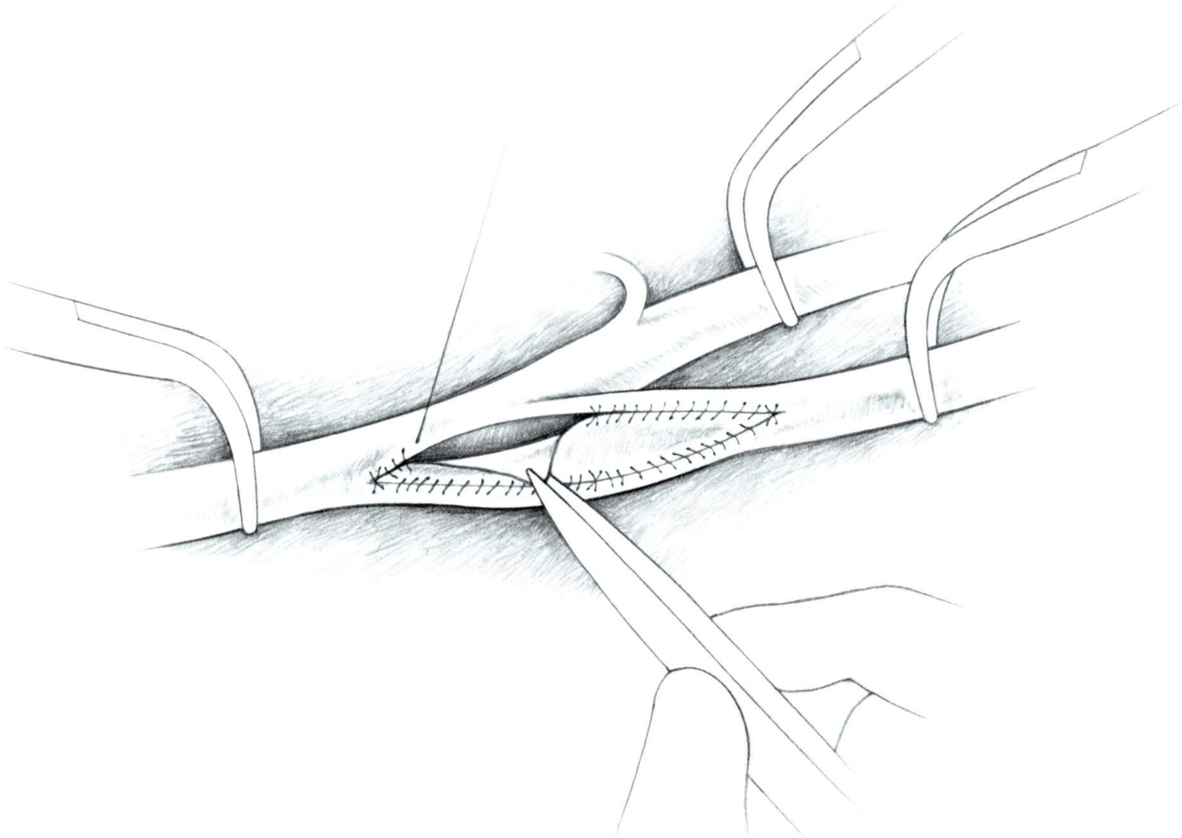

Figure 1.41 Carotid arteriotomy closure with patch.

commonly used. A vein patch may also be used. To practice this, excise a portion of the harvested internal jugular vein and fashion a vein patch. Closure can then be performed as noted in the illustrations (see Fig. 1.41).

- Remember in the clinical situation to forward bleed and back bleed all three arteries and to direct initial blood flow into the external carotid artery for five heartbeats to insure clearance of debris and air.

- Jugular vein: The jugular vein is aberrantly located in 24% of cases. Keep this in mind when performing blind jugular vein catheterization. Therefore, ultrasound-directed jugular vein catheterization technique is recommended.

AXILLARY ARTERY (EXPOSURE FOR AXILLARY ARTERY BYPASS)

INCISION

Transverse incisions parallel to the clavicle may be performed or incise lateral based flaps over the superior pectoralis major muscle (Figs. 1.42 and 1.44).

OPERATIVE PROCEDURE

Identify the fibers of the pectoralis major muscle. Split these fibers 1 cm inferior to the clavicle in line through the entire pectoralis major muscle down to the chest wall. Continue this laterally to the medial edge of the pectoralis minor muscle. The subclavian artery crosses the first rib and becomes the axillary artery. Identify the axillary artery with the axillary vein inferior to it (see Fig. 1.42). Expose the axillary artery from the first rib to the pectoralis minor muscle. This is the first part of the axillary artery. It has two branches, one is the small superior thoracic and the next is the thoracoacromial artery. The second portion of the axillary artery is posterior to the pectoralis minor muscle, and the third portion is the axillary artery beyond the pectoralis minor muscle (Fig. 1.43).

CLINICAL HIGHLIGHTS

- Axilloaxillary artery bypass is done for indications such as proximal subclavian disease resulting in "subclavian steal syndrome." This is rarely done today. Most common is transposition of the subclavian

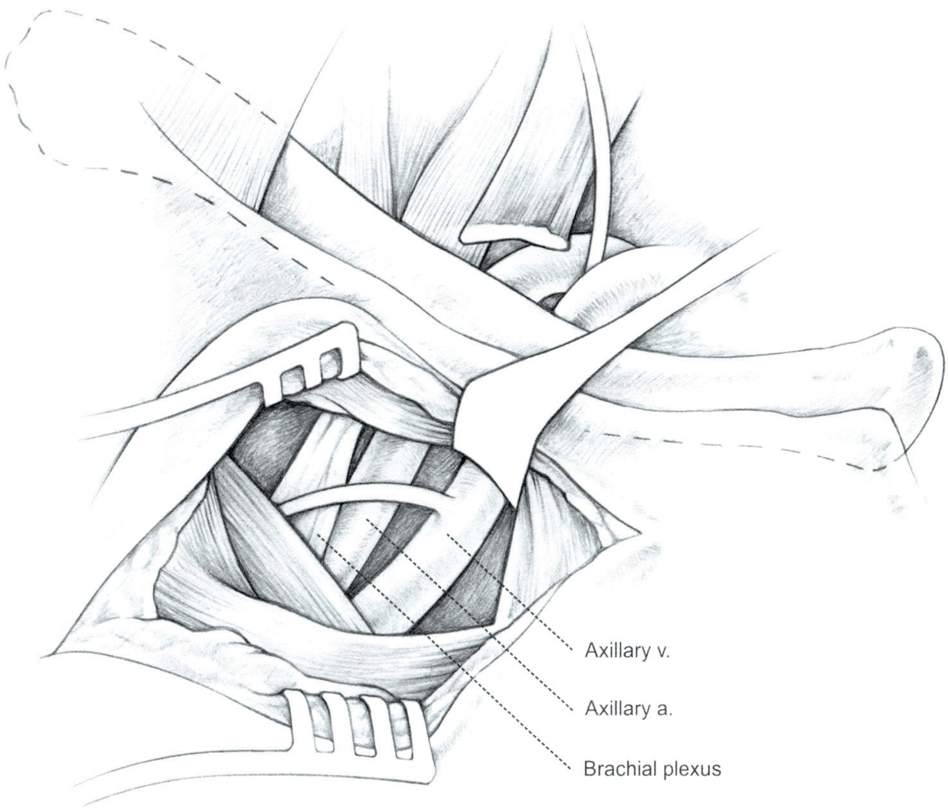

Figure 1.42 Exposure of axillary artery below clavicle.

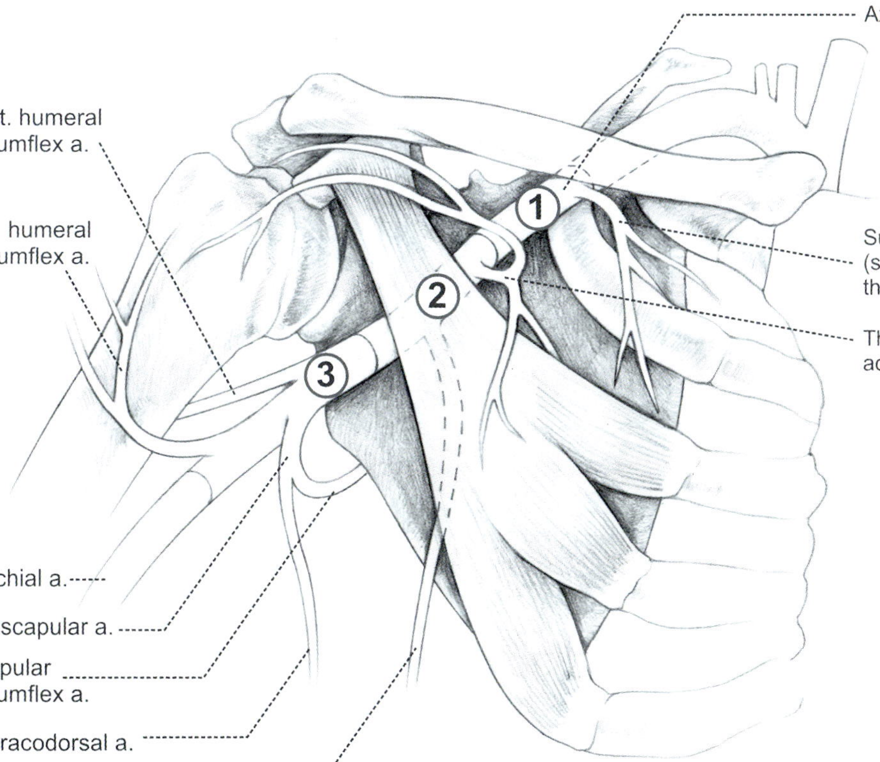

Figure 1.43 Segments and branches of the axillary artery.

artery (prevertebral portion) to the common carotid artery or carotid-subclavian bypass with PTFE (GoreTex) or Dacron.

- Axillobifemoral bypass grafting may be used to improve blood supply to the lower extremities if aortic surgery is contraindicated and for aortoiliac graft infections.

- Access to this region is also useful for treating presence of a seventh cervical rib (cervical rib) leading to subclavian artery aneurysm. Compression of the subclavian artery is caused by firm attachments between the abnormal seventh cervical rib and the first (thoracic) rib causing compression of the subclavian artery and brachial plexus.

BASE OF THE NECK: SUBCLAVIAN ARTERY AND VEIN

INCISION

Reflect all previously made flaps laterally to widely expose the base of the neck.

OPERATIVE PROCEDURE

Excise the clavicles bilaterally. This is done by blunt dissection of the subclavius muscle and soft tissue attachment. Acromioclavicular ligaments and sternoclavicular ligaments can be sharply divided with number 10 blade scalpel. The idea is to gain complete access to the subclavian artery. Upon removal of the clavicle, identify the subclavian vein. Note that it is inferior and anterior to the artery. Replace the clavicle and use a clean needle and syringe to make a subclavian stick. Please note the horizontal alignment needed by the needle and syringe to gain access to the subclavian vein. Aim for the suprasternal notch.

Deep cervical fascia overlies the anterior scalene muscle and continues to the first rib. The phrenic nerve lies under the fascia and on top of the muscle. Divide the fascia and the underlying anterior scalene muscle leaving the phrenic nerve intact. Note the position of the phrenic nerve as it crosses the anterior scalene muscle (Fig. 1.44). Do not divide the transverse cervical or subscapular arteries that lie on the deep cervical fascia as you might during radical neck dissection.

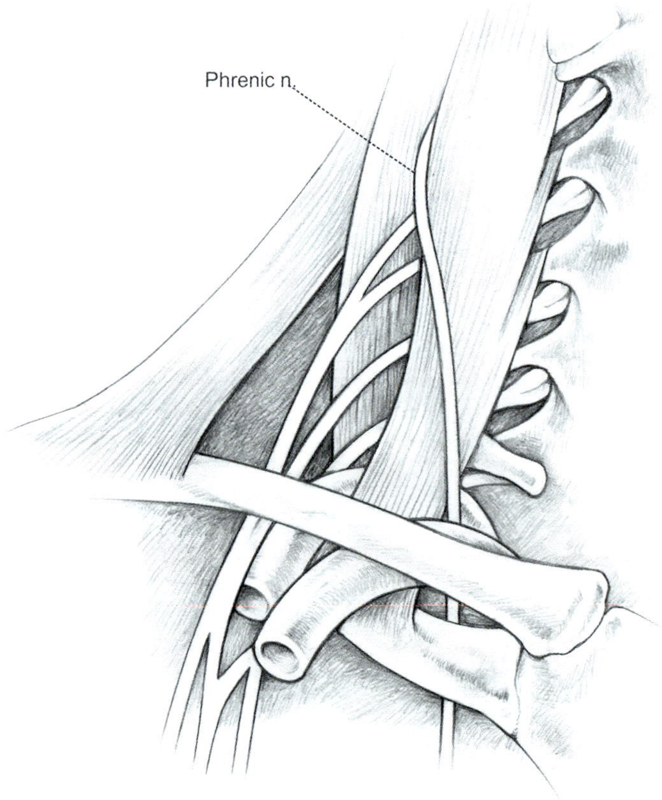

Figure 1.44 Position of the phrenic nerve as it crosses the anterior scalene muscle.

Note the close proximity of the left common carotid and left subclavian arteries. See how a short conduit may be used as a bypass to relieve the subclavian steal syndrome.

Dissect the left and right jugular vein. On the left side, identify the entrance of the thoracic duct at the junction of the jugular vein and subclavian vein. This duct will be a thin wispy collection of lymphatics joining the venous system from a posterior approach (Fig. 1.45). On the right side in a similar position, you may be able to identify the even wispier anteriorly located right lymphatic duct.

After division of the anterior scalene muscle, the entire subclavian artery is exposed (see Fig. 1.47). This dissection may be the only time in your surgical career you ever see the entire subclavian artery. There are three portions of the subclavian artery (Fig. 1.46). The first portion is proximal to the anterior scalene muscle. The second portion is behind the anterior scalene muscle, and the third portion is distal to the anterior scalene muscle. The first portion of the subclavian artery gives rise to three arterial branches: the vertebral, internal thoracic (IMA), and thyrocervical trunk in that order. The second portion gives rise to the singular costocervical trunk. The third portion of the subclavian artery usually gives rise to an artery that pierces through the branchial plexus. That artery is the dorsal scapular (50%), the transverse cervical (20%), *or* the suprascapular artery (10%) (see Fig. 1.46).

CLINICAL HIGHLIGHTS

- Remember subclavius muscle and fascia need to be penetrated during a subclavian venipuncture. A superior-vertical needle track under the clavicle may be used to start, and once under the clavicle, the medial-horizontal direction may be continued to the subclavian vein. Aim for the suprasternal notch.

- The vertebral arteries ascend in the neck to join one another to form the basilar artery, the only place in the body where two arteries join to become one. They bifurcate into the posterior cerebral arteries

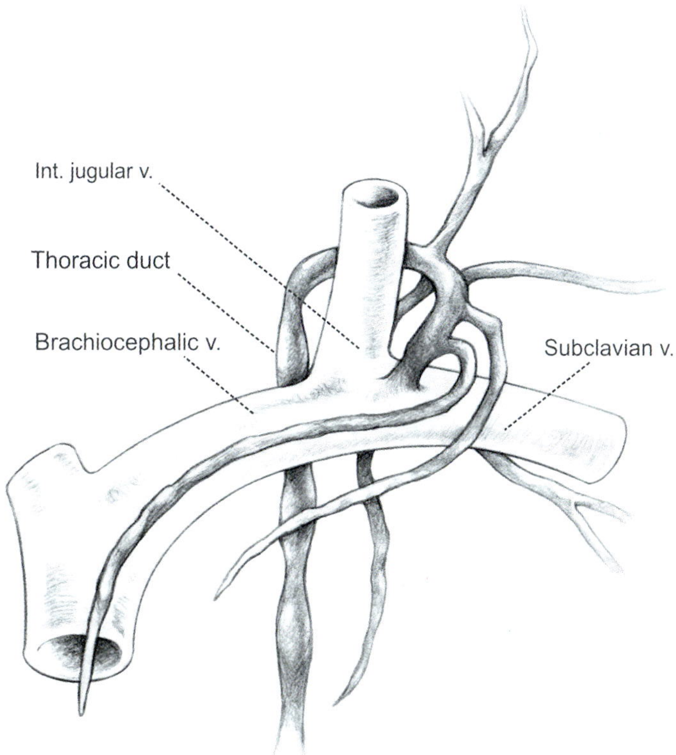

Figure 1.45 Anatomy of the thoracic duct as it drains into the venous system.

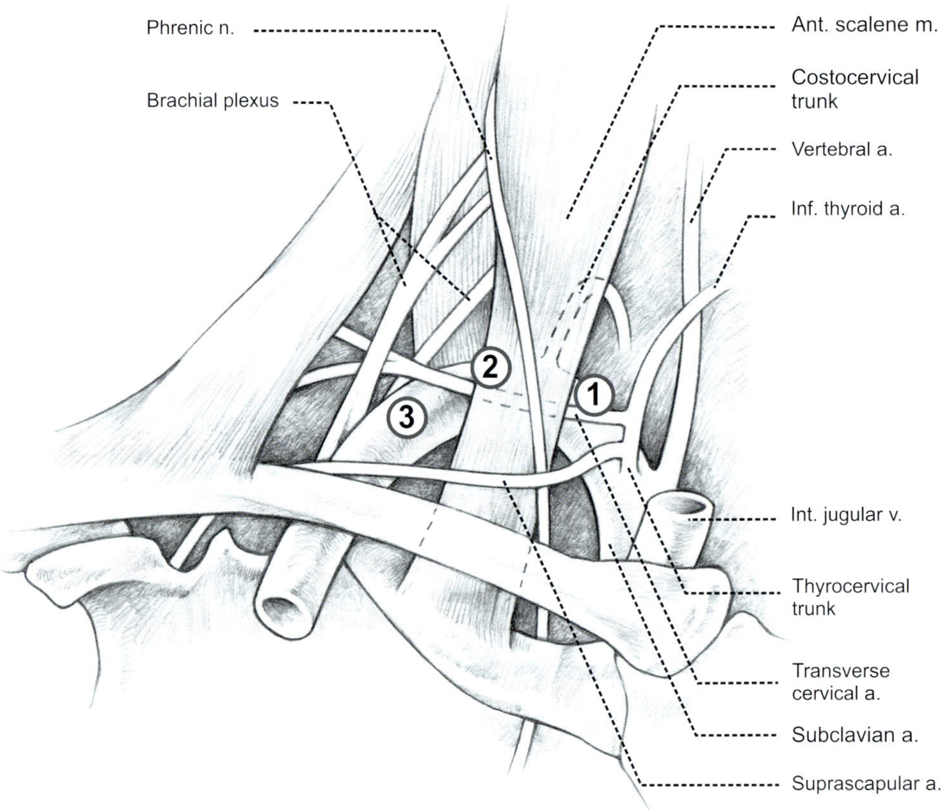

Figure 1.46 Segments and branches of the subclavian artery.

and form the circle of Willis via posterior communicating branches.

- The thyrocervical trunk forms the inferior thyroid artery, transverse cervical artery, and suprascapular artery. This means that the subclavian artery may be ligated in its third portion anywhere distal to the anterior scalene muscle and should allow viability of the upper extremity via the scapular anastomotic network (Fig. 1.47).

- The left common carotid artery and left subclavian artery lie in close proximity. A short bypass is used for subclavian steal syndrome.

- The subclavian artery passes behind the anterior scalene muscle and arches over the first rib, coursing through the scalene triangle (anterior and middle scalene muscles). The subclavian vein does not traverse the scalene triangle, but courses anterior to the insertion of the anterior scalene muscle on the first rib. In thoracic outlet syndrome (TOS), neurologic symptoms are the most common (94%-97%), venous in 2%-3%, and arterial symptoms in 1%-2%.

- To place a subclavian venous line, look at the anteroinferior lay of the subclavian vein relative to the subclavian artery. Place a subclavian venous catheter now under direct vision (see Fig. 1.42).

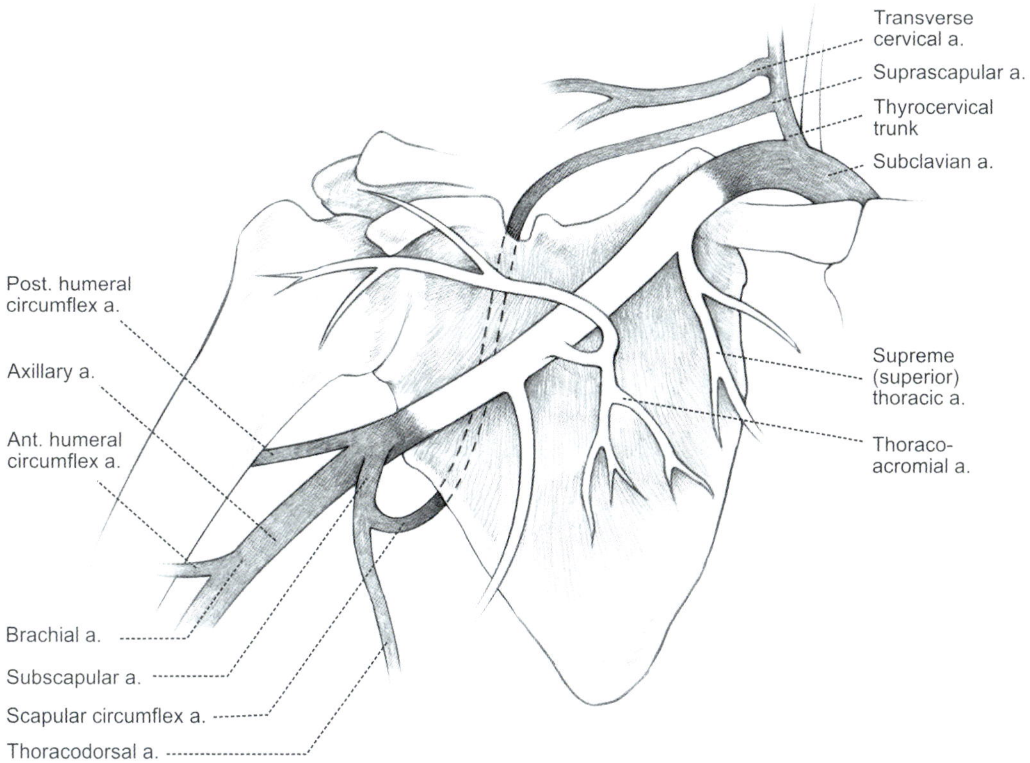

Figure 1.47 Collateral circulation around the shoulder between the subclavian and axillary arteries.

2

Breast and Neurovascular System of the Upper Extremity

Outline

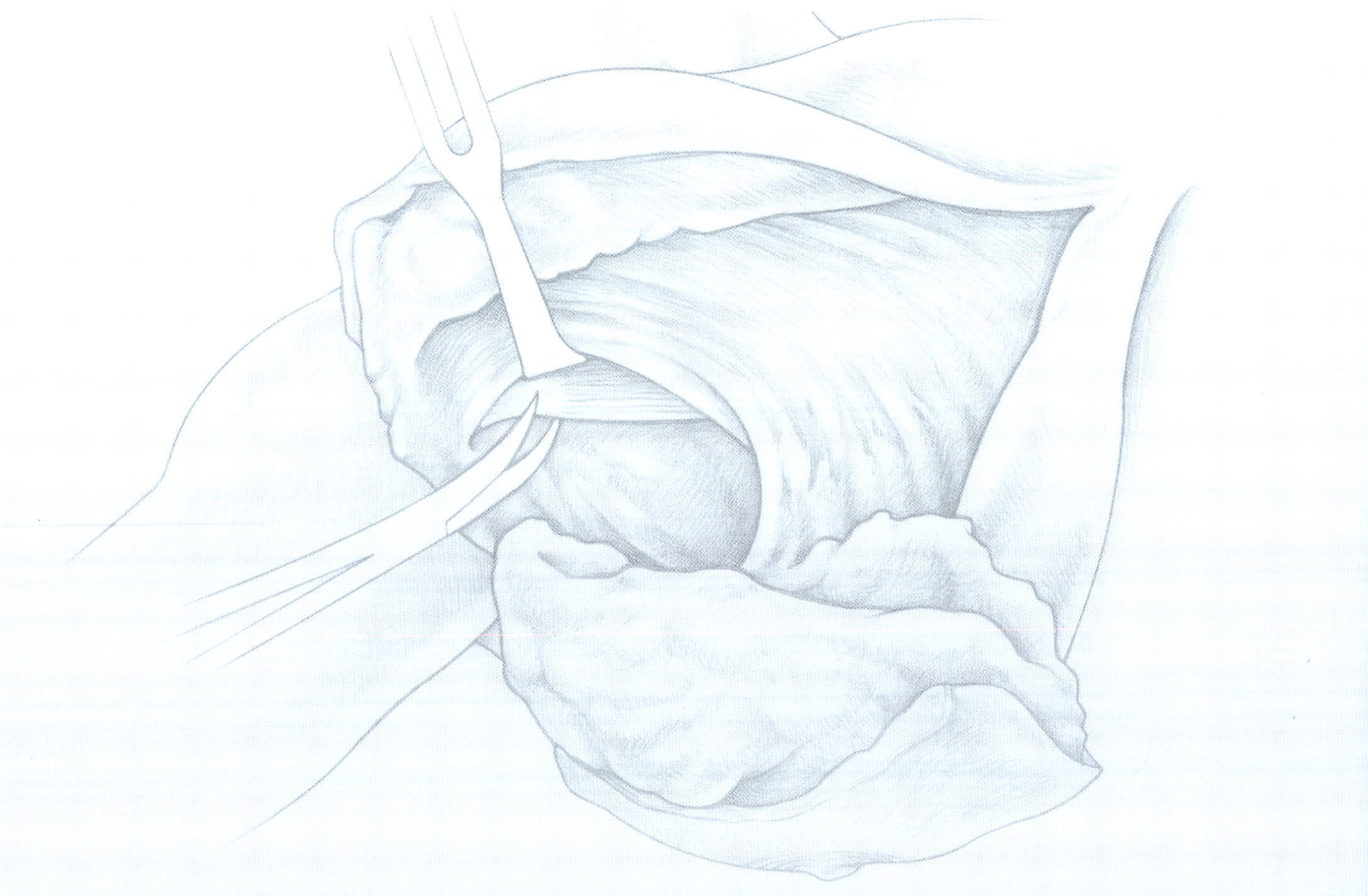

BREAST (MASTECTOMY)

INCISION

Abduct the arm as much as possible and restrain with cheesecloth sling tied to the dissection table (Fig. 2.1). Modified radical mastectomy (mastectomy and levels 1 and 2 axillary lymph node dissection) incision is a horizontally oriented elliptical incision, which includes the nipple-areolar complex and redundant skin. The goal is to remove enough skin to have the healed site lie as smooth and flat as possible without compromising the blood supply to the skin. Skin and nipple-areolar skin–sparing mastectomies are performed for patients undergoing reconstruction. Incise as shown in Figure 2.2.

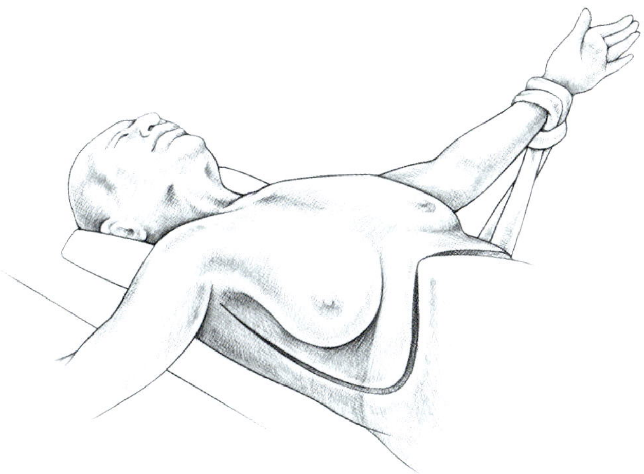

Figure 2.1 Setup for mastectomy and axillary node dissection.

OPERATIVE PROCEDURE

Create thin skin flaps of 0.5 to 1 cm in the plane of the subcutaneous fat-breast tissue interface. You want to preserve the blood supply to the dermal matrix on the skin. Form a superior and an inferior flap. Beginning medially, find the pectoralis fascia, the so-called clavipectoral fascia, and strip the breast along with the fascia from the pectoralis muscle (Fig. 2.3). During this subfascial dissection, approach the breast tissue from the medial aspect and elevate it laterally. This is best performed by blunt and sharp dissection with a knife. One technique is to sweep the blade perpendicular to the muscle.

CLINICAL HIGHLIGHTS

While performing the mastectomy, remember the blood supply to the breast. The major blood supply comes from above (superiorly) (Fig. 2.4).

- The first portion of the subclavian artery gives rise to the internal thoracic artery (internal mammary artery), which gives off transthoracic perforators, second and third perforators generally being the largest. Preserving the branches of the vascular supply that go to the skin are a goal of modern mastectomies.
- The axillary artery gives rise to the following:
  - First portion—superior thoracic artery
  - Second portion—thoracoacromial artery

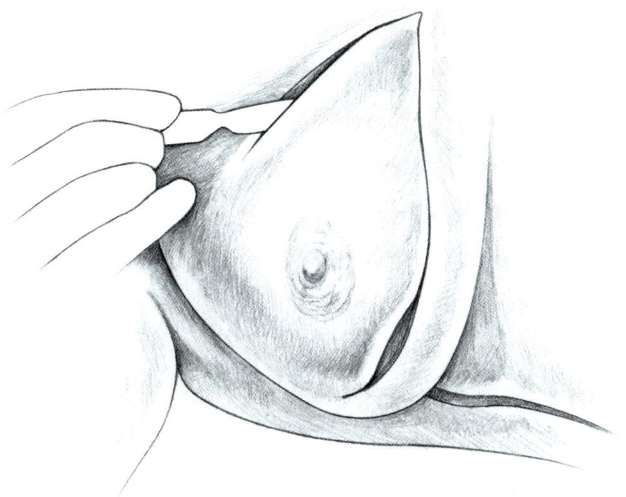

Figure 2.2 Skin incision for modified radical mastectomy.

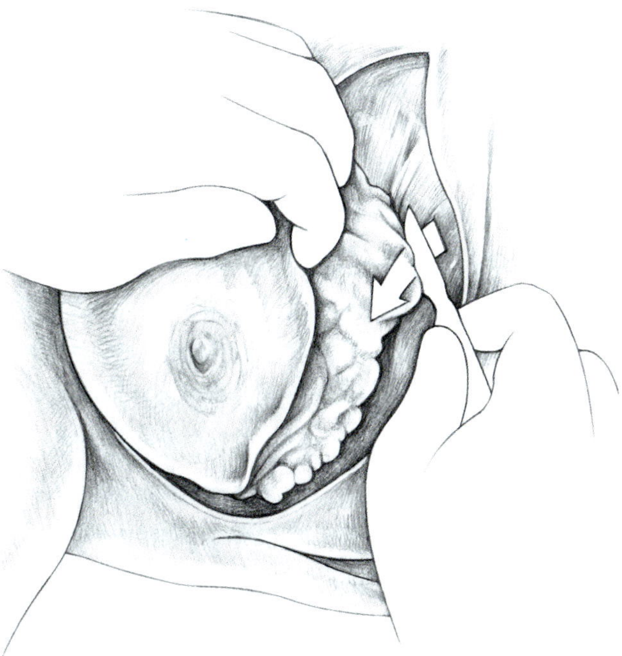

Figure 2.3 Creating skin flaps for mastectomy.

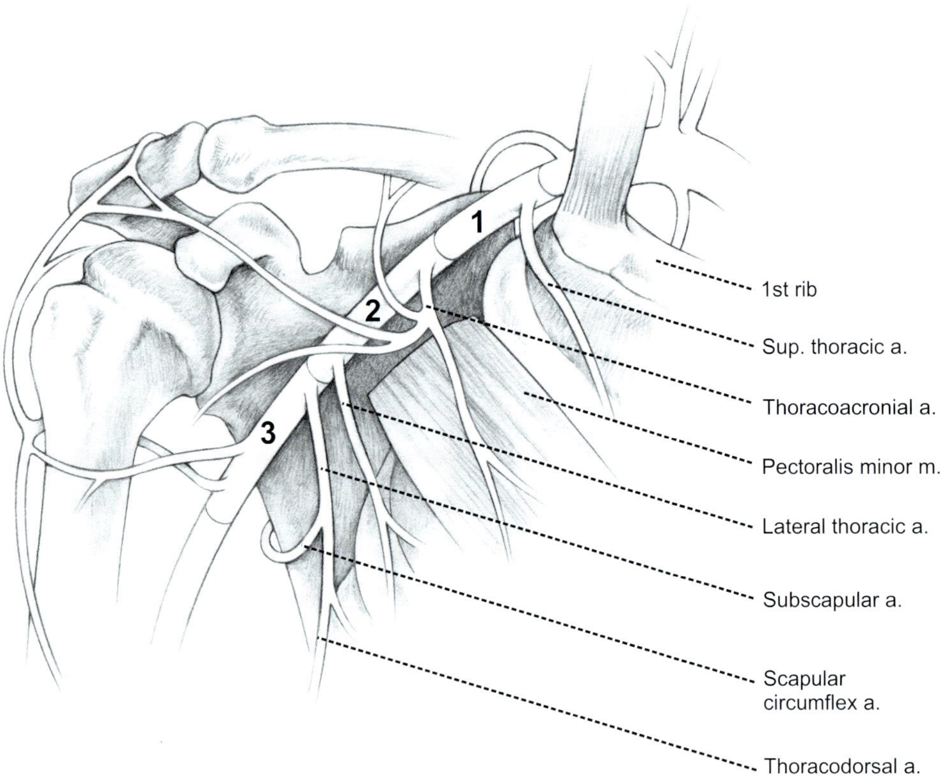

Figure 2.4 Blood supply to the breast and upper extremity.

- Third portion—lateral thoracic artery at the lateral edge of the pectoralis minor sending perforators through and around the edge of the pectoralis major. Subscapular artery completes the scapular anastomosis (see Fig. 1.48)

The minor blood supply of the breast:

- Small branches of the intercostal arteries perforate through to the retromammary space coming through the pectoralis and serratus muscles.

The nipple-areolar blood supply:

- The blood supply to the nipple-areola complex comes from a periareolar plexus supplied by the above-named vessels. Preservation of one of these blood supply sources will prevent necrosis of the nipple after surgical incisions in the area and is very important for nipple-areolar skin–sparing mastectomies. Venous drainage corresponds to arterial supply.

The lymphatic drainage:

- Lymphatic drainage generally follows the arteries. However, 75% of breast lymphatic drainage is to the axilla and the remainder to the internal thoracic nodes. Minor pathway is directly transthoracic to the intercostal lymphatics (Fig. 2.5). Remember that the clinical location of axillary nodes has been referred to levels 1, 2, and 3 from the lowest to the highest on the chest wall (Fig. 2.6). Sentinel node sampling is currently directing more specific node dissection techniques and de-emphasizes the clinical axillary node–level classification.

AXILLA (AXILLARY NODE DISSECTION)

INCISION

Abduct the arm and restrain as previously instructed. Extend the incision from the acromial process of the shoulder down the biceps brachii about halfway to the elbow.

OPERATIVE PROCEDURE

Begin the axillary dissection by peeling the axillary skin and fat inferiorly off biceps and pectoralis major muscles. Clear the tissue down to the lateral edge of pectoralis major muscle. Continue dissection down to the surface of the axillary vein (Fig. 2.7). Typically, you will find a draining tributary vein from the axilla on the inferior margin of the axillary vein. Transect this vein. Move the axillary tissue laterally away from the chest wall to expose the long thoracic nerve on the serratus anterior muscle. Occasionally, it can be palpated as a

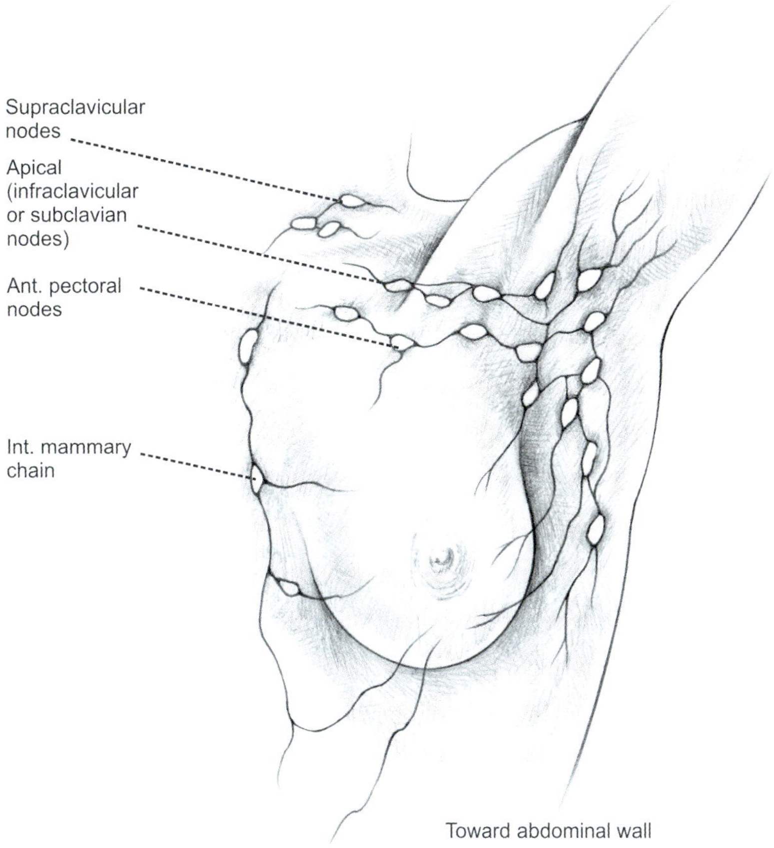

Figure 2.5 Lymphatic drainage of the breast.

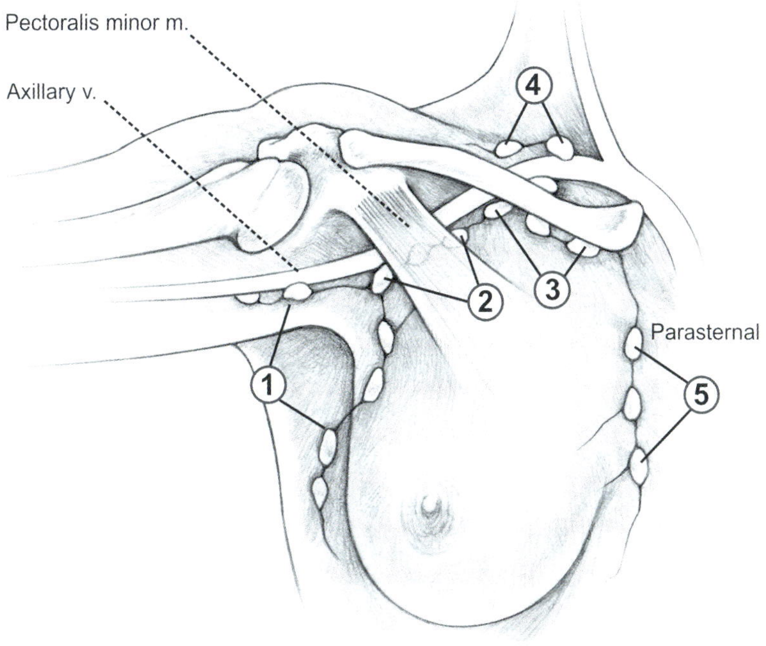

Figure 2.6 Axillary node levels for malignant staging.

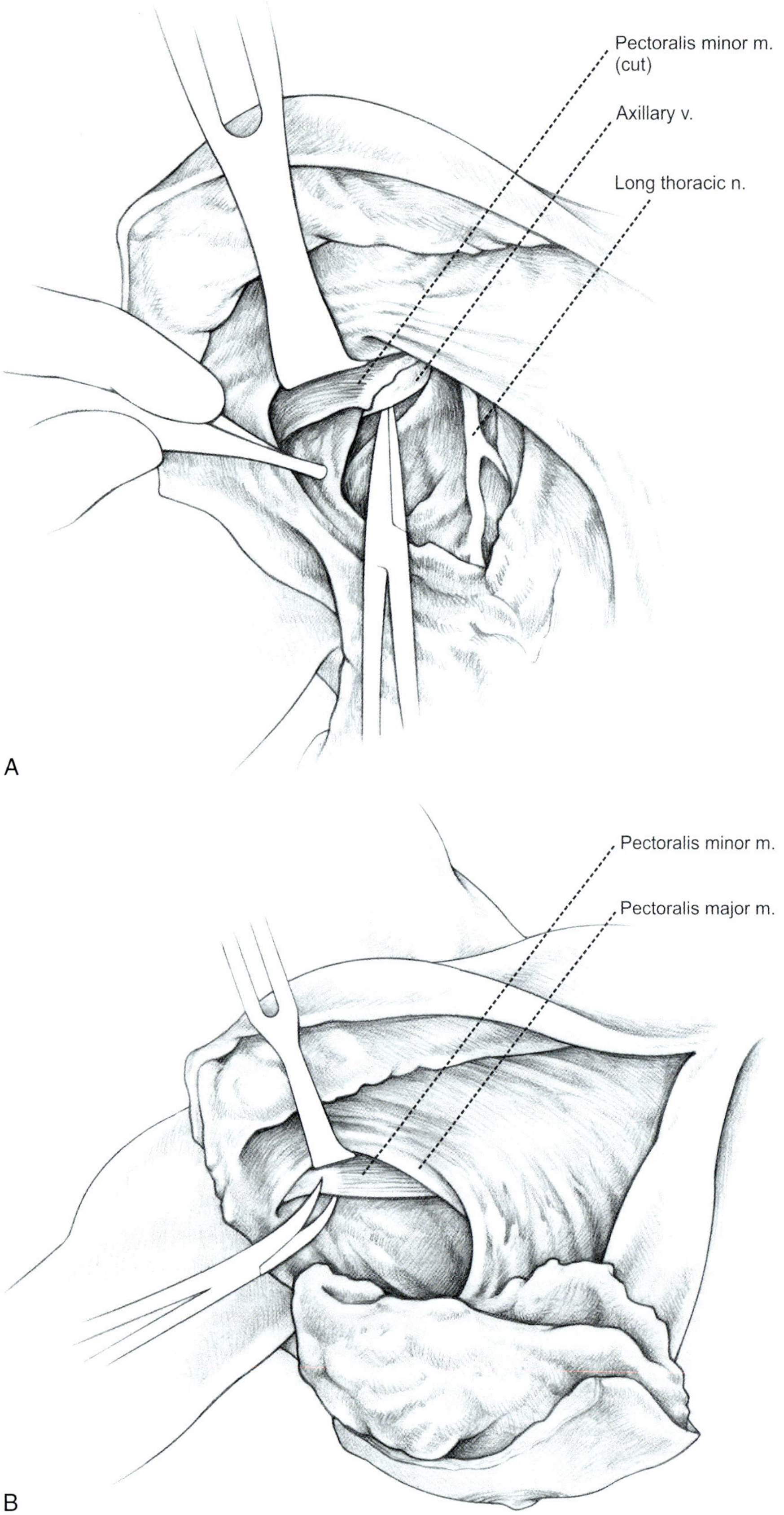

Figure 2.7 **A:** Axillary dissection—aligning procedure on axillary vein. **B:** Axillary dissection—visualize level 2 nodes.

bowstring along the muscle. Posterior and lateral to the transected tributary vein, the thoracodorsal vein, artery, and nerve may be identified (Fig. 2.8). Follow the thoracodorsal nerve down to the medial side of the latissimus dorsi muscle. Next, clear off the inferior and anterior portions of the axillary vein going medially to the chest wall. Intercostal brachial sensory nerves will be encountered crossing the axillary tissue and may be preserved.

To visualize level 2 nodes on a preserved body, you may need to transect the pectoralis minor muscle (Fig. 2.8). Using blunt dissection, clear the medial chest wall posteriorly. With the long thoracic and thoracodorsal nerves identified, you may now clear the axillary contents to complete the axillary dissection (Fig. 2.9).

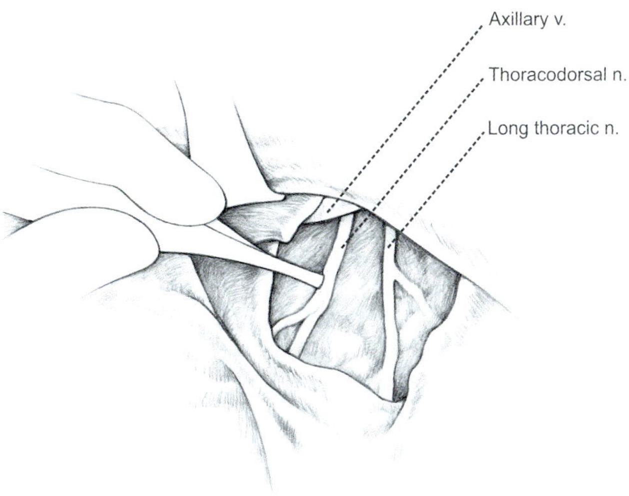

Figure 2.8 Axillary dissection—identify and preserve thoracodorsal nerve and vessels.

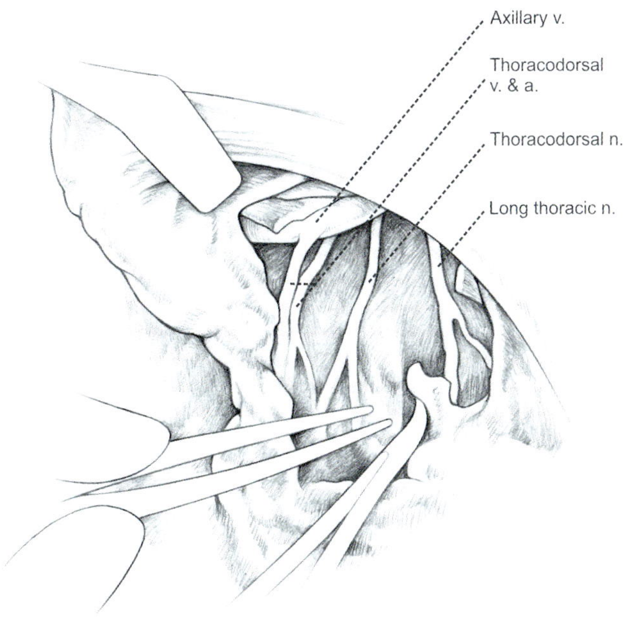

Figure 2.9 Axillary dissection—completing the dissection.

CLINICAL HIGHLIGHTS

- The thoracodorsal blood supply to latissimus is essential to preserve for any breast operation. Latissimus dorsi muscle myocutaneous flaps are commonly used for reconstruction. Thus, all neurovascular supply must be preserved to latissimus dorsi.

- The long thoracic nerve must be preserved to prevent winging of the scapula. This condition is a debilitating upper extremity problem that has very limited clinical rehabilitation. Note that innervation of the long thoracic nerve is C5, 6, and 7. A low cervical spine injury may denervate the serratus anterior muscle.

- During the dissection, preservation of intercostal brachial nerves may be considered. They are cutaneous nerve supply to the skin of the inner arm. This will not compromise survival from breast cancer or completeness of the axillary dissection.

- A standard axillary dissection includes level 1 and level 2 nodes (see Fig. 2.6).
  - Level 1 includes all nodes lateral to pectoralis minor muscle: nodes found below the axillary vein, to the chest wall, and within the axillary fat.
  - Level 2 includes all nodes deep to pectoralis minor muscle; the majority are along the axillary vein. They may also be found between the pectoralis major and minor (Rotter nodes).
  - Level 3 includes all nodes medial to pectoralis minor; the majority are along the axillary vein. There are only a few level 3 nodes, and they are not removed during a standard axillary dissection.
  - Supraclavicular nodes are located on the axillary vein above the clavicle.
  - Internal mammary (internal thoracic) nodes lie along the internal mammary (internal thoracic) vessels. These vessels are found 1 cm lateral to the sternal edge. Internal thoracic nodes are not accessed here because it is very rarely done.

UPPER EXTREMITY (VASCULAR ACCESS)

INCISION

Abduct the arm and restrain as previously instructed. You may need to externally rotate the arm as well. The tissues are stiff but with moderate effort will succumb to rotation. Continue the incision on the brachial artery of the upper arm inferiorly down the upper arm and then down the mid-aspect of the flexor(volar) forearm onto the wrist (Fig. 2.10). Peel back the skin and

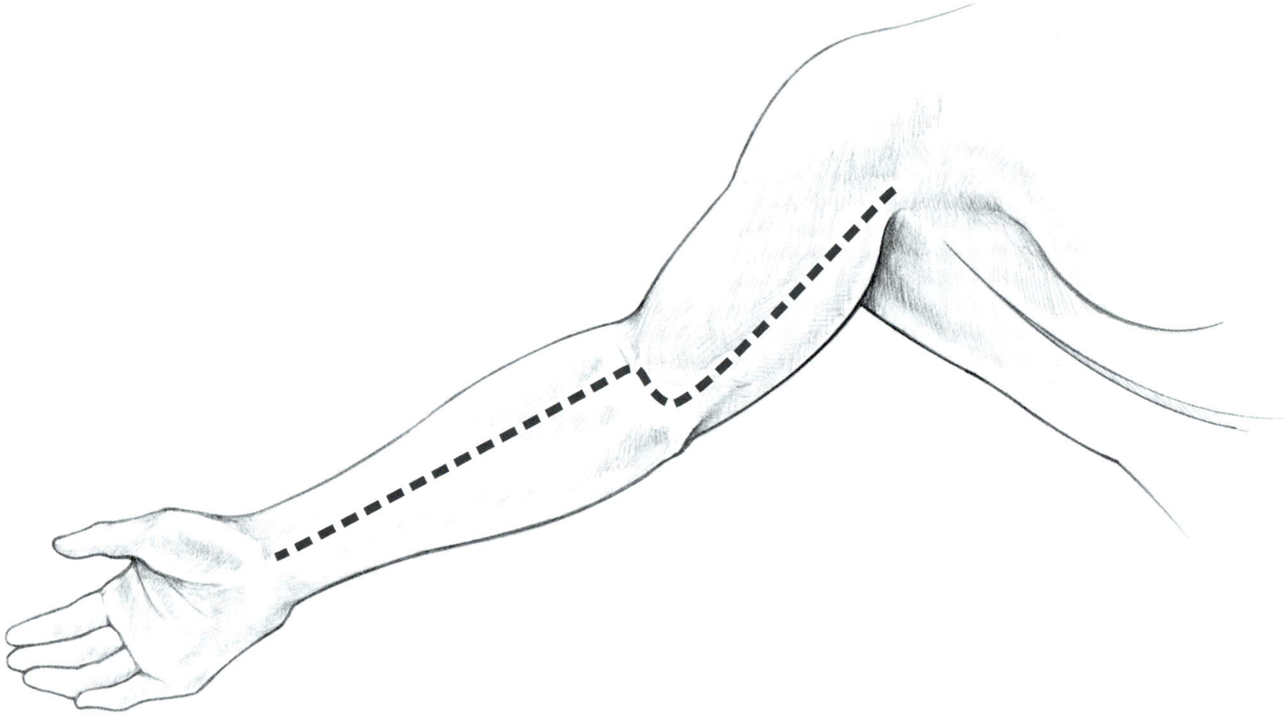

Figure 2.10 Upper extremity incisions for vascular exposure.

subcutaneous tissue down to fascia. Next, turn attention to the axillary sheath and note that the artery, nerves, and veins lie within it. Slit this sheath longitudinally.

OPERATIVE PROCEDURE (VASCULAR EXPOSURE)

Note the three portions of the axillary artery (see Fig. 2.4). The first part of the axillary artery extends from the first rib to the medial portion of the pectoralis minor muscle. The second portion is behind the pectoralis minor muscle. And the third portion of the axillary artery extends beyond the lateral border of the pectoralis minor muscle.

Identify branches of the axillary artery as follows:

- First part—supreme thoracic artery (very small branch)
- Second part—thoracoacromial artery (coming anteriorly) and lateral thoracic artery (descending pattern)
- Third part—subscapular artery (large branch leading to circumflex scapular artery), anterior humeral circumflex (small) and posterior humeral circumflex artery (large)

Continue to dissect the brachial artery inferiorly (Fig. 2.11). Note the radial and ulnar collateral arteries arising from it. Next, direct your attention to the inner aspect of the elbow. Extend the arm as much as possible and open the bicipital aponeurosis (Fig. 2.12). Identify the brachial artery and follow it distally as it divides into the ulnar and radial arteries. The radial artery continues under the brachioradialis muscle (Fig. 2.13).

Open the skin at the wrist and identify the radial artery. The ulnar artery and nerve may be identified with lateral retraction of flexor carpi ulnaris. Finally, identify the carpal tunnel, which is superficial to the median nerve. A carpal tunnel release will be performed in the next section. Once this is open, notice the nerve to opponens pollicis muscle arising from the radial side of the median nerve.

CLINICAL HIGHLIGHTS

- Remember the scapular anastomotic network, which may allow upper extremity viability if the third portion of the subclavian or the first portion of the axillary artery is ligated (see Fig. 1.48).
- Vascular exposure for any upper extremity vessel is possible by using a portion of the incision performed. Note the ready accessibility of the axillary and brachial arteries through medial arm incisions if traumatic injury needs to be rapidly addressed. This access may also be used to treat thromboembolic phenomena.

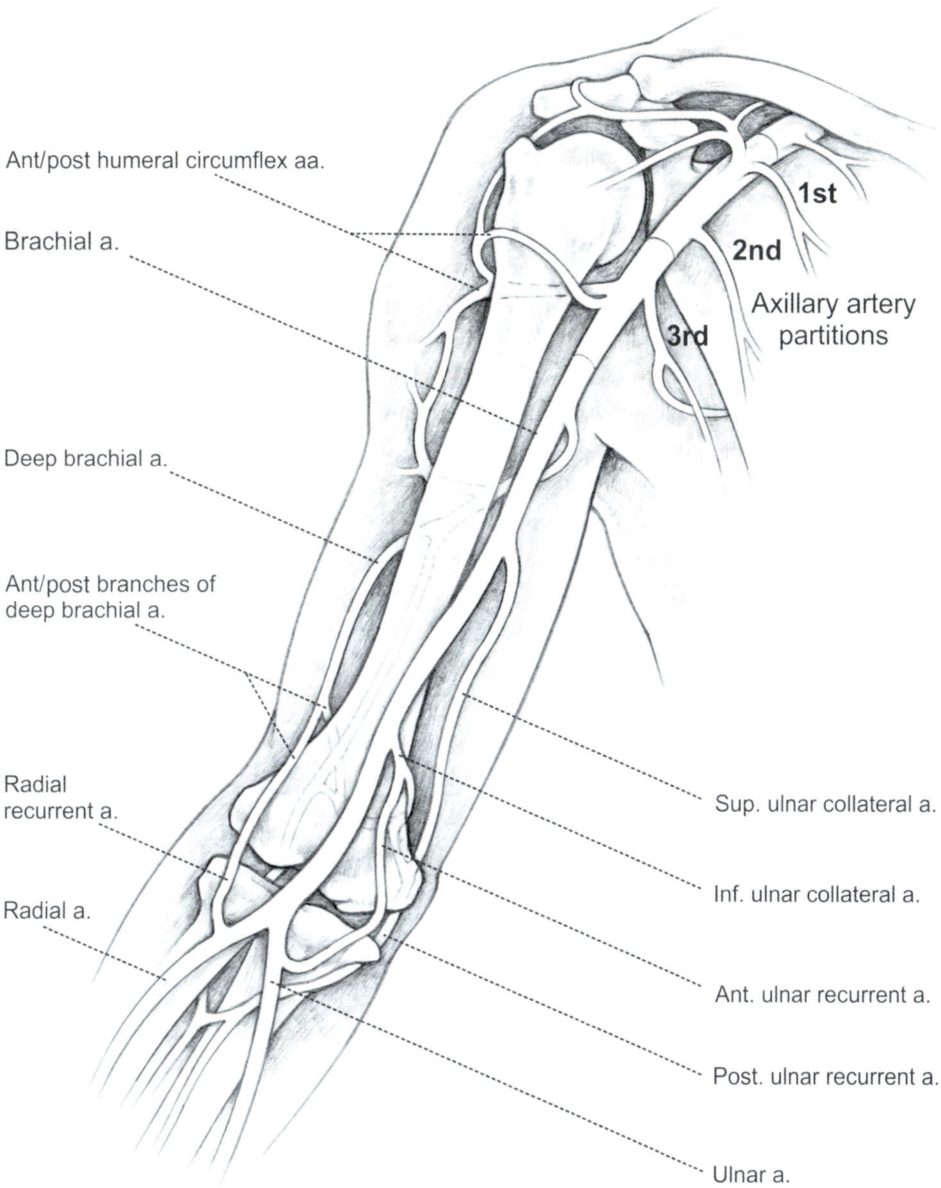

Figure 2.11 Axillary and brachial arteries and collateral circulation around the shoulder and elbow.

- The ulnar collateral arteries descending from above and the ulnar recurrent arteries ascending from below create an anastomotic network about the elbow. Also contributing to this collateral network is profunda brachii artery from above and radial recurrent artery from below (see Fig. 2.11).

- Placement of dialysis access shunts in the forearm requires knowledge of the relationship of the brachial, radial, and ulnar arteries. Dialysis access shunts may be placed with the arterial hood inferiorly on the radial artery or any position up to the brachial artery. Size of the anastomosis is crucial. Note that the ulnar artery should be left undisturbed for this purpose.

- Radial artery access may be necessary for arterial line placement, and this exposure exercise you have just performed familiarizes you with the procedure. Also, Cimino (AV or arteriovenous) fistulas for dialysis access performed at the wrist utilize the radial artery when the Allen test shows a patent ulnar artery and intact palmar arch.

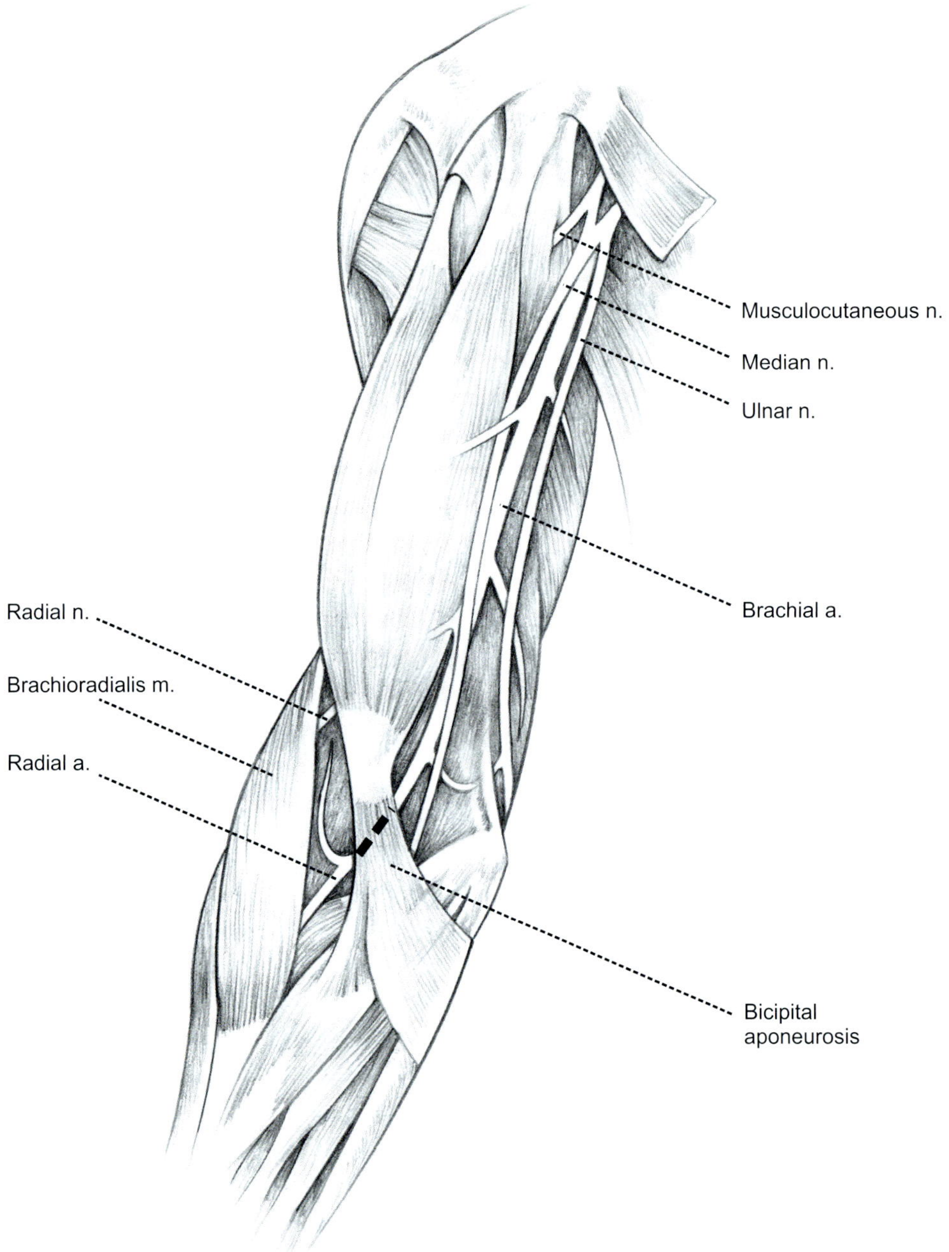

Figure 2.12 Axillary and brachial arteries in situ.

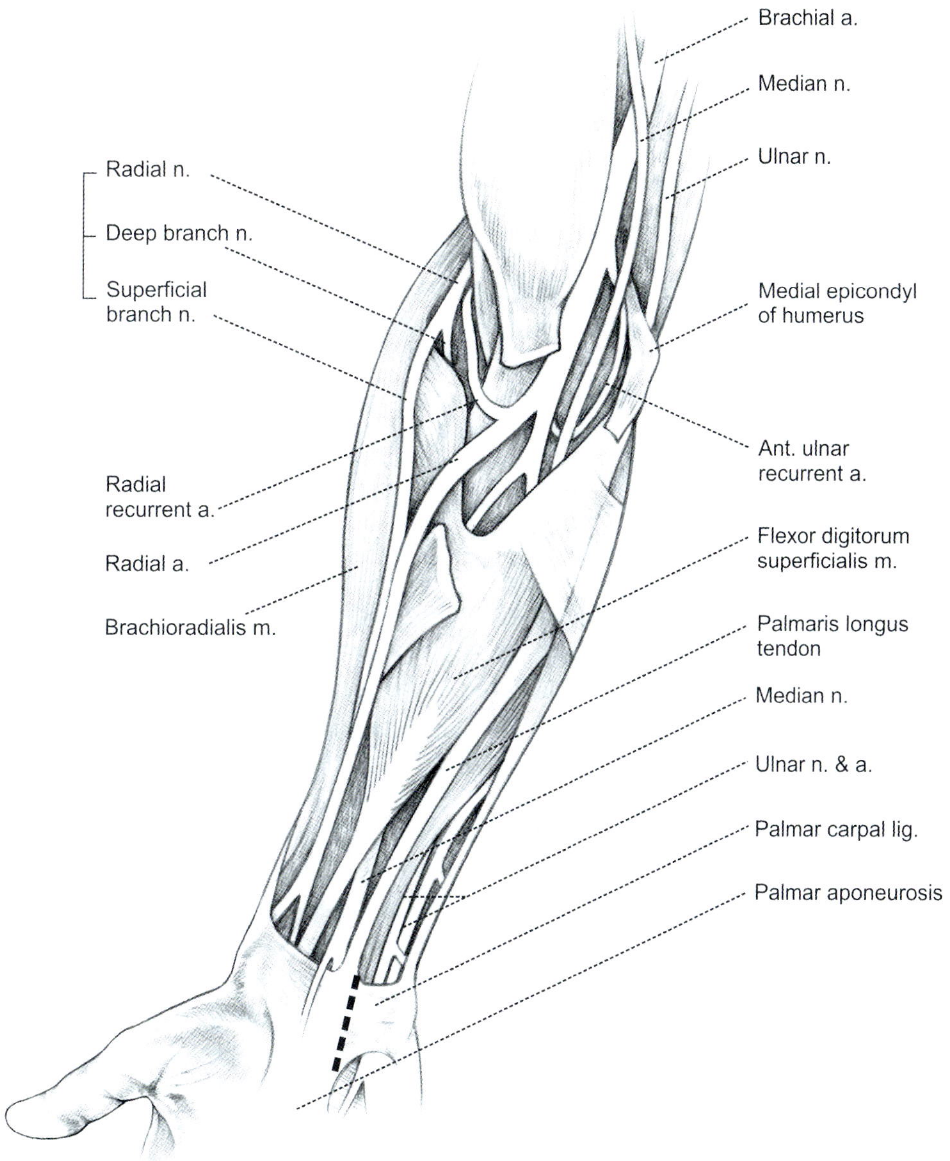

Figure 2.13 Brachial, radial, and ulnar arteries.

WRIST AND HAND (CARPAL TUNNEL RELEASE)

INCISION

The incision in the anatomy lab is done to maximize exposure of relevant structures. Open the wrist in a sigmoid fashion. Begin on the midpalm at the level of the first web space. Continue the incision in the palmar crease, and then continue straight across the palmar wrist. Finish the incision by sweeping to the ulnar side of the palmar wrist (Fig. 2.14). Identify the palmar aponeurosis at the base of the incision. Create skin flaps extending medially and laterally. The palmaris longus tendon should now be identifiable (Fig. 2.15).

The incision is done differently in the OR: a longitudinal incision is aligned in the palmar crease in-line with the radial border of the ring finger. The incision starts just distal to the distal wrist crease and ulnar to the palmaris longus tendon. Complete the distal aspect of the incision at the Kaplan cardinal line.

OPERATIVE PROCEDURE

Incise the palmar aponeurosis longitudinally in-line with its fibers. Retract the medial and lateral aspects of the palmar fascia. Take care along the medial aspect as the ulnar neurovascular bundle lies in close proximity. The radial aspect of the palmaris brevis may overlie the transverse carpal ligament and may require partial elevation medially. Incise the palmar carpal ligament longitudinally along its ulnar aspect (see Fig. 2.15). Adipose tissue is encountered when the transverse ligament release is completed distally. The superficial palmar arterial arch lies within the adipose and must be protected (Fig. 2.16).

Attention is then directed proximally. With a tenotomy scissors, gently dissect superficially and deep to the distal edge of the antebrachial fascia. A groove director, Freer elevator, or similar narrow flat instrument is then placed deep to the antebrachial fascia protecting the median nerve and flexor tendons.

The transverse carpal ligament is then released above the groove director, a short distance across the wrist. If the fascia is thickened and difficult to release, stop. Extend the original incision obliquely across the wrist crease a short distance medially to allow additional visualization and safe release of the fascia. All releases must be performed under direct vision. Blind releases will inevitably result in iatrogenic nerve injury (see Fig. 2.16). With this dissection complete, retract the median nerve and identify the recurrent (motor) branch of the median nerve. The recurrent branch of the median nerve will track radially to the thenar side of the hand (Fig. 2.17).

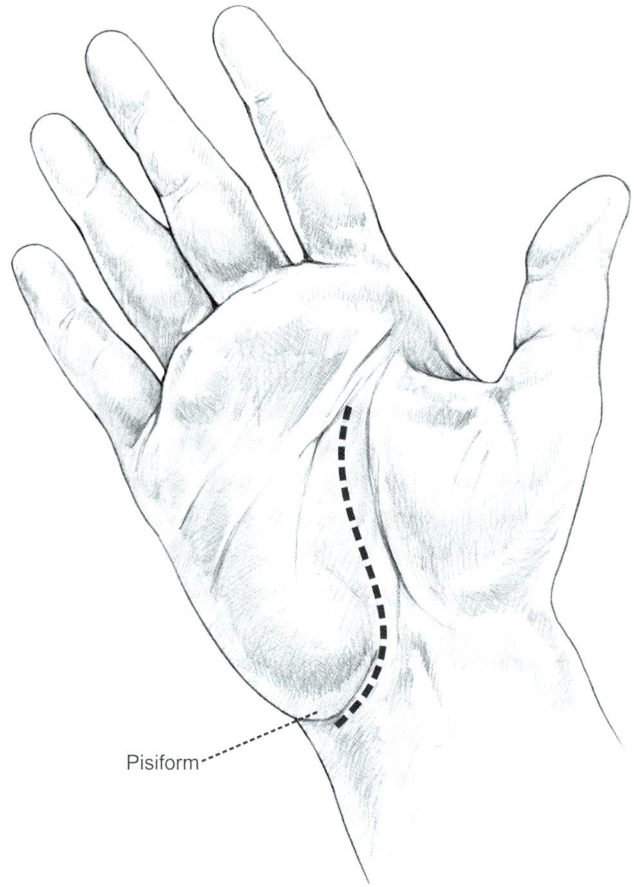

Figure 2.14 Incision for carpal tunnel release.

CLINICAL HIGHLIGHTS

- Remember that the recurrent (motor) branch of the median nerve innervates the flexor pollicis brevis, abductor pollicis brevis, and opponens pollicis.
- The recurrent (motor) branch of the median nerve allows opposition of the thumb, which is crucial to human hand function.
- Remember that the superficial and deep palmar arches are collateral circulation between the radial and ulnar arteries (see Fig. 2.16).

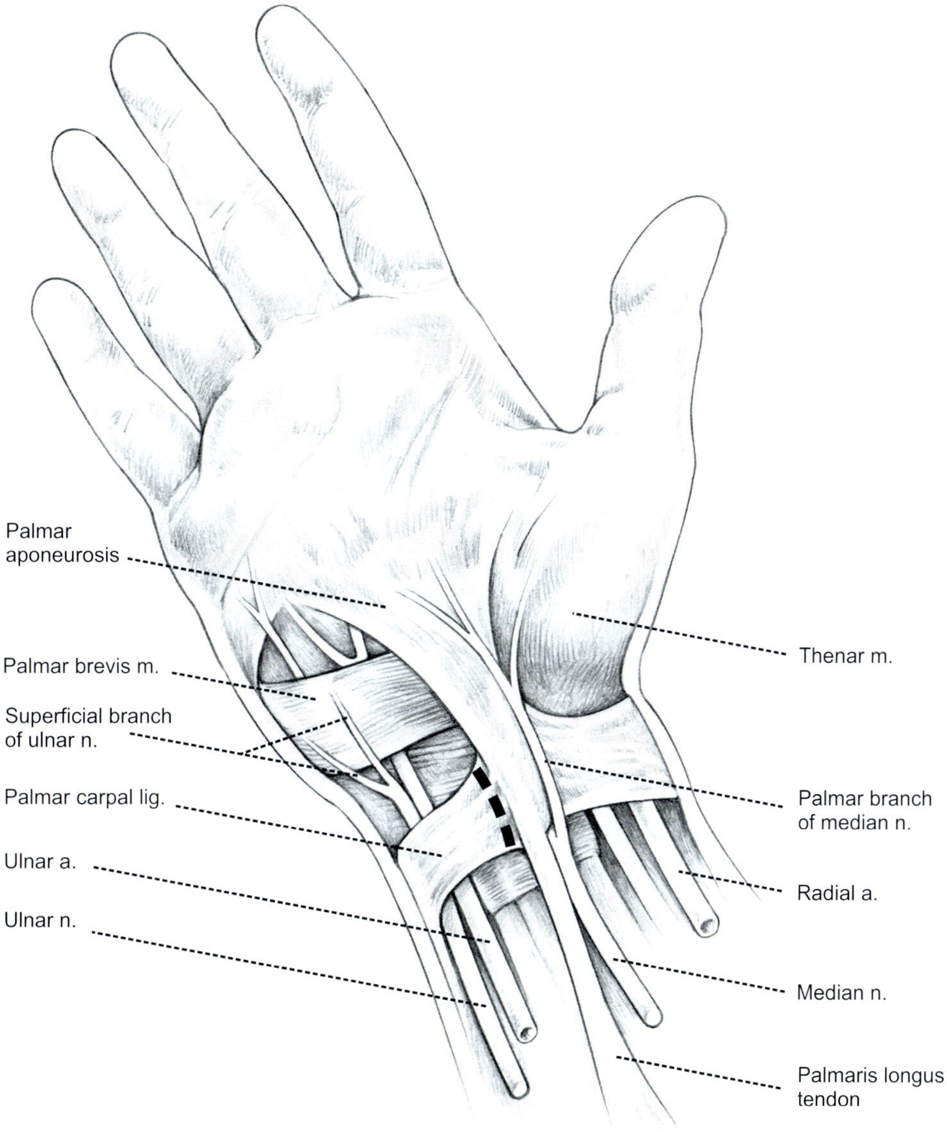

Figure 2.15 Palmar aponeurosis incision.

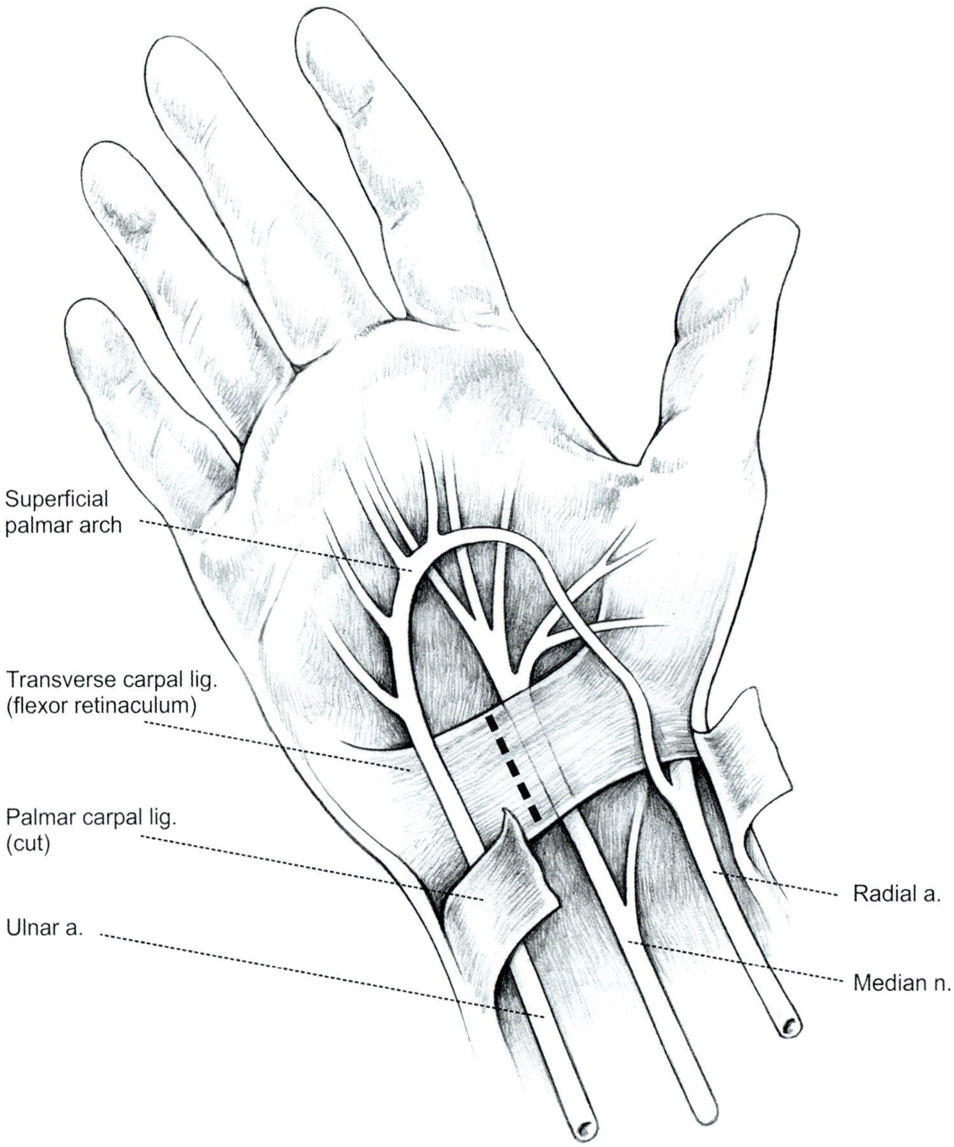

Figure 2.16 Median nerve, and superficial palmar arch, and incision on transverse carpal ligament.

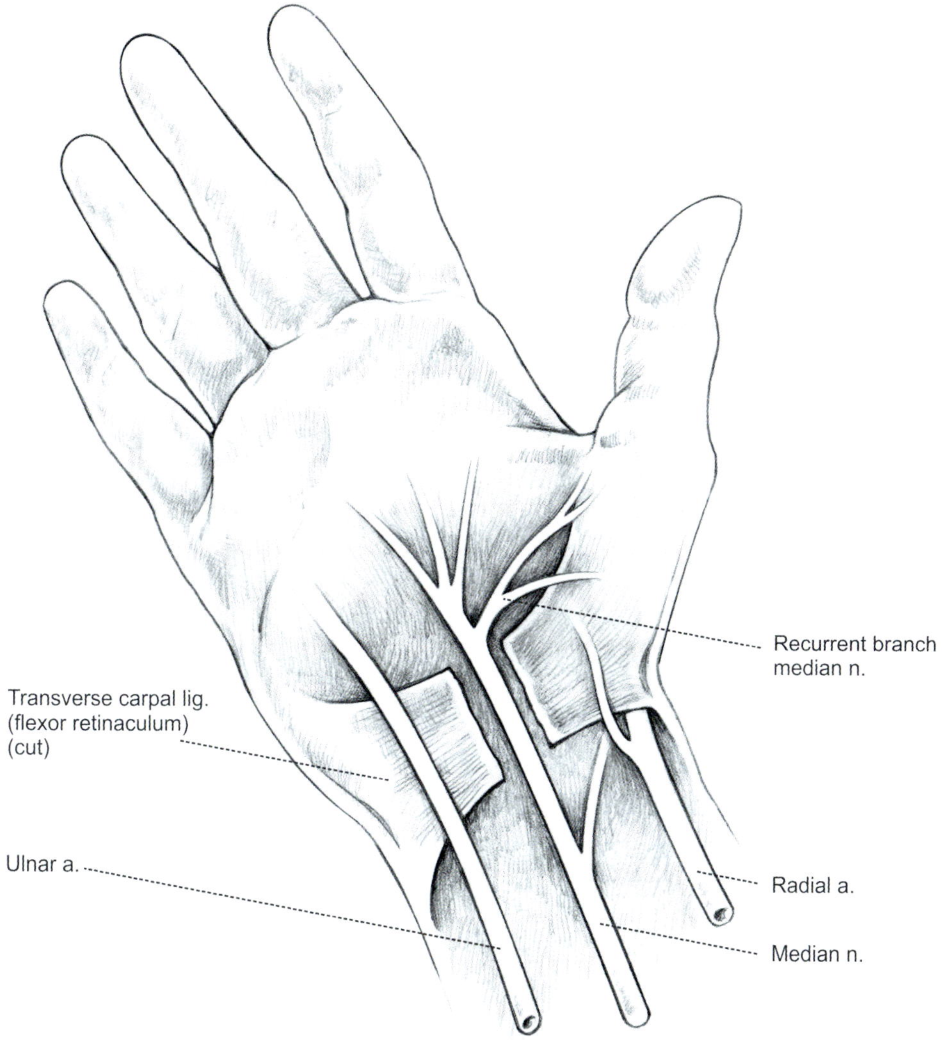

Figure 2.17 Completed carpal tunnel release and location of recurrent median nerve branch.

3

Thorax

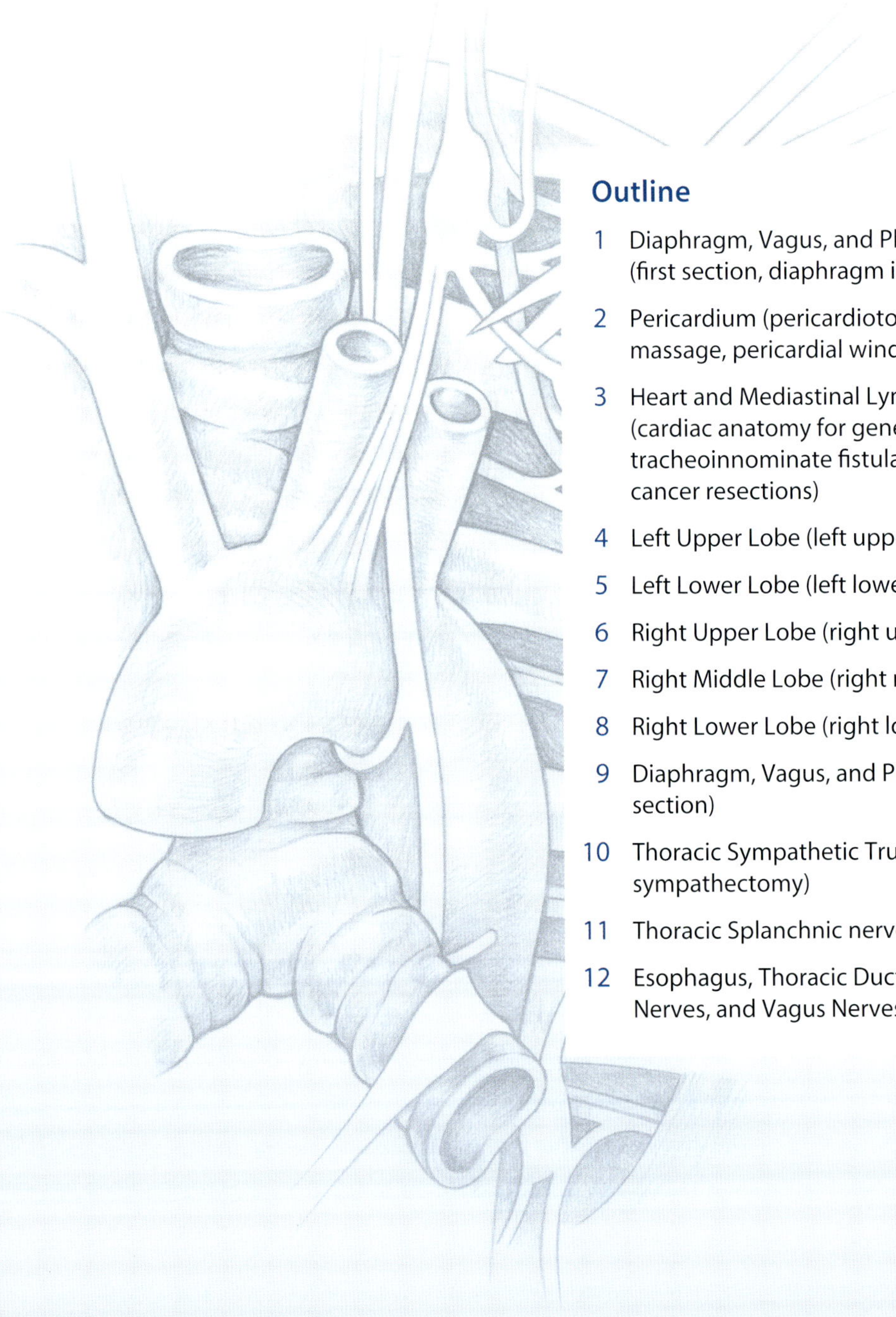

Outline

1 Diaphragm, Vagus, and Phrenic Nerves (first section, diaphragm incisions)

2 Pericardium (pericardiotomy, open cardiac massage, pericardial window)

3 Heart and Mediastinal Lymph Node Stations (cardiac anatomy for general surgeons, tracheoinnominate fistula, nodal stations for cancer resections)

4 Left Upper Lobe (left upper lobectomy)

5 Left Lower Lobe (left lower lobectomy)

6 Right Upper Lobe (right upper lobectomy)

7 Right Middle Lobe (right middle lobectomy)

8 Right Lower Lobe (right lower lobectomy)

9 Diaphragm, Vagus, and Phrenic Nerves (last section)

10 Thoracic Sympathetic Trunk (thoracic sympathectomy)

11 Thoracic Splanchnic nerves (splanchnicectomy)

12 Esophagus, Thoracic Duct, Recurrent Laryngeal Nerves, and Vagus Nerves

DIAPHRAGM, VAGUS, AND PHRENIC NERVES (FIRST SECTION DIAPHRAGM INCISIONS)

PERICARDIAL WINDOW

You may choose to perform a pericardial window to start this chapter. If not, skip to **Incision** (next paragraph). A 6-cm upper midline incision is performed over the xiphoid process. Resection of the xiphoid process will facilitate the procedure. Once the xiphoid has been resected, anterior traction on the sternum will allow dissection of the diaphragmatic surface of the pericardium with electrocautery and blunt dissection. Grasp the pericardium with two Allis or tonsil clamps and, retracting inferiorly, incise the pericardium between the clamps. A rush of pericardial effusion (clear or bloody, depending on the etiology) is expected. Identify an area of the central tendon of the diaphragm, which is contiguous with the pericardium. Using an Allis clamp, grip the diaphragm and make a 2-cm incision to gain access to the pericardial space (Fig. 3.1). A chest tube is then placed through a separate stab incision immediately below the surgical incision and placed into pericardial sac for postoperative drainage. This "pericardial window" is very useful for pericardial effusions such as malignant effusions. A chest tube can be placed through this pericardial window to gain drainage for the effusion causing tamponade. With release of the tamponade fluid, a rather vigorous gush of fluid is generally experienced. Care must be taken not to injure the heart or cardiac vessels.

INCISION

Create a double chevron abdominal incision at the rib margin bilaterally and carry that down to the midaxillary line (Fig. 5.1). At the shoulder area, the skin incision is placed over the pectoralis muscle, which will be divided. The skin incision is carried lower to the midaxillary line. The midaxillary line is then utilized to inferiorly incise the skin down to the abdominal chevron incision. Divide the pectoralis muscle, ribs, and intercostal muscles. Divide the soft tissue attachments of the diaphragm and pericardium on the inner surface of the anterior chest wall. The anterior chest wall can be removed or, as the authors prefer, rotated up using superior tissues as a hinge (Fig. 3.2). The operative procedure below assumes that the neck procedures have been performed. If the neck has not been opened, refer to Chapter 1.

OPERATIVE PROCEDURE

In the left neck, identify the phrenic nerve. It lays on the surface of the anterior scalene muscle (Fig. 1.31). Next, identify the left vagus nerve. Note that the vagus nerve is medial to the phrenic nerve. The phrenic nerves lie on the lateral surface of the mediastinum and

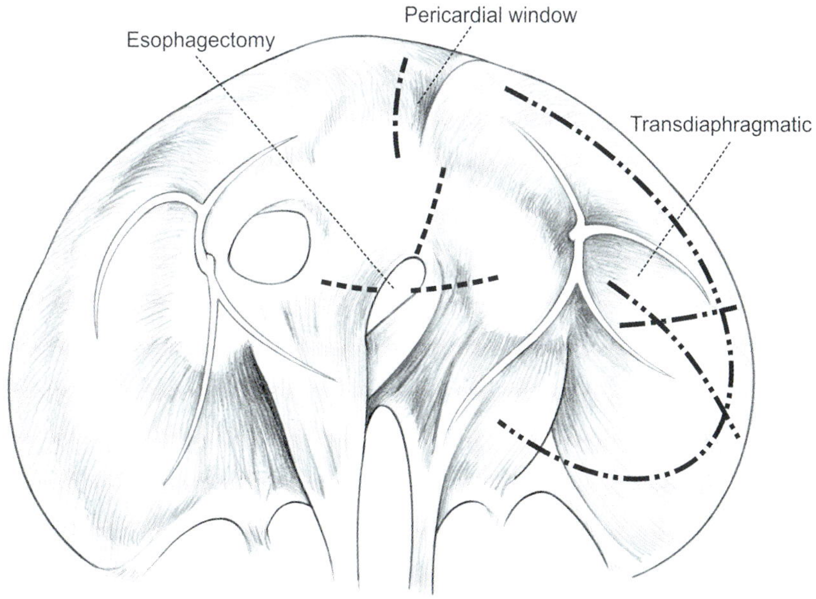

Figure 3.1 Abdominal view of the diaphragm and diaphragmatic incisions relative to phrenic nerve distribution.

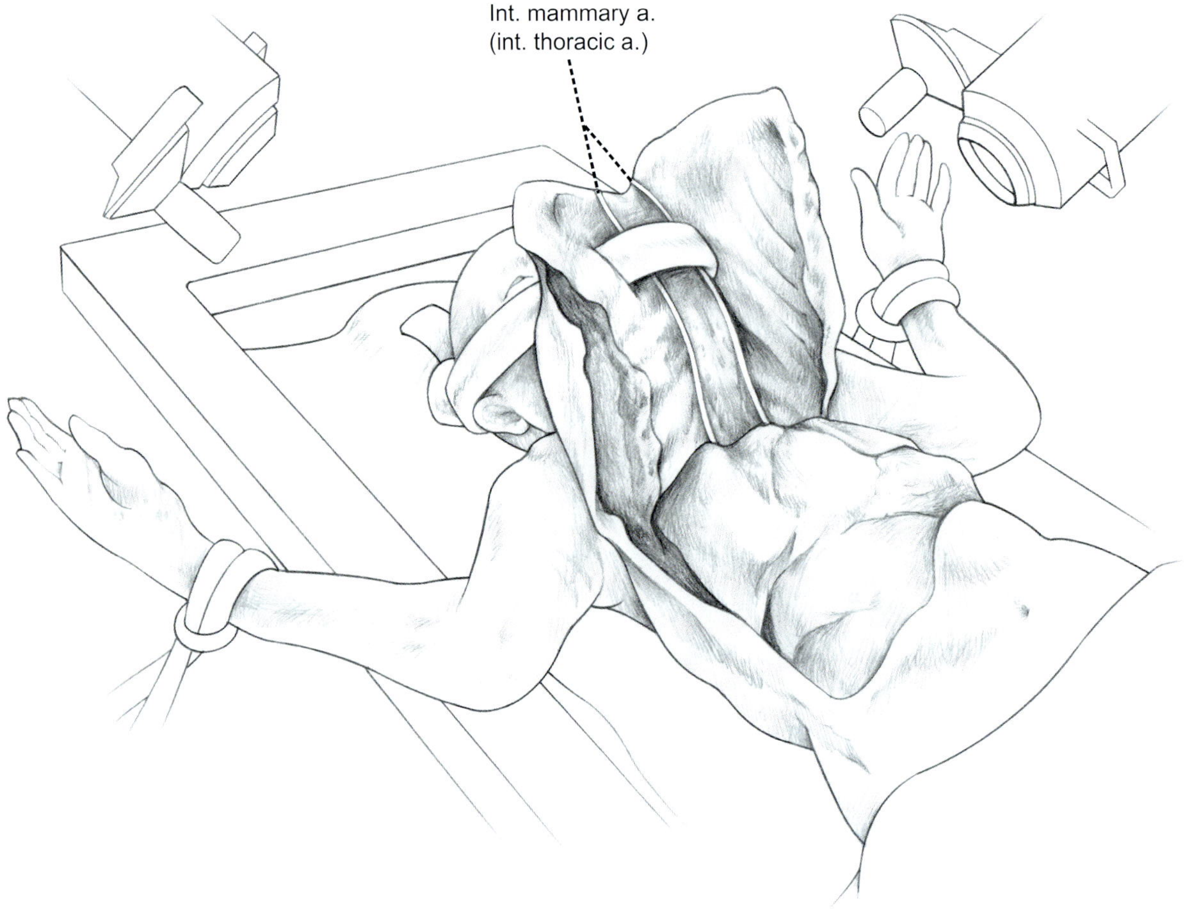

Figure 3.2 Incisions for opening the thorax; note restraints for arms and anterior chest plate.

pericardial sac, anterior to the vagus nerves (Figs. 3.3 and 3.4). The phrenic nerve splays out on the thoracic surface of the diaphragm to innervate the muscle. A major branch pierces the central tendon to enter the abdomen and then splay out in a similar fashion on the abdominal surface of the diaphragm.

On the thoracic side of the diaphragm, note how the phrenic nerve splays out in a radial fashion. To remember this, put both arms down at your sides, hold your hands straight up from your legs at a right angle, and spread your fingers. Think of the phrenic nerve as your arm extending down the side of your body, which is the mediastinum and the phrenic nerve innervation pattern as the fingers and thumb of your hand (Fig. 3.1). Note also that the inferior phrenic artery arising from the aorta on the abdominal side supplies the diaphragmatic muscle and sends a branch ascending through the central tendon alongside the descending branch of the phrenic nerve through the same foramen in the central tendon. Next, identify the vagus nerve adjacent to the carotid artery, which was previously exposed (Figs. 3.3 and 3.4). Follow the vagus nerve inferiorly. Note that it lays medial and inferior to the phrenic nerve in the superior mediastinum. Further dissection will resume after heart and lung sections (Section 9).

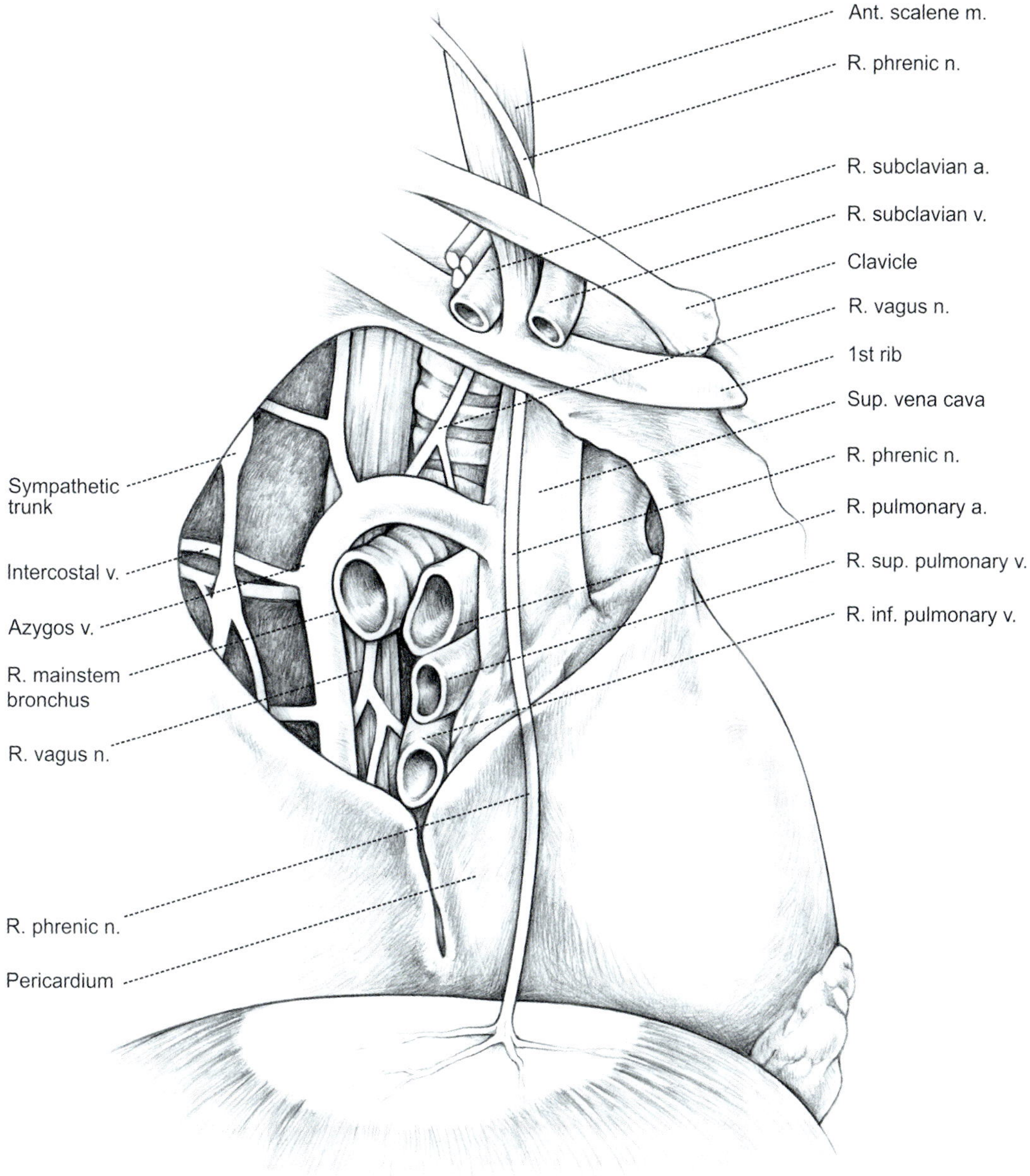

Figure 3.3 Right hemithorax; phrenic, vagus, and sympathetic nerves; azygos vein arching over hilum of the lung.

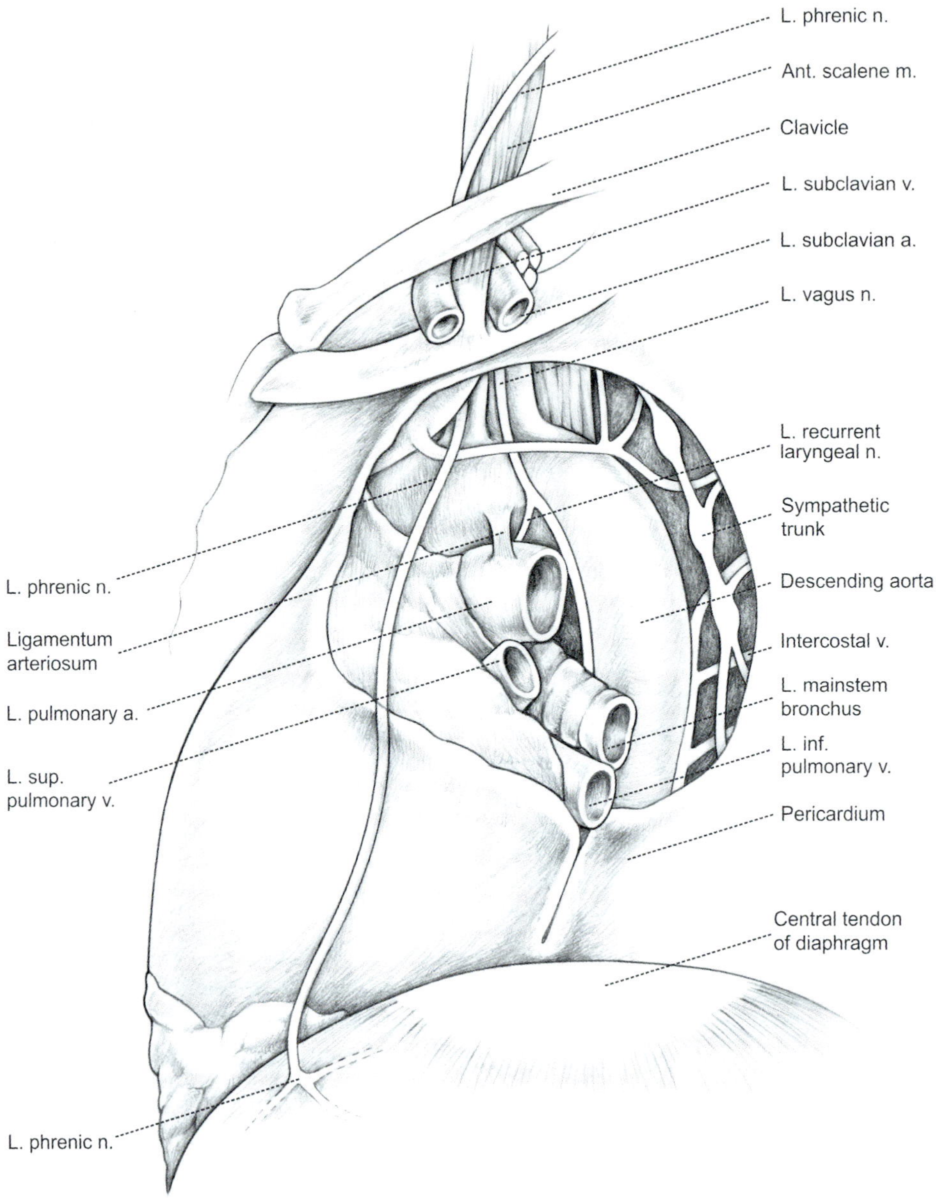

Figure 3.4 Left hemithorax; phrenic, vagus, and sympathetic nerves; ligamentum arteriosum; hilum of the lung.

PERICARDIUM (PERICARDIOTOMY, OPEN CARDIAC MASSAGE, PERICARDIAL WINDOW)

INCISION

Elevate the chest wall on its superiorly based hinge. Remember an anterior or emergency thoracotomy is performed through the left seventh interspace. The breast crease is a useful landmark when time is critical. A rib retractor or an assistant using both hands is put in place to gain access to the left chest. With the anterior chest wall elevated, note the flexibility of the chest wall. Count down to the seventh interspace to orient relative to the heart.

OPERATIVE PROCEDURE

Identify the left phrenic nerve, which descends close and anterior to the left lung hilum (Fig. 3.4). To perform an emergency pericardiotomy, remember that the pericardial slit should be oriented in an inferior to superior fashion parallel to the phrenic nerve (Fig. 3.5, dotted line). To avoid damage to the phrenic nerve, make the pericardiotomy large enough to admit your hand. Palpate the heart with the fingers and palm behind the heart and thumb on the anterior wall of the heart. Notice the thick nature of the left ventricle and the very thin nature of the right ventricle. **Care must be used when the thumb is on the anterior surface of**

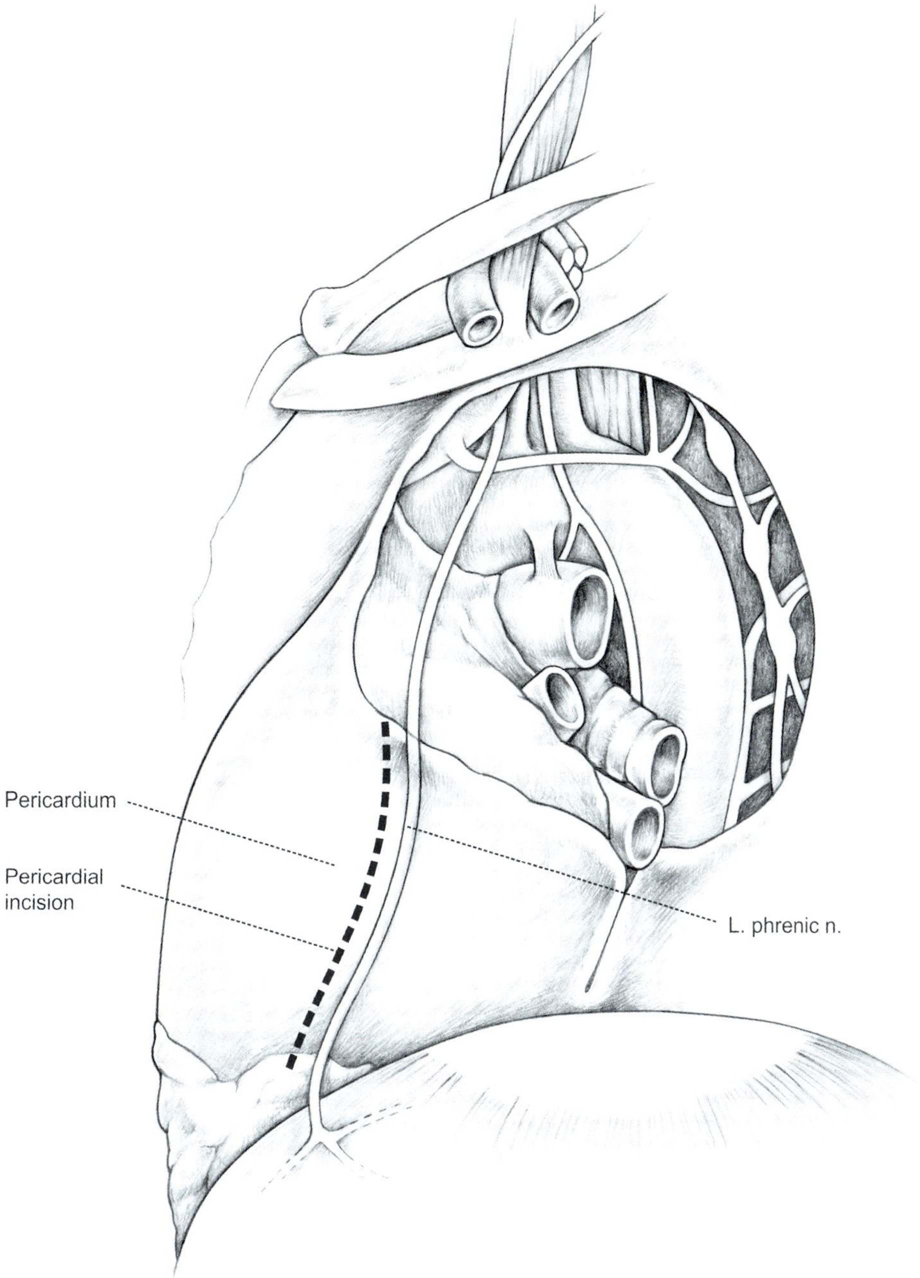

Figure 3.5 Pericardiotomy for open heart massage.

the heart. The heart may be encircled with the hand and compression performed *with care* to avoid right ventricular perforation and injury from excessive pressure from the operator's thumb. Alternatively, open cardiac massage with the operator's hand behind the heart and pushing anteriorly to the sternum may be used (Fig. 3.6). Practice this now by dropping the anterior chest wall down such that open cardiac massage can be performed by pressing the heart anteriorly into the sternum.

Slit the pericardium up to the base of the heart. Cardiac herniation can occur with a large pericardiotomy that is not closed. If open cardiac massage is successful, the surgeon may consider a trip to the operating room to close the pericardiotomy and thoracotomy with proper lightning and instrumentation.

Pericardial window, see Section 1. If not yet done, replace the pericardium in its normal position. Put the anterior chest wall down over it. Using Allis or Pean

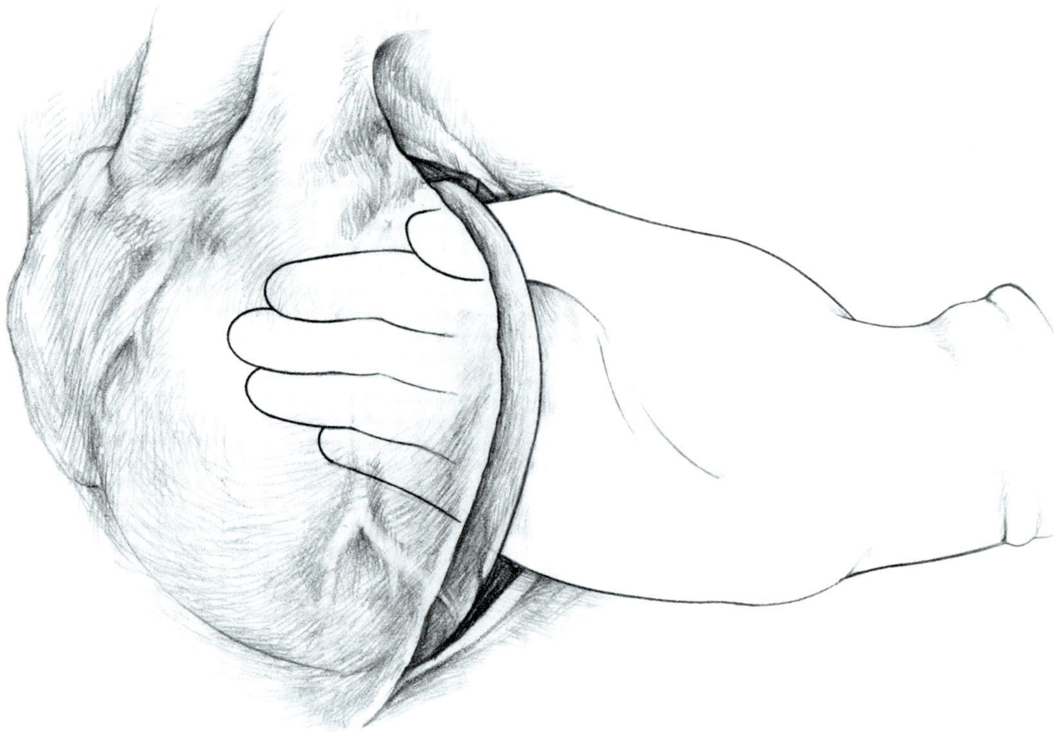

Figure 3.6 Hand placement for open heart massage.

clamps, stretch the diaphragm anteriorly. Identify an area of the central tendon of the diaphragm, which is contiguous with the pericardium. Using an Allis clamp, grip the diaphragm and make a 2-cm incision to gain access to the pericardial space (Fig. 3.1).

HEART AND MEDIASTINAL LYMPH NODE STATIONS (CARDIAC ANATOMY FOR GENERAL SURGEONS, TRACHEOINNOMINATE FISTULA, NODAL STATIONS FOR CANCER RESECTIONS)

INCISION

Elevate the anterior chest wall on its superiorly based hinge. Open the anterior pericardium in a vertical fashion to expose the heart.

DISSECTION

Notice the intrapericardial aortic root. Palpate the anterior and posterior heart remembering that the anterior chamber of the heart is the right ventricle and the posterior chamber of the heart is the left ventricle. Identify the superior and inferior vena cava and transect them. Identify the ascending aorta, pulmonary arteries, and pulmonary veins and transect them. Remove the heart from the pericardial sac. Understand that this is similar to the cardiectomy performed for organ donation and cardiac transplantation. Identify the aortic valve and examine for any sign of calcification (Fig. 3.7). Calcification is common in the aortic position in the elderly due to the high-pressure nature of the left side of the heart. You may also find a bicuspid aortic valve, which is a congenital condition that occurs with regularity in the general population. Dissect out the left anterior descending, circumflex coronary and right coronary arteries (Fig. 3.7). Notice that these are epicardial vessels with perforators descending into the mass of the cardiac muscle. Think of this relationship when considering left ventricular hypertrophy as a risk for endocardial infarction secondary to aortic stenosis or hypertension. Identify the ascending nodal branch off the proximal right coronary artery going to the sinoatrial node.

Next, open the right atrium, identify the tricuspid valve, and slide an index finger into the right ventricle. Using scissors, open the right ventricle and notice the thin nature of the ventricular wall (Fig. 3.8). Also, notice that the ventricle is anteriorly located and crescent shaped. Place a finger through the aortic valve

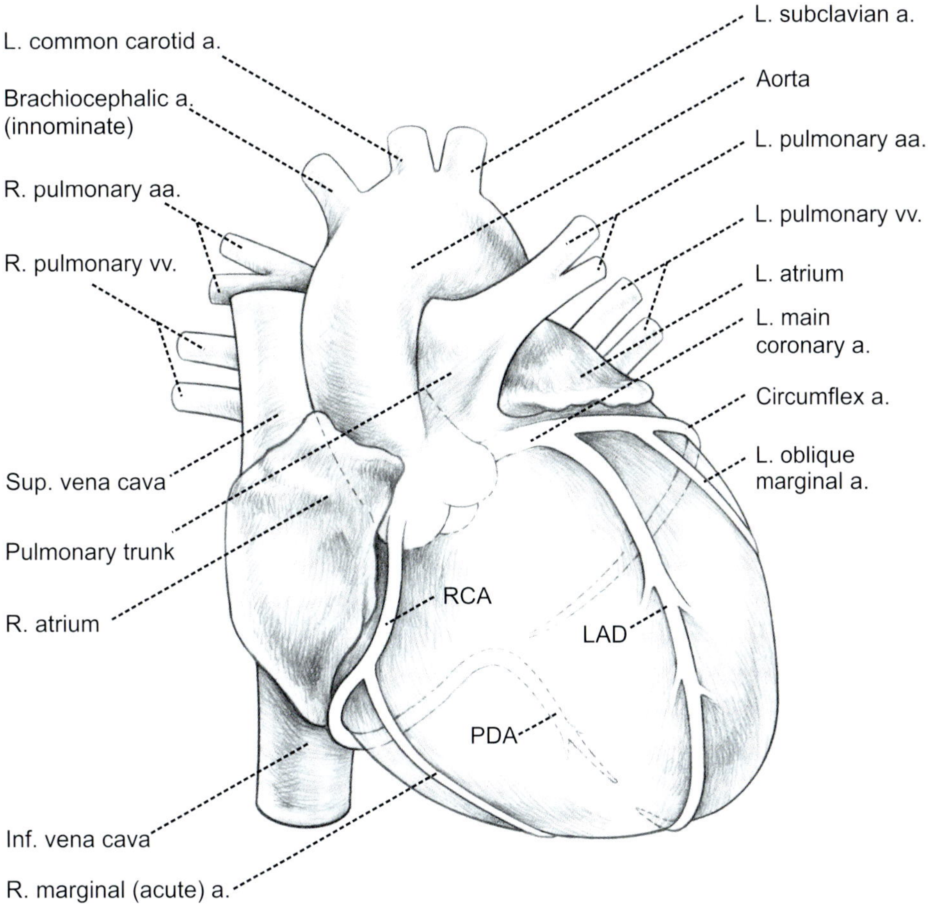

Figure 3.7 Composite view of the heart and great vessels.

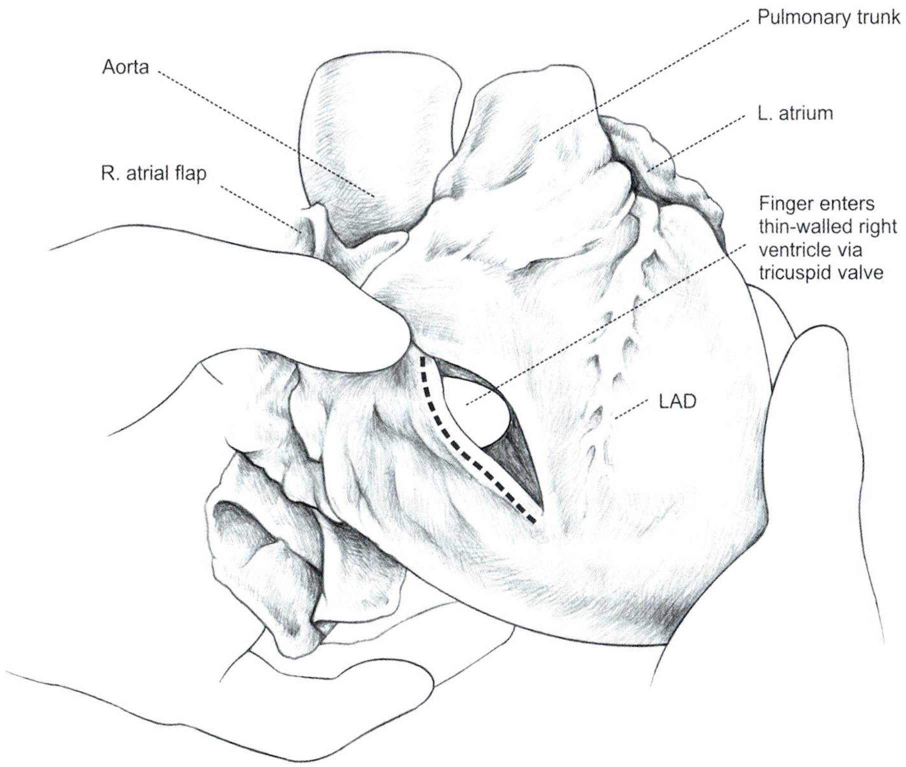

Figure 3.8 Finger placed to demonstrate thin walled right ventricle.

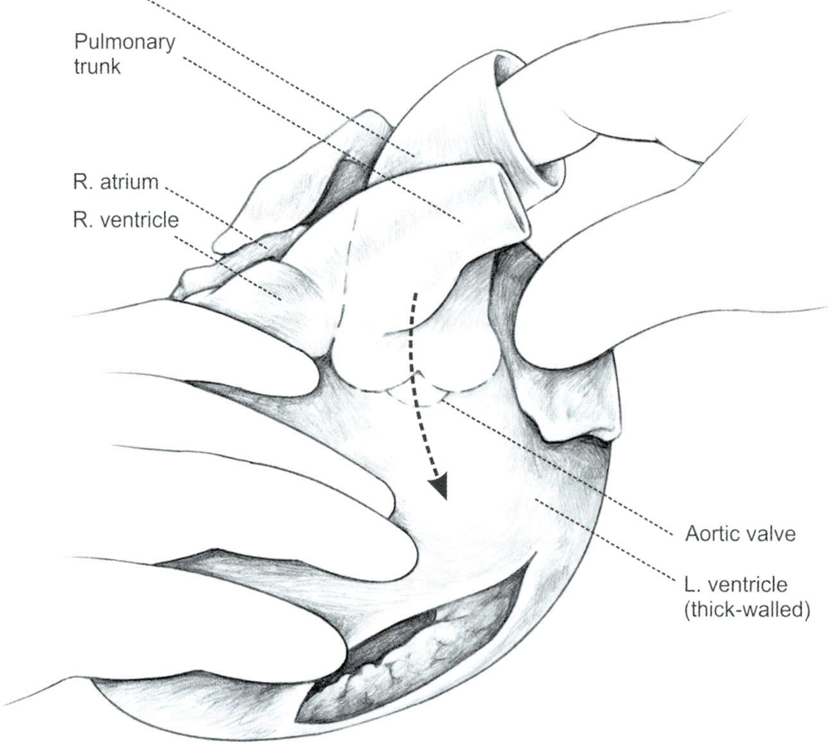

Figure 3.9 Finger placed to demonstrate the aortic valve location and thick-walled left ventricle.

and use a scissors to open the left ventricle (Fig. 3.9). Notice the mitral leaflets hanging like curtains from the fibrous skeleton of the heart and the contiguous nature of the aortic and mitral valves and the central fibrous body. Now reorient the heart in its normal anterior and posterior position and place the index finger down the aorta to the aortic valve. Using the left hand, open the right atrium and palpate medially. Notice the very close proximity of your right and left index fingers. This should help you appreciate the proximity of the AV node represented by the left index finger and the aortic valve represented by the right index finger. Aortic valve replacement can lead to conduction problems postoperatively.

Dissect the ascending aorta, brachiocephalic (innominate) artery, and right carotid artery in the anterior mediastinum (Fig. 3.10). Identify and dissect the brachiocephalic vein as it crosses the superior mediastinum. Note the relationship between the innominate vein and the manubrium. Consider this close relationship when thinking about the sternal saw used in median sternotomy to gain access to the mediastinum, especially relevant during a "redo" sternotomy.

See Chapter 1, Section 3 for tracheostomy. For purposes of dissection and relational anatomy, enlarge this tracheostomy to admit the surgeon's finger. Place a finger into the trachea, have an assistant elevate the anterior chest wall and with the opposite hand palpate the innominate (brachiocephalic) artery. Note their intimate relationship at the thoracic outlet with the trachea and the close relationship to the superior vena cava superiorly. A high-pressure tracheal balloon can erode through the trachea into the soft innominate artery, which is intimately related anatomically. This may initiate a tracheoinnominate fistula.

Refer to the diagram of clinical peritracheal node stations used for thoracic cancer staging purposes (Fig. 3.11A, B).

Removing mediastinal tissue, identify these peritracheal nodes in stations 1, 2, 3, and 4 (Fig. 3.11A). Identify the aorticopulmonary station 5 nodes and the aortic station 6 nodes (Fig. 3.11B). Next, from the right chest, rotate the hilum to see the tracheal bifurcation and station 7 nodes (you may choose to perform this after the lungs have been resected) (Fig. 3.11C). Alternatively,

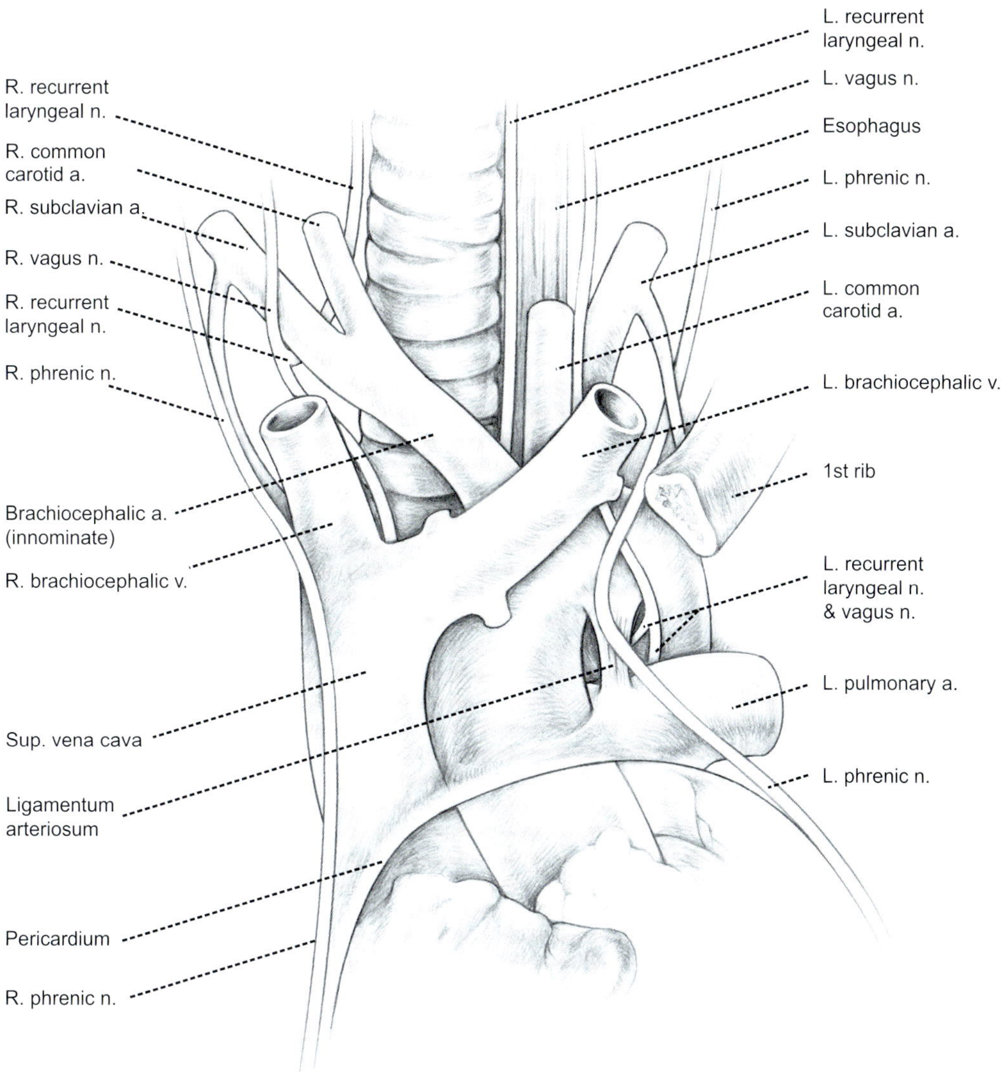

Figure 3.10 Detailed view of the great vessels in the superior mediastinum.

open the posterior pericardium and identify the distal trachea and its bifurcation. The subcarinal nodes in this position are station 7 (Fig. 3.11C). Station 8 and 9 nodes are in the inferior pulmonary ligament.

Identify the station 5 nodes in the aorticopulmonary window and remove them. The ligamentum arteriosum, a remnant of the ductus arteriosus, should now be visible. Identify the recurrent laryngeal nerve just lateral to the ligamentum arteriosum. Note the relationship between the ligamentum arteriosum and the left subclavian artery relative to the descending aorta. Consider these relationships when thinking about a powerful deceleration injury in a head-on motor vehicle collision (Fig. 3.11B).

CLINICAL HIGHLIGHTS

- The anterior chamber of the heart is the right ventricle. Open cardiac massage must be performed with care to avoid right ventricular perforation if the thumb is in the anterior position. My preference is to have the thumb behind the heart so the whole hand presses the heart anteriorly against the sternum.

- Degenerative aortic valve disease is common in elderly Americans in the 8th and 9th decades of life. Congenital bicuspid aortic valves occur in one to two percent of the general population and result in clinically significant aortic valve stenosis in the 4th and 5th decades of life.

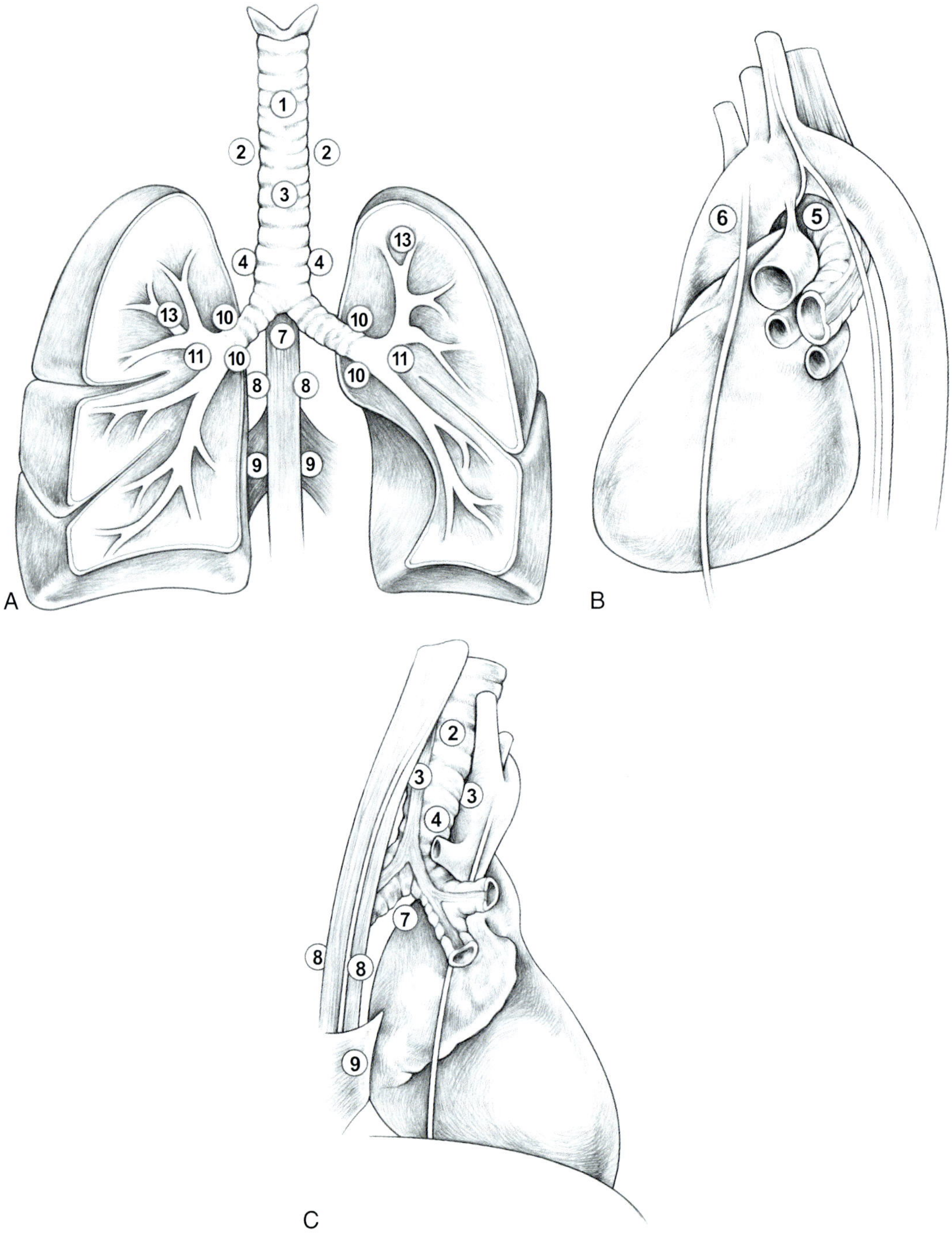

Figure 3.11 **A–C:** Lymph node stations used for lung cancer resections; anterior, left, and right chest views, respectively.

- Aortic stenosis can lead to left ventricular hypertrophy. Note that the surface coronary vessels with perforators must reach through this hypertrophied coronary muscle. This is how inner cardiac wall muscle ischemia can occur.

- Aortic stenosis may present with shortness of breath, exercise intolerance, or sudden death. Preload and vascular tone are very important in aortic stenosis. Induction of anesthesia typically vasodilates and decreases the blood pressure. This is particularly dangerous in the presence of aortic stenosis. Remember that chest compressions and closed cardiac massage may be useless in the presence of aortic stenosis. Therefore, it is critical to listen for a systolic murmur in the preoperative physical exam on all patients preparing for an anesthetic. If a holosystolic, crescendo-decrescendo, harsh murmur is heard, consider a preoperative cardiac echo.

- The atrioventricular node and left bundle are at risk during aortic and mitral valve replacement.
- Remember the proximity of the brachiocephalic (innominate) vein and manubrium when using a sternal saw for a median sternotomy.
- If a herald bleed of a tracheoinnominate fistula is observed, hyperinflate the tracheal cuff and prepare for emergency surgery.
- The aorta is tethered in three spots in the chest: (1) the aortic valve, (2) the ligamentum arteriosum, and (3) the diaphragm. Traumatic deceleration injury may occur at any of these locations. The descending aorta beyond the ligamentum arteriosum and left subclavian artery is a long segment of untethered vessel, and traumatic rupture just distal to the ligamentum is the most common location for traumatic injury. This is why a tear may occur in this position during a major deceleration injury such as a head-on motor vehicle collision.
- The internal thoracic artery (internal mammary artery, IMA) is the preferred conduit for coronary artery bypass grafting.
- The epigastric arteries are extensions of the IMAs and are used as a basis for a TRAM musculocutaneous flap for breast reconstruction after surgery for breast cancer.

LEFT UPPER LOBE (LEFT UPPER LOBECTOMY)

INCISION

It is most useful if cheesecloth or heavy suture is used to tie to the table once it is flipped up. This will improve visualization of the apices of the lung and further dissection of the lungs (see Fig. 3.2).

Prior to the left upper lobectomy, identify the hilum of the lung and the phrenic nerve. Notice the close association of the hilum and the phrenic nerve. It is vital to remember that the phrenic nerve needs to be identified and dissected free of the hilum prior to becoming involved with the lobectomy. This is particularly true for tumors that have caused distortion of the hilum. So, we will begin the dissection with mobilization of the phrenic nerve.

While examining the anterior chest plate that is now flipped up, identify the right and left internal thoracic artery (internal mammary arteries). Arising from the subclavian artery, these clinically useful arteries have several features. First, remember that they are located 1 cm lateral to the sternal margin (Fig. 3.2). Also, note proximally how the IMA dives off the chest wall near the subclavian vein. This is the limit of proximal dissection when harvesting the IMA. When performing a lateral thoracotomy, the intercostal portion of the incision must be extended anteriorly and posteriorly, but remember that the anterior extension should not divide the internal thoracic (mammary) vessels. Second, the internal thoracic arteries continue down the posterior rectus sheath as the superior epigastric arteries, which will continue inferiorly as the inferior epigastric arteries.

Examine the left upper lobe and left lower lobe by elevating the upper lobe at the major fissure. Look at the diagram of the bronchopulmonary segments (Figs. 3.12 A, B and 3.13 A, B). See that the anterior segment of the upper lobes is quite large. The apicoposterior segment is the other portion of the superior division of the **left** upper lobe. From the diagram, you can see that the apical and posterior segments of the **right** upper lobe are separate. Again, on the **left** side, the lingula is divided into a superior and inferior set of segments, while on the **right** side, the middle lobe is divided into lateral and medial segments. Finally, notice that the superior segment of the lower lobes, the "B6" segments, rises quite high in the posterior thoracic cavity. The major fissures on each side are aligned rather vertically, and the minor fissure on the right side is arranged horizontally. Keep this in mind when viewing a chest x-ray.

OPERATIVE PROCEDURE

Identify the vascular structures at the hilum of the lung. Open left upper lobectomy classically begins by retracting the apex of the lung inferiorly and identifying and incising the pleura overlying the pulmonary artery as it exits the mediastinum centrally in a curvilinear fashion running anteriorly to posteriorly. Incise the pleura and identify the superior pulmonary vein and pulmonary artery (Fig. 3.14). On the pulmonary artery, notice that there is a plane next to the pulmonary artery, which opens into an areolar space. This areolar space or plane can be used to dissect the pulmonary arteries into the

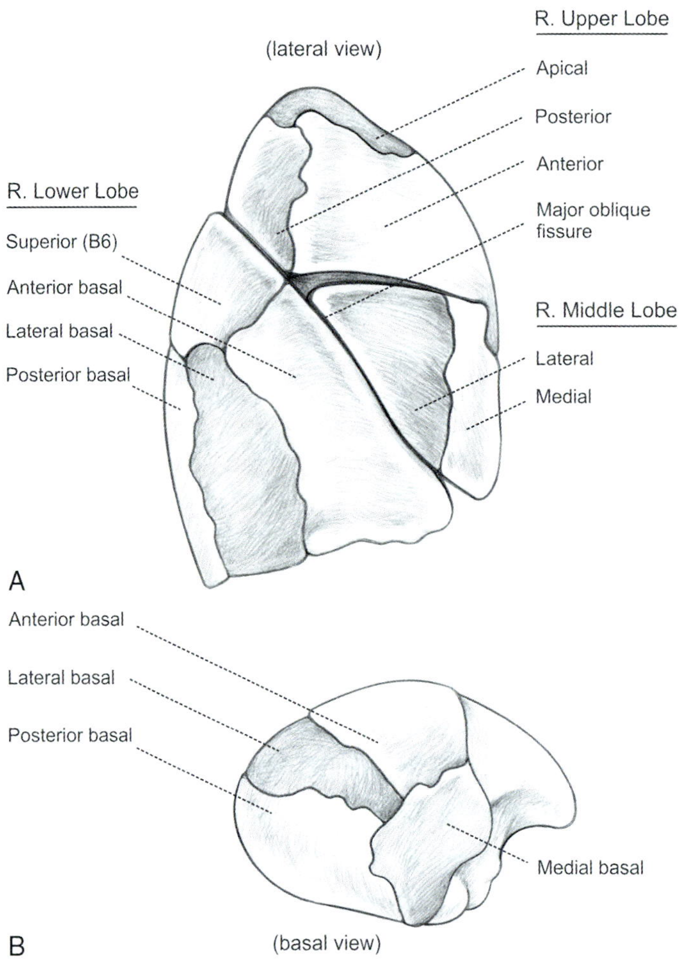

Figure 3.12 Right lung segments. **A:** Lateral view **B:** Basal view.

parenchyma of the lung. This is particularly useful when performing a segmentectomy anywhere within the lung. Use that areolar plane next to the pulmonary artery to identify the first upper lobe branch. Typically, this is the apical posterior trunk. These arteries may arise individually. Once you have followed these vessels in the areolar plane out into the upper lobe to affirm their course, you may then divide them.

The author's operative technique when handling pulmonary arteries is as follows: 2-0 silk ligature suture on the proximal and distal segment of the vessel followed by 3-0 silk stick tie on the patient side of the vessel. Use a right angle under the vessel to elevate and spread the vessel. This isolates the vessel from underlying tissues and prevents past-point of underlying structures. A no. 15 blade on a long handle is then used to divide the vessel. This technique is a safe, secure method for division of any vessel anywhere in the body. Alternatively, branches of the pulmonary artery may be stapled, clipped, or tied. Pulmonary arteries have a "slinky"-type configuration and are delicate. They can come apart in rings if excessive traction or close tumor proximity is encountered. This nasty little feature of pulmonary arteries can be handled quite well with horizontal mattress suture and Teflon pledgets for vessel repair. Any technique that is effective for handling pulmonary arteries will be useful for pulmonary veins as well.

Elevate the upper lobe and look into the major fissure. Notice that the major fissure runs in a rather vertical line. Remember that the upper and lower lobes tend to be anterior and posterior lobes. The left upper lobe and lingula are a single anatomic unit. Remember that the (left) lingula and the right middle lobe are analogous structures. While looking in the major fissure, notice the superior segment of the lower lobe. This superior segment is often referred to as the "B6" segment. Divide the pleura in the midportion of the major fissure and identify the pulmonary artery and

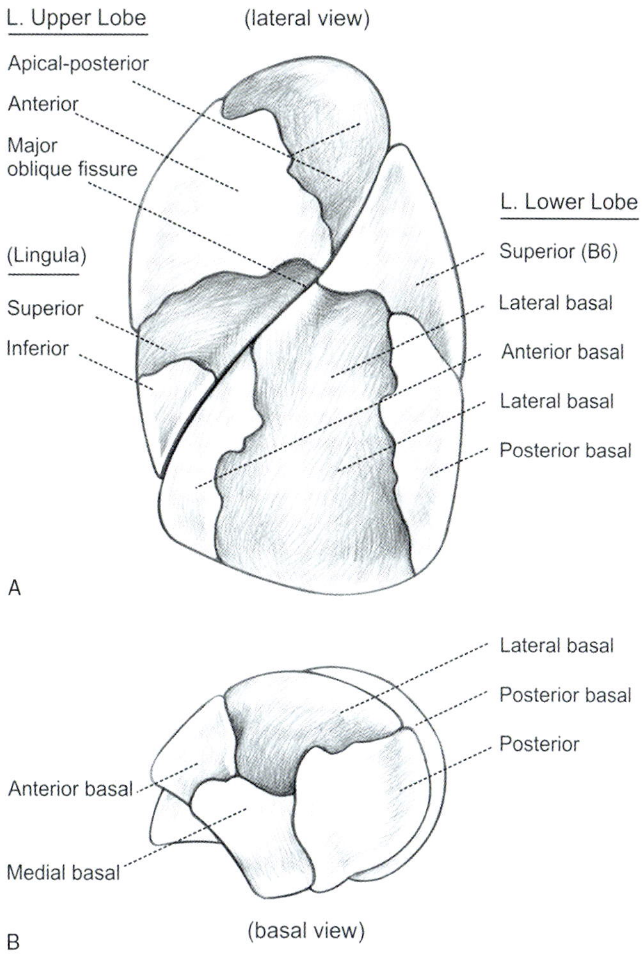

Figure 3.13 Left lung segments. **A:** Lateral view **B:** Basal view.

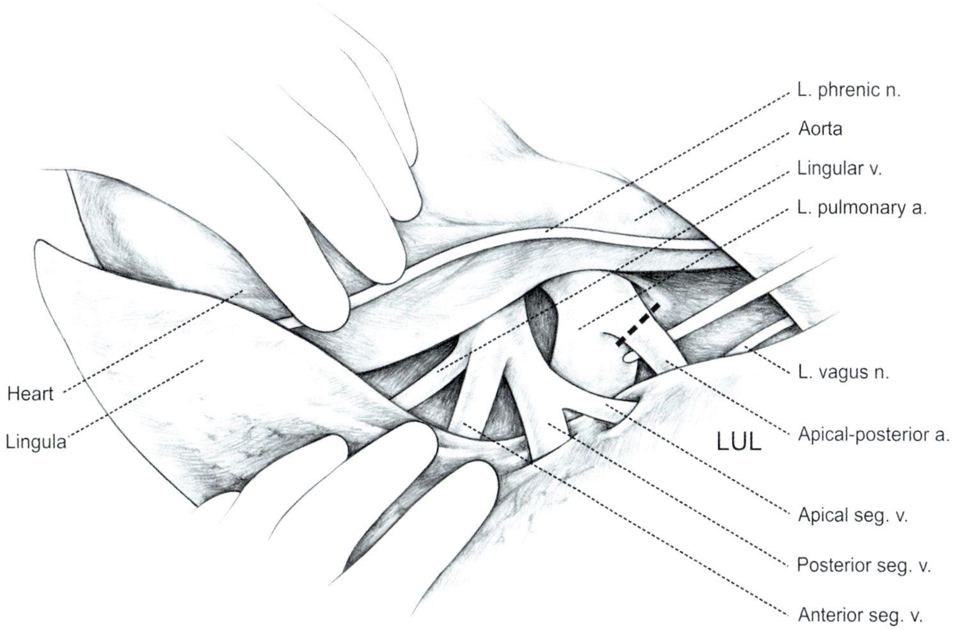

Figure 3.14 Left upper lobectomy; division of pulmonary artery segmental branches, view at superior aspect of the hilum.

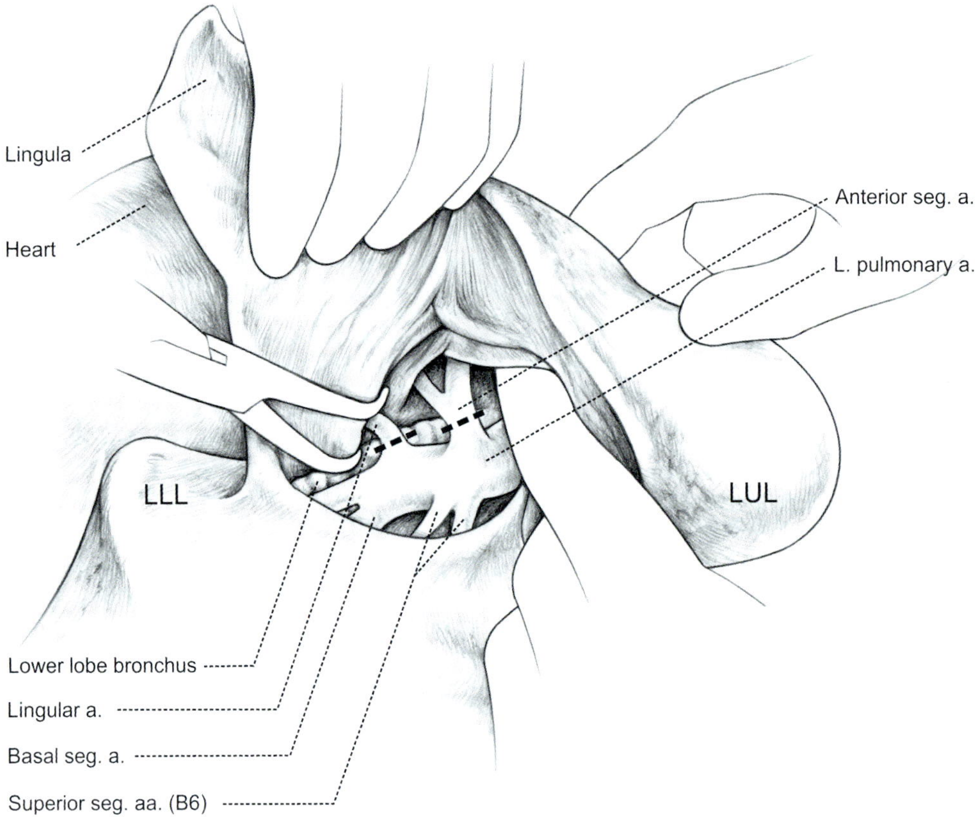

Figure 3.15 Left upper lobectomy; division of pulmonary artery segmental branches, view in the major fissure.

its branches (Fig. 3.15). Identify the main trunk of the pulmonary artery as well as the lingular branch and the anterior segmental branch. Notice that the lingular branch comes off the main pulmonary artery opposite the B6 segmental artery. This landmark of the opposing vessels is useful for orientation within the fissure. Divide the lingular and upper lobe segmental vessels. Now, examine the pulmonary artery for any remaining branches. Examine the pulmonary artery both from the major fissure and from the anterior and superior portions of the hilum. Segmental arterial supply from the pulmonary artery is highly variable and experienced pulmonary surgeons address anatomical anomalies with the following phrase "there is always another pulmonary artery branch." One anomaly you may want to look for on the left pulmonary artery is lingular artery branch descending inferiorly from the trunk of the main pulmonary artery. This proximal takeoff of the lingular artery can only be accessed from the anterior view of the left pulmonary artery (Fig. 3.14). Also note that pulmonary arteries rarely cross fissures, while pulmonary veins do with regularity and may need to be divided when developing an incomplete fissure.

Once you believe that all pulmonary artery branches are divided, you may touch your fingertips as shown (Fig. 3.15). Then identify and clean off the superior pulmonary vein. Again, trace the venous branches into the parenchyma a short way. This will give the surgeon a good picture of the proximal venous branch pattern of the left upper lobe. Once these venous branches have been identified and isolated, divide each of them (Fig. 3.16). All vascular connections should be now clearly divided from the left upper lobe. Identify the left mainstem bronchus as it courses under the left main pulmonary artery. Follow the upper lobe and lingular lobe bronchial segments into their respective areas of parenchyma for a short distance. This will give the surgeon a good picture of the bronchial distribution into the left upper lobe. Divide these segmental bronchi and remove the left upper lobe (Fig. 3.17). A pleural flap to cover the bronchial stump may be used here and for any lung resection.

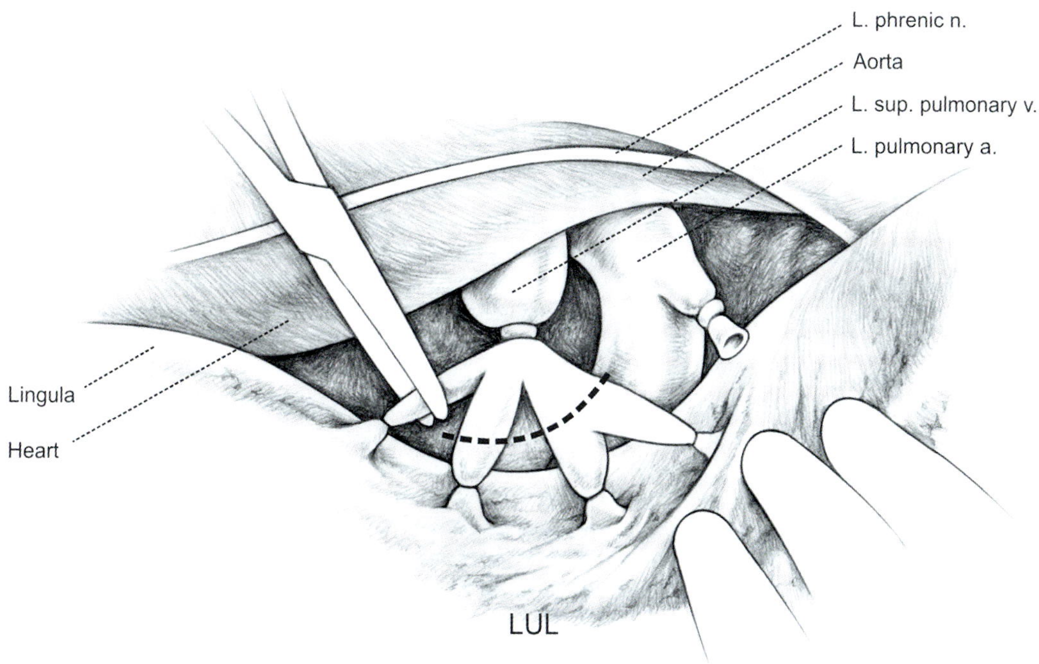

Figure 3.16 Left upper lobectomy; division of superior pulmonary vein.

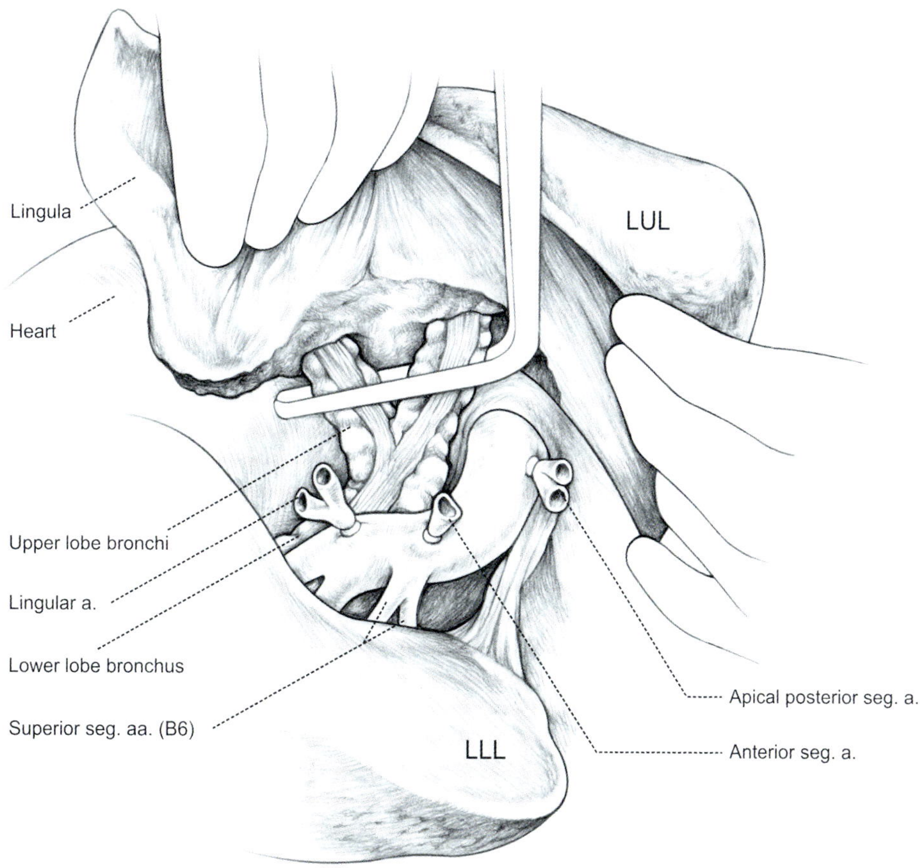

Figure 3.17 Left upper lobectomy; clamp across upper lobe bronchi.

LEFT LOWER LOBE (LEFT LOWER LOBECTOMY)

OPERATIVE PROCEDURE

Retract the left upper lobe anteriorly. Incise the pleura overlying the pulmonary artery and identify the areolar plane surrounding the left pulmonary artery (Fig. 3.18). Expose segmental branches to the lingula and to the superior segment of the lower lobe (B6 segment). Dissect the pulmonary artery to expose branches to the rest of the basilar segments. Note that the vessels to the B6 segment may come off proximal to the lingular artery branches. This is important to remember when doing a left lower lobe. The B6 segmental arteries may need to be taken individually to preserve blood supply to the lingula. Divide the segmental branches of the pulmonary artery to the lower lobe.

To identify the inferior pulmonary vein, it will be necessary to divide the inferior pulmonary ligament. This ligament usually is found with a pleural reflection extending from the lower lobe to the inferior most portion of the mediastinum. This ligament may be divided sharply and bluntly in an ascending fashion up to the inferior pulmonary vein. Dissect the left inferior pulmonary vein to identify the branches to the inferior lobe (Fig. 3.19). Divide these branches. The next step will be to identify the bronchus to the left lower lobe.

Return to the major fissure and again retract the upper lobe medially if it still remains. Note that the pulmonary artery is anterior to the bronchus on the left side. The left pulmonary artery drapes over the bronchus in its course from its origin superior to the bronchus and ending up anterior to the bronchus as you see before you. Dissect the lower lobe bronchi proximally and distally into the lung parenchyma. Note that the B6 segmental bronchus may arise at the level of the lingular segmental bronchus. This may necessitate dividing or stapling the B6 segmental bronchus and inferior lobar bronchus separately to avoid encroachment on the lingular bronchus (Fig. 3.20).

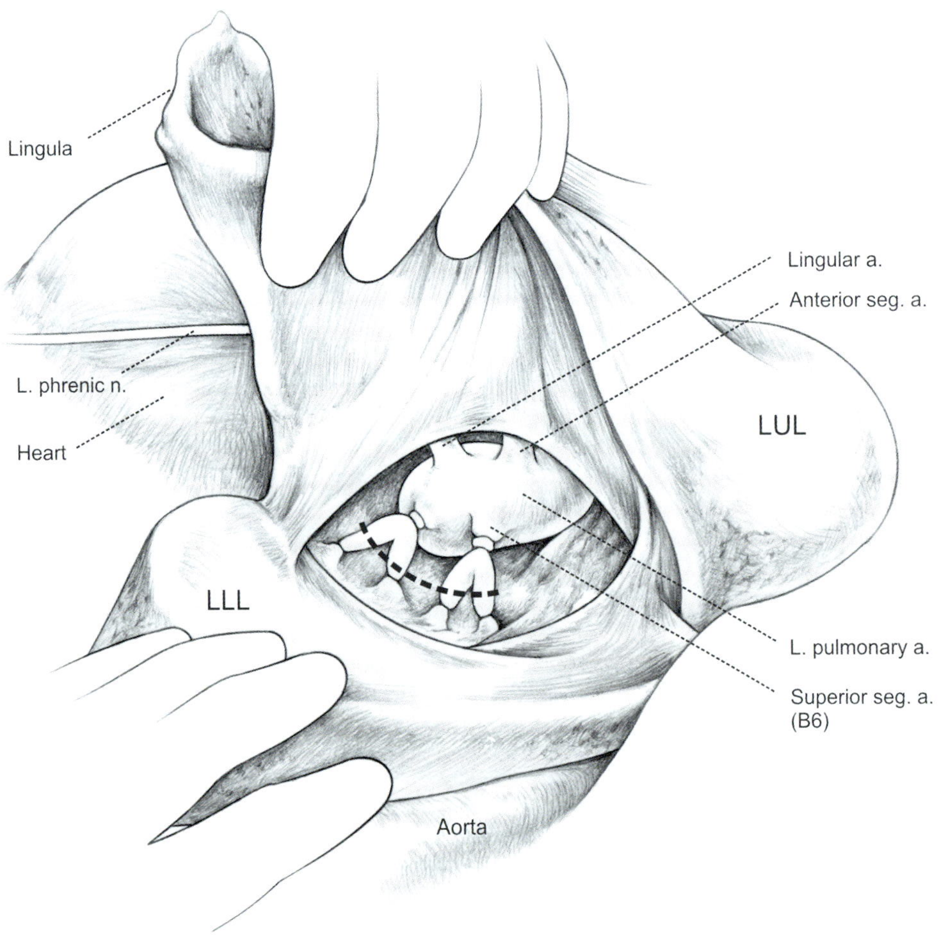

Figure 3.18 Left lower lobectomy; division of lower lobe segmental branches, view in the major (oblique) fissure.

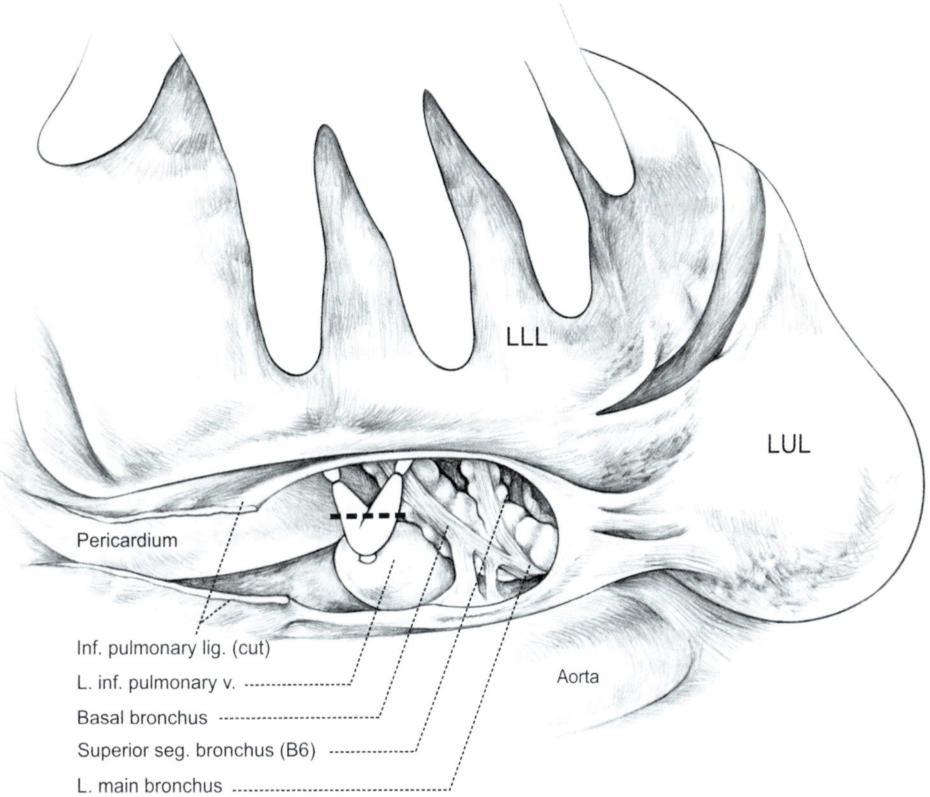

Figure 3.19 Left lower lobectomy; division of inferior pulmonary vein, view from the posterior hilum with the lung retracted forward.

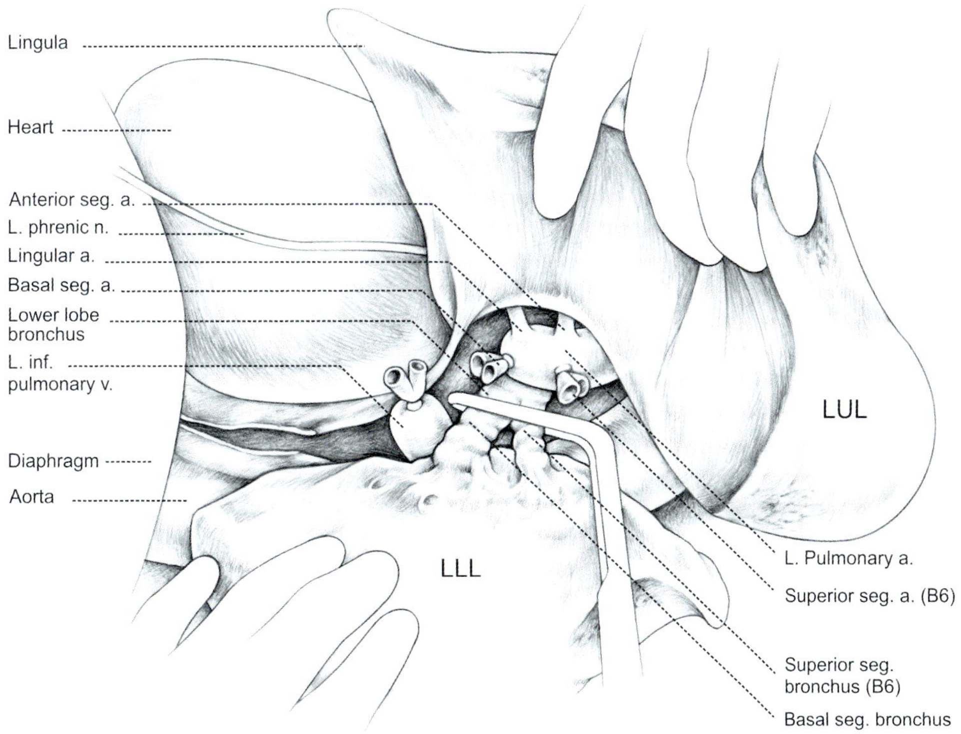

Figure 3.20 Left lower lobectomy; clamp across lower lobe bronchi.

Upon division of the bronchi, the lower lobe may now be removed. At this point, if the upper lobe has already been removed, note the relationships of the left pulmonary artery to the left mainstem bronchus. The artery originates above the bronchus and drapes over the bronchus ending up in an anterior location. The pulmonary veins are typically located in an anterior and inferior position relative to the bronchus for the superior and inferior pulmonary veins, respectively. After division of the bronchus, some surgeons advocate elevation of local pleura to cover the bronchial stump.

RIGHT UPPER LOBE (RIGHT UPPER LOBECTOMY)

INCISION

Notice the location of the right upper lobe in its relation to the right middle and lower lobes. As on the left side, the major fissure is oriented in an oblique plane, which causes the upper and lower lobes to be anterior and posterior in their locations. The minor fissure is a horizontally situated division between the upper and middle lobes. Any of these fissures may be very complete, somewhat complete, or incomplete. If the fissure is incomplete, it may be opened with gentle use of a Kittner in the operating room. If the fissure consists of lung parenchyma rather than filmy adhesions, sharp and blunt dissection may be necessary to open a fissure to gain access to the underlying vascular structures. Once the underlying vascular structures have been identified and controlled, a GIA-type stapler may be used on the remaining lung parenchyma to divide and seal the lung. Stapling will minimize subsequent postoperative air leaks.

OPERATIVE PROCEDURE

Identify the phrenic nerve as it courses down the superior vena cava and pericardium. Elevate the phrenic nerve anteriorly. Divide the overlying pleura and identify the superior pulmonary vein and the right pulmonary artery (Fig. 3.21). Identify the apical-anterior artery, find the areolar plane surrounding the artery, and follow the course of the apical and anterior branches of the artery into the lung parenchyma. These vessels may come off individually from the main pulmonary artery, or they may come off as a common trunk and bifurcate later. Once these vessels are clearly identified, they may be divided.

Elevate the right upper and middle lobes and incise the pleura in the center of the confluence of the major and minor fissures (Fig. 3.22). Identify the main pulmonary artery and its branches as shown in the figure. Notice that the "B6" segmental artery (superior segmental artery of the lower lobe) is given off opposite the segmental artery to the middle lobe. Identify the posterior segmental artery, find the areolar plane surrounding

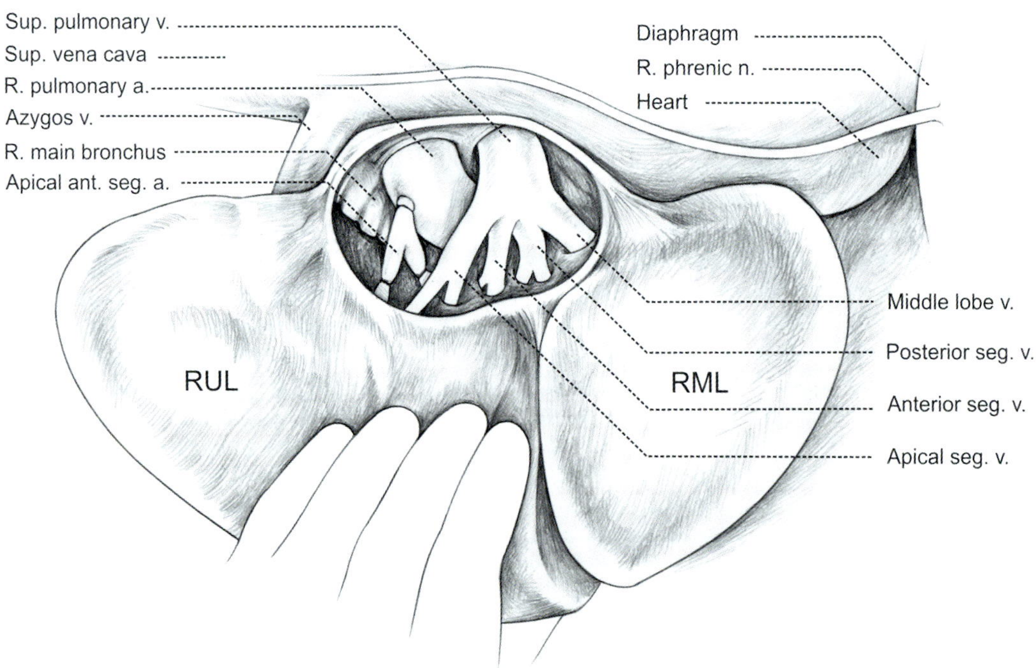

Figure 3.21 Right upper lobectomy; division of pulmonary artery segmental branches, view at superior aspect of the hilum.

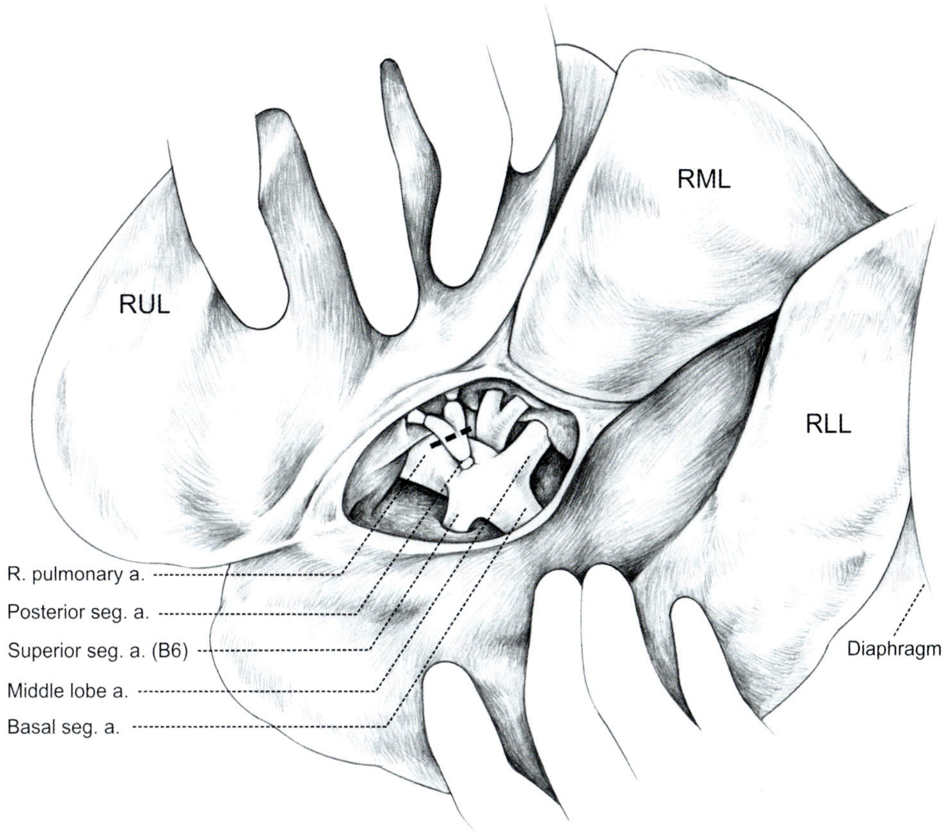

Figure 3.22 Right upper lobectomy; division of pulmonary artery segmental branches, view in confluence of major and minor fissures.

the pulmonary artery, and follow it out into the parenchyma a short way. Once clearly identified, divide the posterior segmental artery to the upper lobe. At this point, as you would in the operating room, clear off the pulmonary artery and ensure that no remaining segmental arteries are intact. Use your fingers to assure no further arterial branches remain as done on the left side (Fig. 3.15). If found, identify these arteries and divide them.

Identify the superior pulmonary vein and its branches in the anterior portion of the right hilum (Fig. 3.23). While doing a right upper lobectomy in a patient that cannot tolerate loss of the right middle lobe, one point is essential to remember. The middle lobe vein arises from the superior pulmonary vein. Less frequently, it arises from the inferior pulmonary vein. The vein to the right middle lobe must be preserved if the right middle lobe is to remain viable. If this vein is damaged and cannot be repaired, the right middle lobe must also be excised. With that in mind, the surgeon may now begin to identify and divide the veins to the right upper lobe. The segmental veins to the posterior and anterior segments may be discovered posteriorly where they arise from the superior pulmonary vein trunk.

Having preserved the vein to the right middle lobe and having divided the segmental veins to the right upper lobe, pull the right upper lobe forward to begin identifying the bronchus (Fig. 3.24). Note that the right main pulmonary artery lies anterior to the bronchus on the right side. Remember that the **left** pulmonary artery traversed posterior to the bronchus.

Follow the segmental bronchi into the parenchyma to clarify their branching patterns. Notice the proximity of the azygos vein to the right upper lobe bronchus. Keep this in mind during dissection of the right upper lobe in the operating room. Notice the very short stubby nature of the right upper lobe bronchus. In fact, the segmental bronchi are stapled immediately distal to their take off in the operating room to avoid encroachment on the right mainstem bronchus and bronchial stenosis. Divide these segmental bronchi individually to view their relationship to the mainstem bronchus (Fig. 3.24).

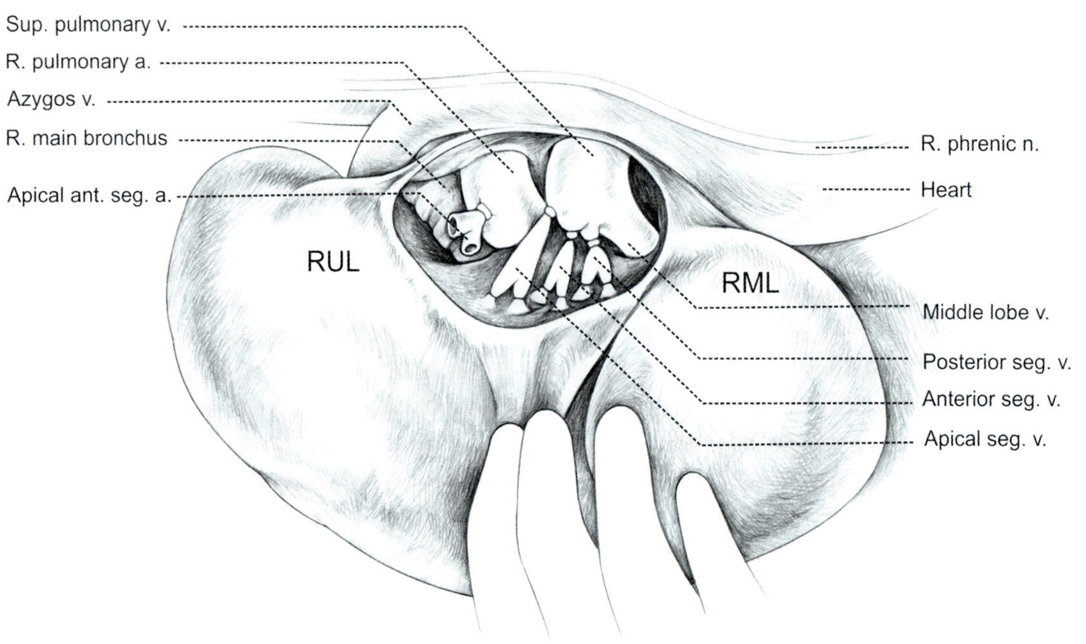

Figure 3.23 Right upper lobectomy; division of superior pulmonary vein, ensure patency of middle lobe vein.

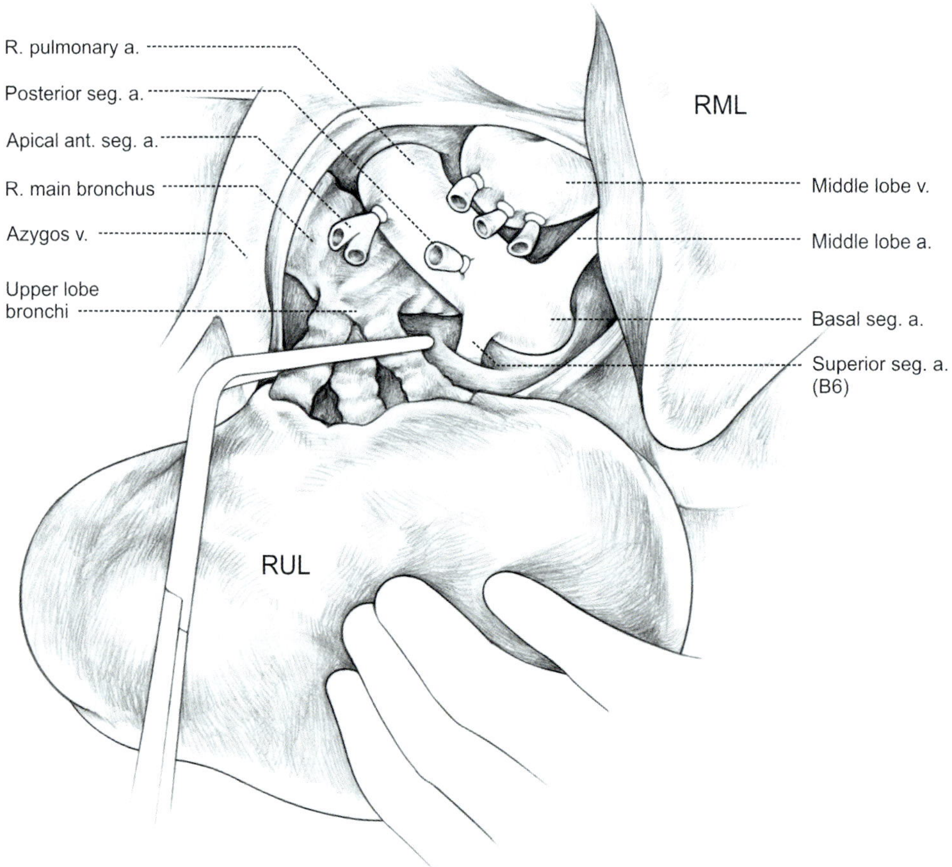

Figure 3.24 Right upper lobectomy; clamp across upper lobe bronchi.

RIGHT MIDDLE LOBE (RIGHT MIDDLE LOBECTOMY)

INCISION

If the upper lobe has not been removed, enter the major fissure and retract the right upper lobe anteriorly.

OPERATIVE PROCEDURE

Identify the pulmonary artery within the major (oblique) fissure and identify the areolar plane surrounding the pulmonary artery. In this areolar plane, clean off the pulmonary artery and its branches to the upper lobe and middle lobe. One, two, or three branches to the middle lobe may be identified. Also, identify the branches to the B6 segment (superior segment of the lower lobe). Note that often the segmental branch arteries to the middle lobe arise opposite the B6 segmental arteries from the main pulmonary artery. Trace the middle lobe vessels into the parenchyma a short distance to gain length on the vessels. Divide the pulmonary artery branches to the middle lobe (Fig. 3.25).

From an anterior position, identify the right middle lobe vein arising from the superior pulmonary vein. This typically arises as a single trunk. Once this vein is identified and dissected into the lung parenchyma a short distance to gain length, it may be divided (Fig. 3.26). Next, identify the middle lobe bronchus. This bronchus arises from bronchus intermedius and ascends to the right middle lobe. A descending pattern to the vessels and an ascending pattern of the bronchus may be somewhat disorienting until bronchial anatomy is remembered. Clean surrounding tissues from the right middle lobe bronchus. Trace this bronchus back to bronchus intermedius. Remember this bronchus must be divided distal enough to ensure that encroachment on the lower lobe bronchus does not occur. Divide the middle lobe bronchus (Fig. 3.27).

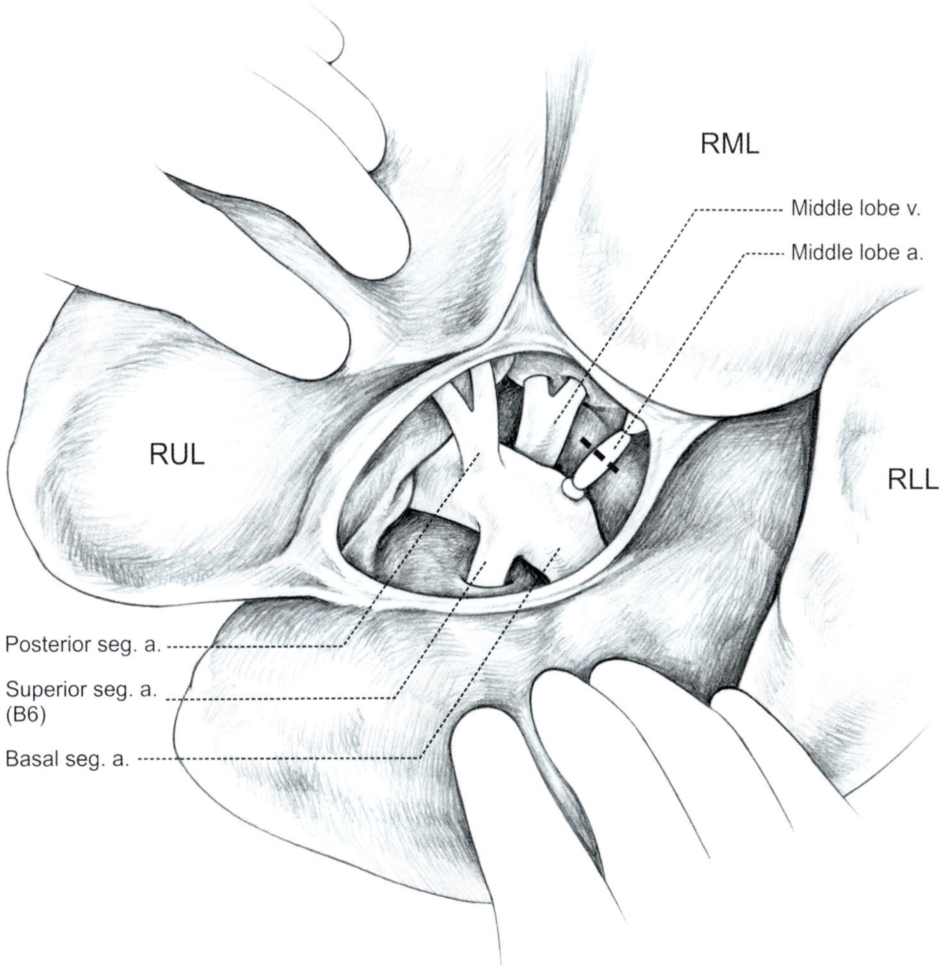

Figure 3.25 Right middle lobectomy; division of pulmonary artery segmental branches, view in confluence of major and minor fissures.

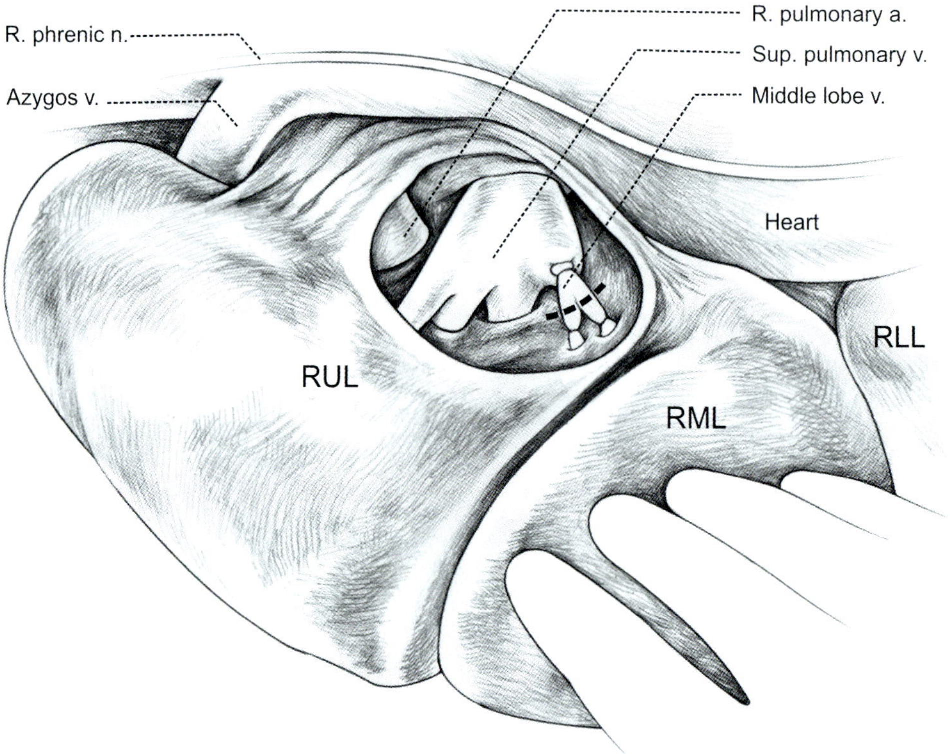

Figure 3.26 Right middle lobectomy; division of pulmonary vein segmental branches, view in confluence of major and minor fissures.

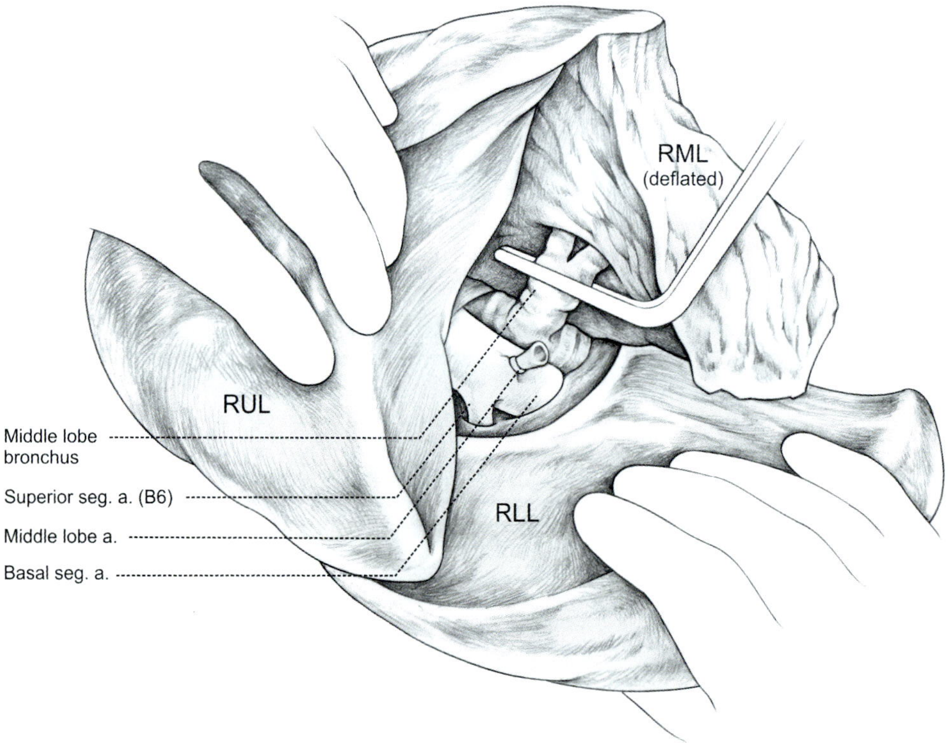

Figure 3.27 Right middle lobectomy; clamp across middle lobe bronchus, view in confluence of major and minor fissures.

The minor fissure is usually not complete. In the operating room, the middle lobe may be separated from the upper lobe using blunt and sharp dissection, or preferably GIA staplers to avoid postoperative air leaks. Inflation of the RUL reveals where to position the stapler in the OR to separate RML from RUL. In the anatomy laboratory, attempt to identify a plane between the upper and lower lobe and using blunt and sharp dissection as previously described, separate the minor fissure and remove the middle lobe. Raw surface of remaining lung tissue may be sutured with absorbable suture in the OR to control postoperative air leaks.

RIGHT LOWER LOBE (RIGHT LOWER LOBECTOMY)

INCISION

If the upper and middle lobes have not been removed, reflect them anteriorly to expose the major (oblique) fissure.

OPERATIVE PROCEDURE

Enter the major fissure and identify the pulmonary artery. In the areolar plane surrounding the right pulmonary artery expose the branches of the pulmonary artery to the B6 (superior) segment of the lower lobe. Typically, these vessels will originate opposite the right middle lobe branches from the pulmonary artery. Clean off the branches to the B6 segment and the basilar segments of the lower lobe. Once identified, these vessels may be divided (Fig. 3.28).

The inferior pulmonary ligament is taken down by dividing the ligament at its inferior most extent. This is done by retraction of the lower lobe laterally to expose the ligament. Sharp and blunt dissection may be used to divide this ligament. Lymph nodes identified in this position are identified as station 9 (Fig. 3.11C). Continue dissection of the right inferior pulmonary ligament up to the right inferior pulmonary vein. Clean off the vein and identify the branch vessels into the segments of the right inferior lobe. Divide these segmental branches (Fig. 3.29). Posterior to the pulmonary

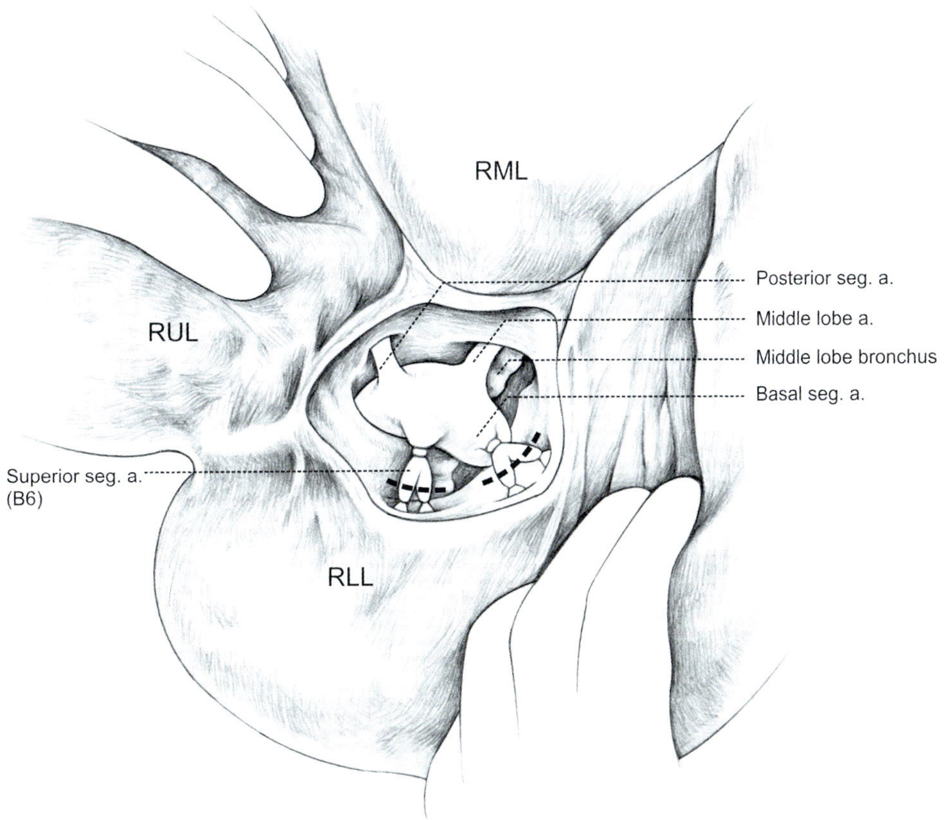

Figure 3.28 Right lower lobectomy; division of pulmonary artery segmental branches, view in confluence of major and minor fissures.

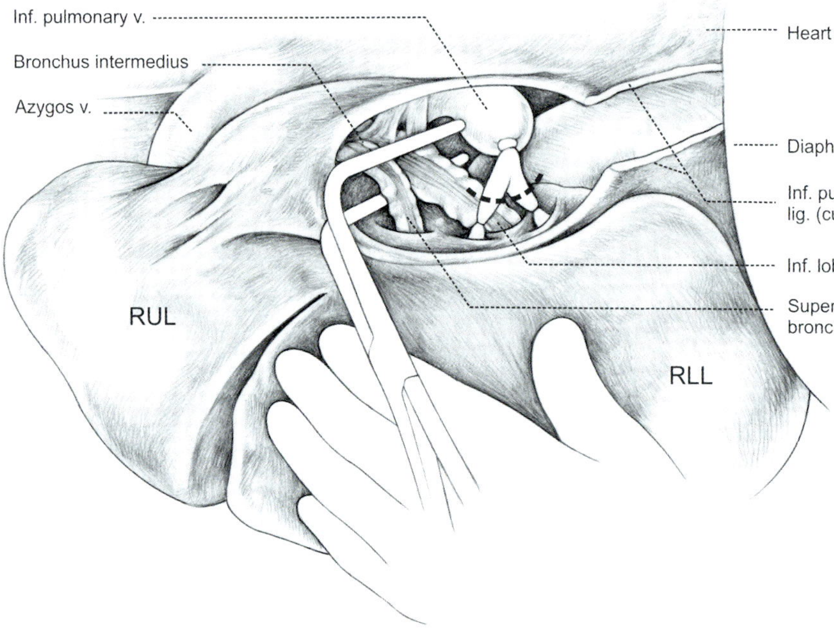

Figure 3.29 Right lower lobectomy; division of pulmonary vein segmental branches, clamp across lower lobe bronchi, view in confluence of major and minor fissures.

veins, identify the bronchus intermedius. Dissect the bronchus to identify the B6 segmental bronchus as well as the segmental bronchi to the basilar segments of the lower lobe and divide them (Fig. 3.29). The right lower lobe may now be removed. If the upper and middle lobes have been removed, now examine the right hilum (Fig. 3.3). Note that the pulmonary artery and its branches to the upper, middle, and lower lobes lie anterior to the upper lobe bronchus and bronchus intermedius. The bronchus to the right middle lobe courses anterior to the pulmonary artery (Fig. 3.27). Remember this picture to avoid confusion during lobectomy. Pulmonary veins on the right side lie anterior to the artery and bronchi. Once this general pattern is understood, anatomical variations encountered during lobectomy are recognized and dealt with safely.

Technical experience hand sewing a bronchus may be gained in the anatomy lab. Using a mainstem bronchus, place sutures originating posteriorly in the membranous portion of the bronchus extending through the cartilaginous anterior portion of the bronchus. These sutures should be placed like spokes on wheel in a radial pattern (Fig. 3.30). The reason for this radial pattern of suture placement is that the pliable soft membranous portion of the bronchus is drawn anteriorly to the rigid U-shaped cartilage to mate securely in an airtight fashion. Most surgeons use the available surgical staplers for this purpose. However, you may encounter a circumstance where bronchial suture technique will be needed.

CLINICAL HIGHLIGHTS (PULMONARY)

- Bronchoscopy is standard of practice prior to any lung resection.
- Remember that the RUL bronchus may take off directly from the trachea. This may present with a

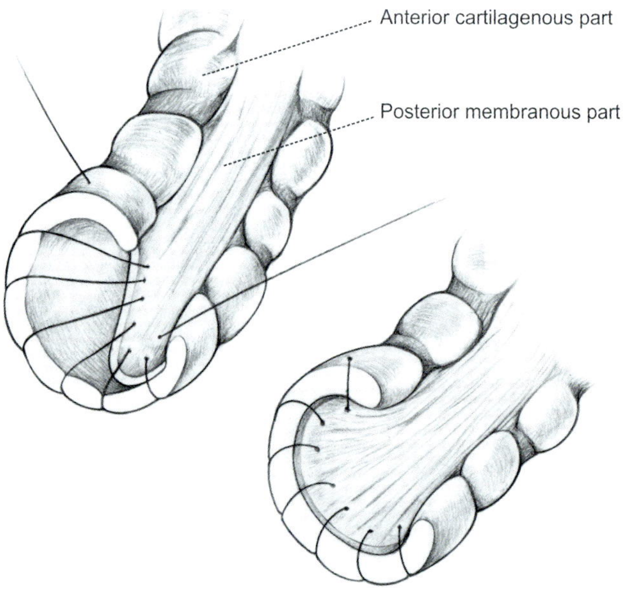

Figure 3.30 Technique to suture a bronchial stump using spokes-in-a-wheel pattern.

chest x-ray in an intubated patient showing severe atelectasis of the RUL. Retracting the endotracheal tube with bronchoscopy will diagnose this bronchial variant.

- To begin lobectomy, be sure to identify and preserve the phrenic nerve when dividing the pleura over the hilum.
- Dual lumen intubation using one lung ventilation markedly improves visualization during thoracotomy. It is essential during thoracoscopy.
- Jet ventilation of the nonventilated lung may be used if needed without compromising visualization in that hemithorax.
- Pulmonary arteries are high-volume thin-walled structures, which may be ligated with stick tie ligatures or between ties and clips for secure closure. During VATS (Video Assisted Thoracic Surgery) lobectomy, surgical staplers are used.
- Pulmonary artery variability is prevalent. So, during lobectomy once the pulmonary artery branches have been cut and tied, experienced surgeons will look for "the last branch still present."
- The right middle lobe vein *must* be identified and preserved to ensure viability of the lobe.
- Bronchial anatomy may be confirmed during lobectomy by clamping the bronchus with the TA stapler. Prior to firing of this stapler, reinflate the lung to confirm bronchial patency to nonresected lobes.
- Lymphatic drainage is generally in a medial and superior direction. Drainage typically remains in the ipsilateral hemithorax with two exceptions.
  - The left lower lobe may drain to the left or right hilum and subcarinal nodes (station 7).
  - Lymph nodes become obstructed when filled with tumor. This may cause aberrant drainage patterns.
- Chest tubes after lobectomy must be placed to drain air at the apex and blood or fluid posteriorly.
- Postoperative chest x-ray should show complete re-expansion of the lung, shift of the mediastinum to the operated side, and elevation of the ipsilateral diaphragm.
- Left pneumonectomy is easier than right. The left pulmonary artery is longer and drapes over the left mainstem bronchus. The right pulmonary artery frequently divides in the mediastinum and is shorter.

DIAPHRAGM, VAGUS, AND PHRENIC NERVES (LAST SECTION)

OPERATIVE PROCEDURE

At the aortic arch, the vagus nerve continues its downward course but sends off the important left recurrent laryngeal nerve to loop underneath the arch and reascend (recur) in the tracheoesophageal groove. Identify the left recurrent laryngeal nerve (Fig. 3.4).

In a similar fashion on the right side, identify the phrenic and vagus nerves and follow them in their descending courses (Fig. 3.3). Note a similar three-dimensional relationship on the right side that you found on the left side. On the right side, the recurrent laryngeal nerve loops underneath the right subclavian artery and ascends (recurs) in the tracheoesophageal groove. Further dissection of the recurrent laryngeal nerves on the right and left side will wait for dissection of the esophagus.

To perform a transthoracic vagotomy through the left chest, mobilize the distal esophagus and pass a finger around the esophagus 5cm above the esophageal hiatus. The right vagus trunk runs on the anterolateral wall and should be palpable. Identify the right and left vagal trunks (Fig. 3.31A, B). Each vagal trunk will have a web of anastomotic nerve branches and esophageal vascular branches associated with it. Identify and transect these branches to free each trunk. The vagus is a single trunk: left (anterior), 96% at *4 cm above* the diaphragm, and right (posterior), 92% *at the level of* the diaphragm. The vagal trunks and branches may be excised. Remember that the thoracic duct lies to the posterior right of the esophagus at the diaphragm, although variability does exist (Fig. 3.32).

CLINICAL HIGHLIGHTS

- The phrenic nerve arises from the ventral rami of the C3, C4, and C5. High cervical transections or lesions due to trauma or tumor may interfere with diaphragmatic function and may require pacing of the diaphragm muscle.
- Remember that diaphragmatic irritation may produce pain referred to the top of the shoulder. In these cases, shoulder joint motion does not elicit shoulder pain.
- Operative incisions of the diaphragm, for instance, to gain access to the adjacent body cavity are performed in such a way that muscle function is preserved upon

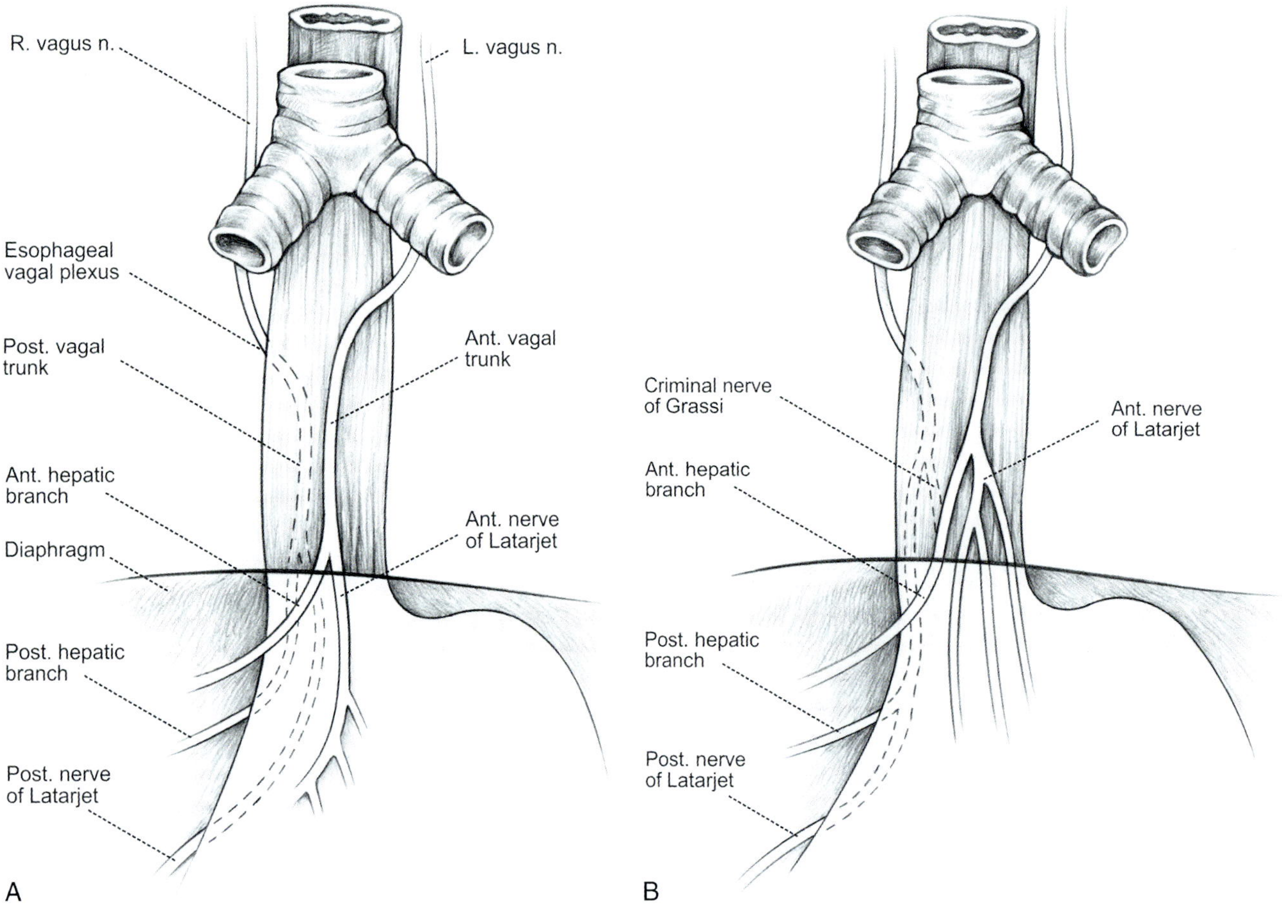

Figure 3.31 **A, B:** Patterns of vagal structure at the esophageal hiatus.

repair of the diaphragm. Incisions starting from the periphery of the diaphragm should be made radially into the central tendon. An alternative is to perform a peripheral incision 1-2 cm from the insertion of the diaphragm at its margin. These incisions maximize function of the diaphragm upon repair (Fig. 3.1).

- A stab wound to the heart may require emergency anterolateral thoracotomy through the 6th interspace. Remember that the pericardial incision to gain access to the heart must be parallel in line with the phrenic nerve.

- To perform open CPR through an emergency (7th interspace) anterolateral thoracotomy and vertical pericardial incision, remember to place your entire hand behind the heart and compress anteriorly toward the sternum. Alternately, the heart may be encircled with the hand and compression performed *with care* to avoid right ventricular perforation and injury from excessive pressure from the operator's thumb.

- Transdiaphragmatic lymphatic channels may allow accumulation of a pleural effusion from the abdomen such as ascites, subdiaphragmatic abscess, or Meigs syndrome.

- There is a dual innervation of the diaphragm. Remember that the posterior rim of the diaphragm has sensory innervation by local intercostal nerves. The central portion of the diaphragm has sensory innervation by the phrenic nerve.

- Transthoracic vagotomy is used to complete an abdominal vagotomy if previous surgery has created dense adhesions (a "hostile abdomen") or if an incomplete vagotomy is documented.

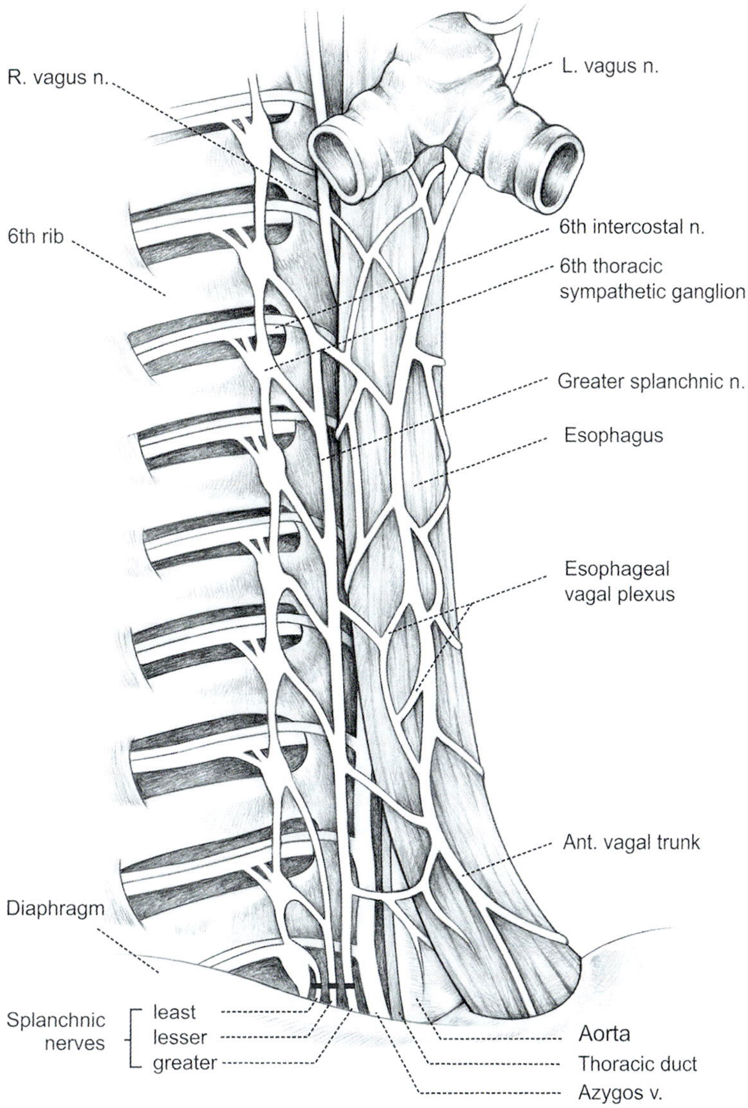

Figure 3.32 Greater, lesser, and least splanchnic nerves, thoracic duct, and vagus nerve on the right side at the diaphragm.

THORACIC SYMPATHETIC TRUNK (THORACIC SYMPATHECTOMY)

INCISION

Flip the anterior chest plate open to expose the entire chest. Remove the lungs that have been previously resected. On the right and left sides, identify the heads of the ribs as they articulate with the lateral vertebral processes. At this point on each side, the sympathetic chain will be visible.

OPERATIVE PROCEDURE

To identify the sympathetic chains, remove the overlying parietal pleura to expose the sympathetic trunks (Fig. 3.4). Exposure and dissection of the cervicothoracic (stellate) ganglion at the apex of the chest may be difficult. If necessary, an assistant can pull on the anterior chest wall to aid exposure. Typically, a vein is identified crossing the sympathetic trunk between the second and third ganglia. During an operative procedure, this vein may be controlled with clips to avoid bleeding. Expose the entire sympathetic chain from the thoracic

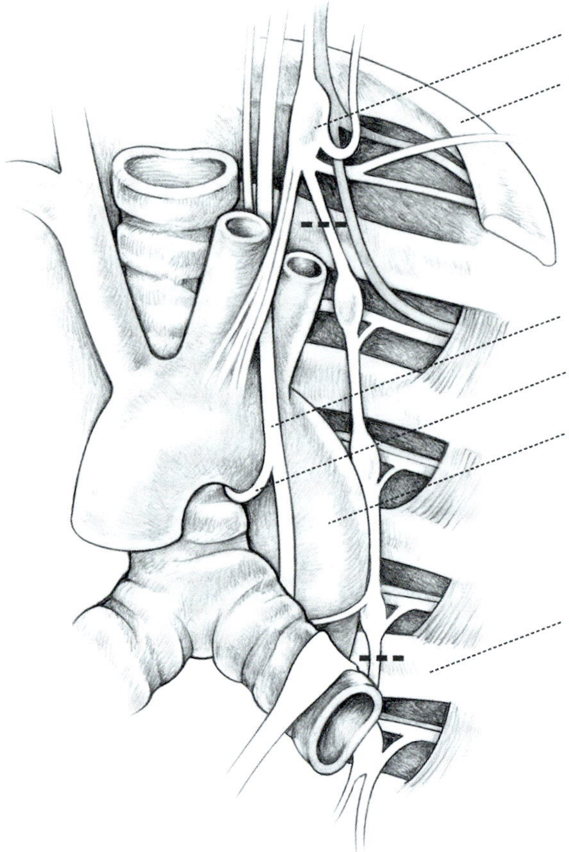

Figure 3.33 Thoracic sympathectomy beginning at T5 on the left side.

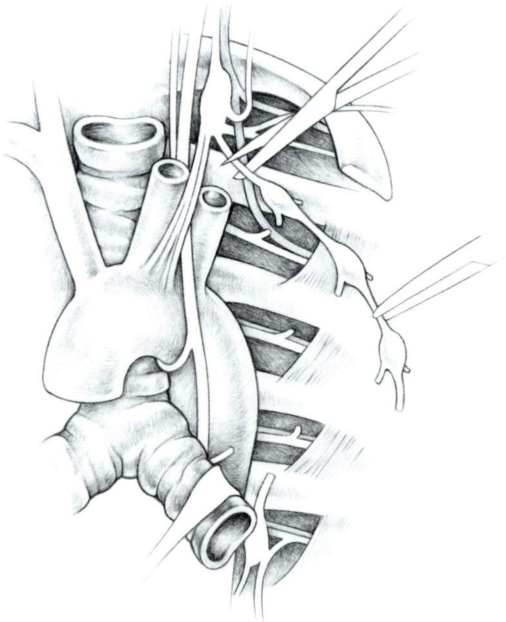

Figure 3.34 Completion of thoracic sympathectomy on the left side.

inlet to the diaphragm. At the level of T5 (Fig. 3.33), transect the sympathetic ganglion and begin a cephalad dissection. Note the lateral contributions to the intercostal nerves (Fig. 3.4). Continue dissection to the stellate ganglion. Note the bulbous nature of this ganglion. You may dissect and excise the lower one-third of this ganglion in continuity with the sympathetic trunk. However, remember that many surgeons in the United States transect the sympathetic chain below the stellate ganglion to avoid Horner syndrome (Fig. 3.34).

THORACIC SPLANCHNIC NERVES (SPLANCHNICECTOMY)

OPERATIVE PROCEDURE

Identify the thoracic sympathetic chain in the lower chest. Remove the pleura overlying the distal thoracic chain and identify the branches that form the greater splanchnic nerve medially (Fig. 3.32). The surgeon will also find branch connections to the vagal plexus surrounding the esophagus. Dissect the greater splanchnic nerve distally to the level of the diaphragm. Lateral to the greater splanchnic nerve the lesser splanchnic and least splanchnic nerves will be identified. These preganglionic fibers will synapse at the celiac ganglion. Interruption of the splanchnic nerves or the celiac ganglion will achieve the same result. These procedures are performed to alleviate chronic pain from pancreatic cancer or chronic pancreatitis. Note that the greater splanchnic nerve originates from the sympathetic ganglia from T5 to T9, the lesser splanchnic nerve from T10 to T11 and the least splanchnic nerve from T12.

CLINICAL HIGHLIGHTS

- The lower extent of thoracic sympathectomy is at T4 for vascular conditions. Buerger disease (thromboangiitis obliterans), Raynaud phenomenon, and collagen vascular diseases are all indications for thoracic sympathectomy in properly selected patients.
- T5 is the lower extent of thoracic sympathectomy commonly used for hyperhidrosis. Long-term relief of palmar hyperhidrosis results from sympathectomy T2-T5. This relieves the excessive sweating of the hand that may result in occupational or social problems.
- Remember that Horner syndrome comprises unopposed parasympathetic input causing ptosis, meiosis, and anhydrosis.
- Splanchnicectomy can be done minimally invasively to interrupt the greater (T5-T9), lesser (T10-T11), or least (T12) splanchnic nerves or interruption of the celiac plexus. These procedures are done to relieve chronic visceral pain typically from unresectable pancreatic cancer or chronic pancreatitis.
- Some surgeons advocate transthoracic vagotomy along with splanchnicectomy for pancreatic pain relief.

ESOPHAGUS, THORACIC DUCT, RECURRENT LARYNGEAL NERVES, AND VAGUS NERVES

INCISION

(This section needs to be coordinated with trans-hiatal esophagectomy in Chapter 6.) Create a window in the posterior pericardium to allow visualization of the underlying esophagus and aorta. Identify the left vagal trunk as it transits to the anterior esophagus (Fig. 3.31A, B). Also, identify the right vagus as it rotates from a lateral to posterior position relative to the esophagus.

OPERATIVE PROCEDURE

Trace the vagus nerves from the thoracic inlet to the diaphragm. Note that the right vagus nerve gives off the right recurrent laryngeal nerve, which loops under the subclavian artery. On the left side, the recurrent laryngeal nerve divides from the vagal trunk and tracks below the arch of the aorta to become recurrent and ascend in the tracheoesophageal groove.

On the right side, note that the cupola of the right hemithorax has the brachial plexus adherent to the parietal pleura. Identify the inferior portion of the brachial plexus as it drapes over the first rib. On the left side in a similar fashion, identify the subclavian vein and inferior brachial plexus. Note their relationships to the apex of the lung.

Continuing down the esophagus, identify the vagal plexus (Fig. 3.32). The right and left vagus nerves interconnect via the vagal plexus. Note that two vagal structures will traverse the esophageal hiatus 88% of the time (Fig. 3.31A), four vagal structures 7% of the time, and more than four vagal structures 5% of the time (Fig. 3.31B).

The thoracic duct is located in the right hemithorax just above the diaphragm (Fig. 3.32). Carefully dissect the esophagus away from surrounding structures. The thoracic duct will be identifiable between the esophagus and the azygos vein. Remember that it is simply a lymphatic structure so expect a thin white friable structure in the fat located between the esophagus and azygos vein. This is the location of a thoracoscopic approach to treatment of chylothorax. Trace the thoracic duct superiorly. This may be a very challenging dissection and may have limited clinical application. Note that the thoracic duct traverses to the left side at about the level of T5 where it ascends on the left side of the esophagus. The thoracic duct then tracks anteriorly to enter the venous system at the junction of the subclavian and jugular veins (Fig. 1.46).

Esophagectomy and use of the stomach as a conduit for either Ivor-Lewis or transhiatal esophagectomy will be performed in Chapter 6, Section 7.

CLINICAL HIGHLIGHTS

- Thoracic arch aneurysms may present with hoarseness from stretch of the left recurrent laryngeal nerve.
- The thoracic aorta is anchored by the aortic valve, left subclavian artery, ligamentum arteriosum, and

the diaphragm. High-velocity trauma may disrupt the aorta due to these solid anchors. The most common is at the ligamentum arteriosum.

- The vagus commonly is a plexus until the esophageal hiatus. Transthoracic vagotomy is done at the esophageal hiatus for this reason.
- The left vagal trunk rotates anteriorly and the right vagal trunk rotates posteriorly to enter the abdomen from the chest. Typically, the posterior vagus is a single large trunk in the abdomen. The anterior vagus may have additional branches at the esophageal hiatus.
- "Criminal nerve of Grassi" may be a high posterior branch traversing the esophageal hiatus separately. This nerve may be the cause of an incomplete vagotomy performed for peptic ulcer disease. Remember this detail when performing a truncal vagotomy.
- The vagus is a single trunk: left (anterior), 96% at *4 cm above* the diaphragm, and right (posterior), 92% *at the level of* the diaphragm.
- Thoracic outlet syndrome may present with neurologic, venous, or arterial symptoms. Most common is a neurologic presentation of ulnar distribution of symptoms in the fourth and fifth fingers.
- Pancoast tumors of the apex of the lung may also present with neurologic symptoms of the upper extremity. These symptoms are caused by tumor in growth into that portion of the brachial plexus, which lies on the cupola of the hemithorax.
- In the OR, the thoracic duct is difficult to identify and isolate. To address chylothorax operatively, mass ligature in the area described above is used for those patients who fail medical therapy (Fig. 3.32).

4

Abdominal Wall

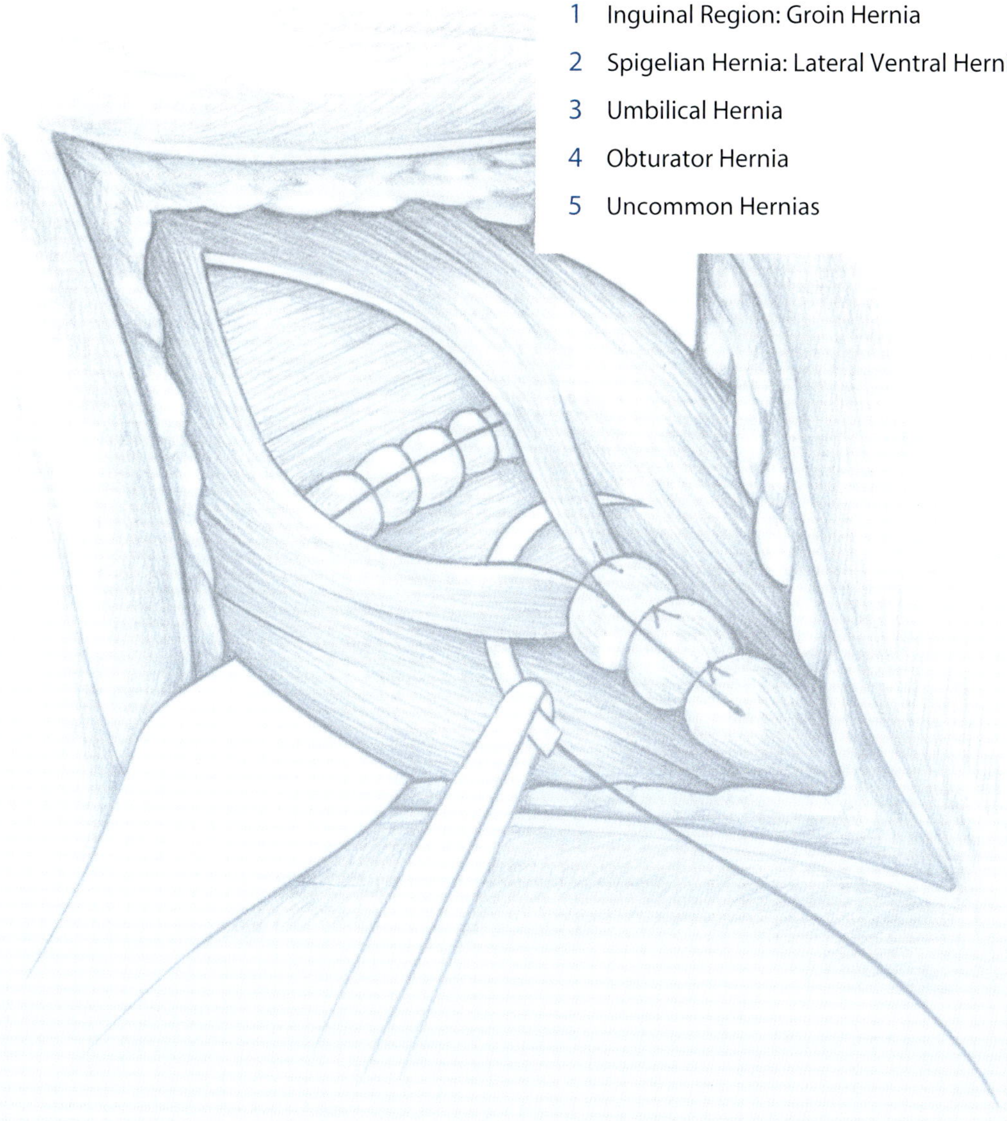

Outline

INGUINAL REGION: GROIN HERNIA

INCISION

Prior to making an incision, think about the location of the internal and external rings through which the spermatic cord runs. The cord drapes over the pubic ramus near the pubic tubercle. The inguinal canal and spermatic cord run in a slanted vertical alignment (Fig. 4.1). For inguinal herniorrhaphy, you may be surprised that an incision placed 30° lateral to straight vertical will align your incision with the directional run of the spermatic cord and inguinal canal. Begin the incision at the pubic tubercle and run superiorly. This incision not only improves visibility of the open repair but also minimizes superficial cutaneous nerve damage.

Make the incision described above ~10 cm in length in the anatomy lab. Note that during open herniorrhaphy in the OR, an incision approximately half this length will suffice. Carry the incision down to the external oblique fascia (see Fig. 4.1).

OPERATIVE PROCEDURE

Clean off the external oblique fascia inferiorly to the inguinal ligament, which is also called the iliopubic tract or Poupart's ligament. Externally, clean off this ligament down to the femoral fascia.

When doing a pediatric hernia, the inferior transverse abdominal fold is used for a transversely aligned incision. The Scarpa fascia in children may mimic the external oblique fascia and mislead the surgeon. This error is easily avoided if Poupart's ligament is identified by blunt dissection as done in this cadaveric adult example. The same technique can be used in adults when distorted anatomy due to obesity or large hernias is encountered.

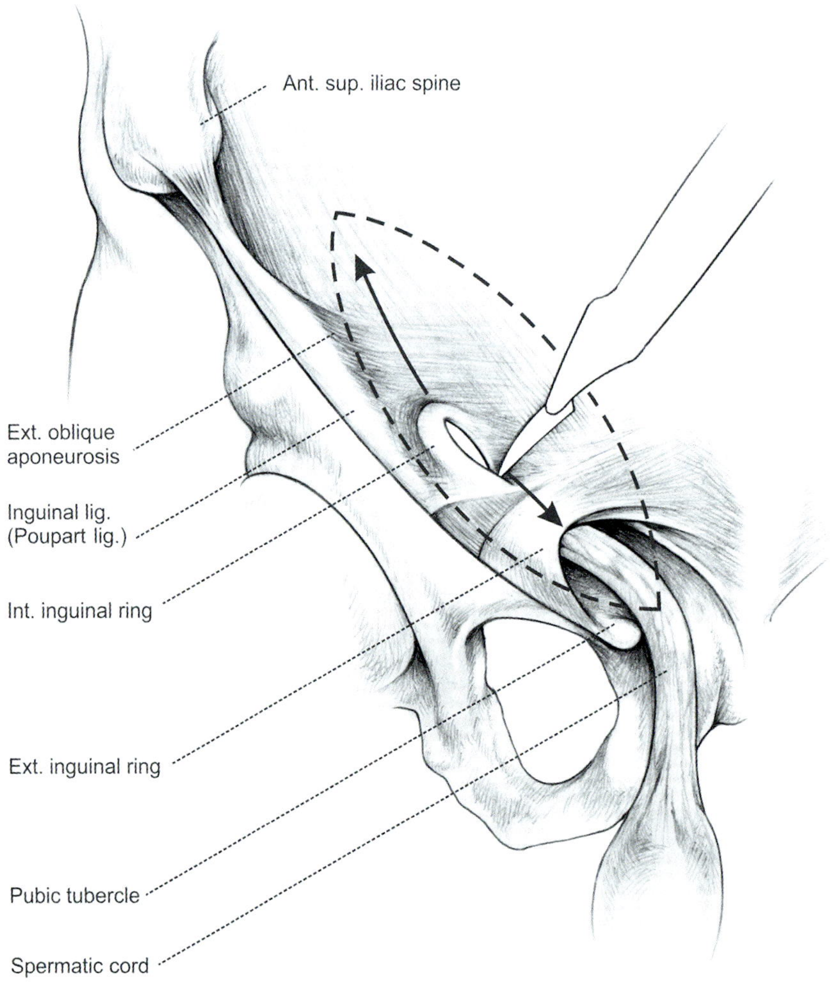

Figure 4.1 Incision alignment of the skin and external oblique fascia.

Note: In female cadaver specimens, use the round ligament instead of the spermatic cord, here described in males.

Clear off and identify the external ring formed by the external oblique fascia (see Fig. 4.1). A Weitlaner self-retaining retractor may be used to hold the skin and fat open to facilitate the dissection. Approximately 2-3 cm above the ring use a scalpel to incise the external oblique fascia for ~5 mm (see Fig. 4.1). Elevate the incised edges of the external oblique with forceps and use a Metzenbaum scissor to spread under the fascia. In line with the fibers, incise the external oblique fascia to the external ring. Spreading under the fascia with scissors is a technique used to avoid injury to the ilio-inguinal nerve running alongside the spermatic cord. Extend the lateral portion of the fascial incision in line with the external oblique fibers to completely identify the entire spermatic cord up to the internal inguinal ring (Fig. 4.2). Kelly or Allis clamps may be used on the edges of the external oblique fascia to hold it open as you may do during an open herniorrhaphy.

Identify the ilioinguinal nerve and separate it from the spermatic cord. Some surgeons will place the nerve behind the Kelly clamp holding the external oblique fascia. The nerve is protected and out of the way. Our illustrations will show the nerve remaining on cremaster muscle.

Elevate the entire spermatic cord by bluntly dissecting the cord off the pubic tubercle. During herniorrhaphy, a Penrose drain may be used to control the spermatic cord. Clear off the fat lateral to the pubic tubercle. The medial fascia thus identified is the transversalis fascia in Hesselbach triangle. Hesselbach triangle is bounded by the inferior epigastric vessels laterally, the rectus sheath medially, and the inguinal (Poupart's) ligament inferiorly (Fig. 4.3). This is the location of direct inguinal hernias and also the "medial recurrence" in recurrent inguinal hernias (Fig. 4.4).

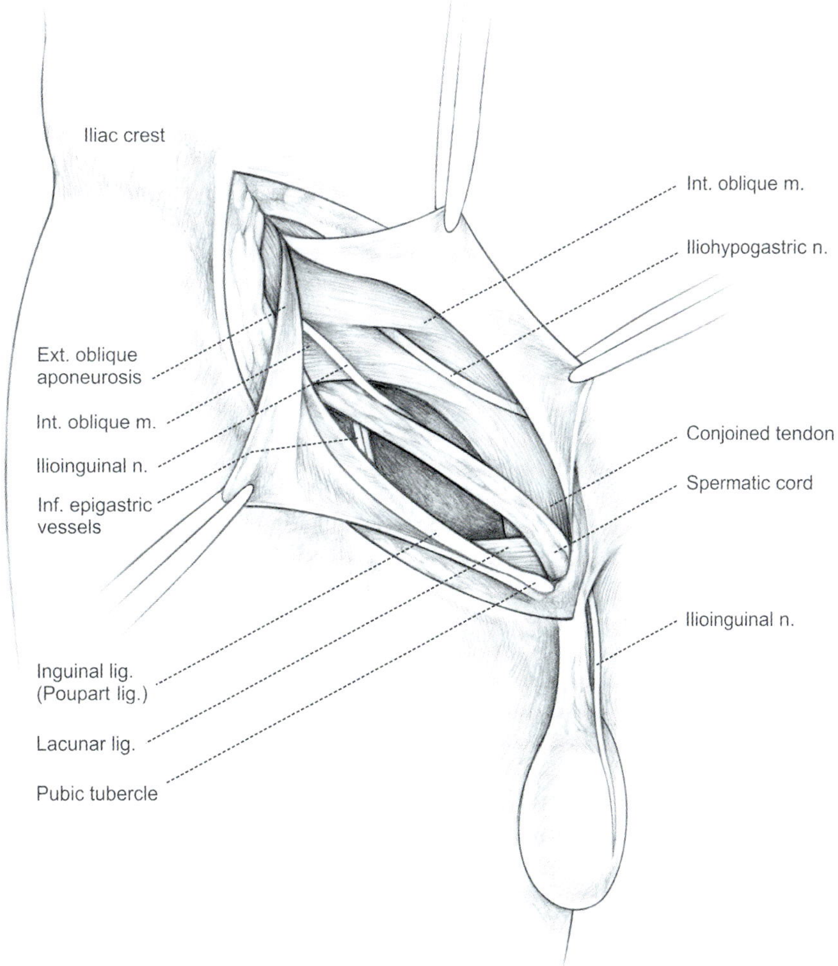

Figure 4.2 Inguinal canal after incision of external oblique fascia.

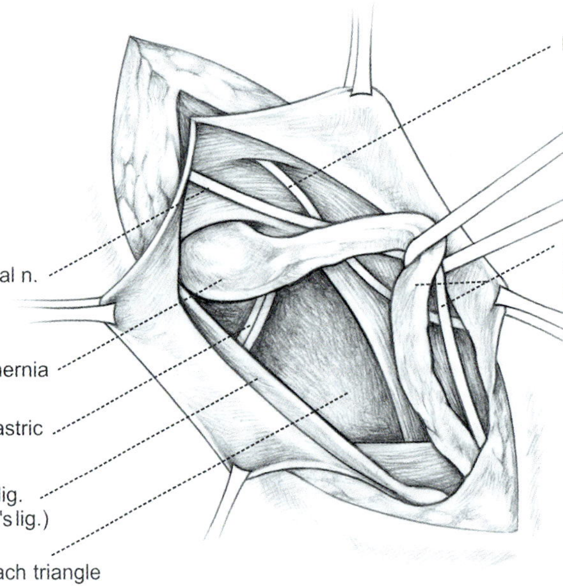

Figure 4.3 Hesselbach triangle and location of indirect inguinal hernia.

Incise the cremaster muscle fibers in line with the direction of the spermatic cord. Carry the incision laterally and widely open the cord to identify the vas deferens, testicular artery, and testicular vein. Identify the internal ring at the superior lateral extent of the spermatic cord. Identify the *anteromedial* area of the cord (see Fig. 4.3). This is the most common location of an indirect inguinal hernia.

Move the spermatic cord to the superior margin of the wound and clean off the inguinal (Poupart's) ligament. Find the empty space medial to the femoral vein that traverses under the ligament. With blunt dissection, open this space and place a finger through it. This is the area where femoral hernias are located (Fig. 4.5A). Place an index finger through this "empty space" and flex the distal phalanx onto the superficial aspect of the external oblique fascia. Clinically, femoral hernias can traverse under the inguinal ligament and flip back up onto the external oblique fascia in just this fashion to mimic an indirect or a direct inguinal hernia. Remove your finger and identify the lacunar ligament (see Fig. 4.5A). Slit the lacunar ligament medially to the pubic tubercle. Note that this is how the surgeon who encounters an incarcerated femoral hernia safely enlarges the orifice to reduce the femoral hernia.

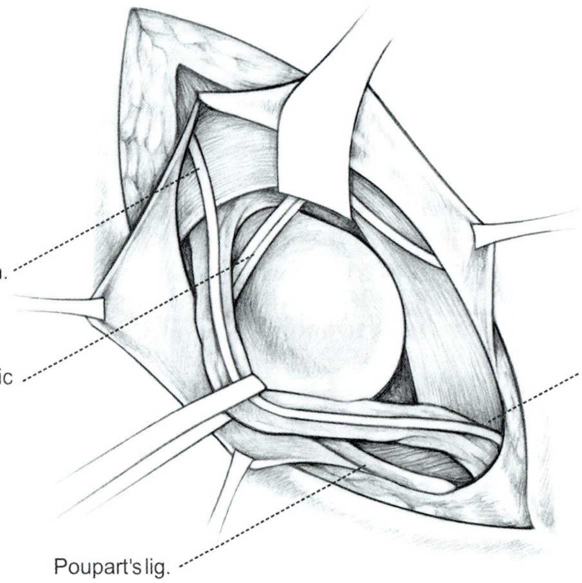

Figure 4.4 Direct inguinal hernia.

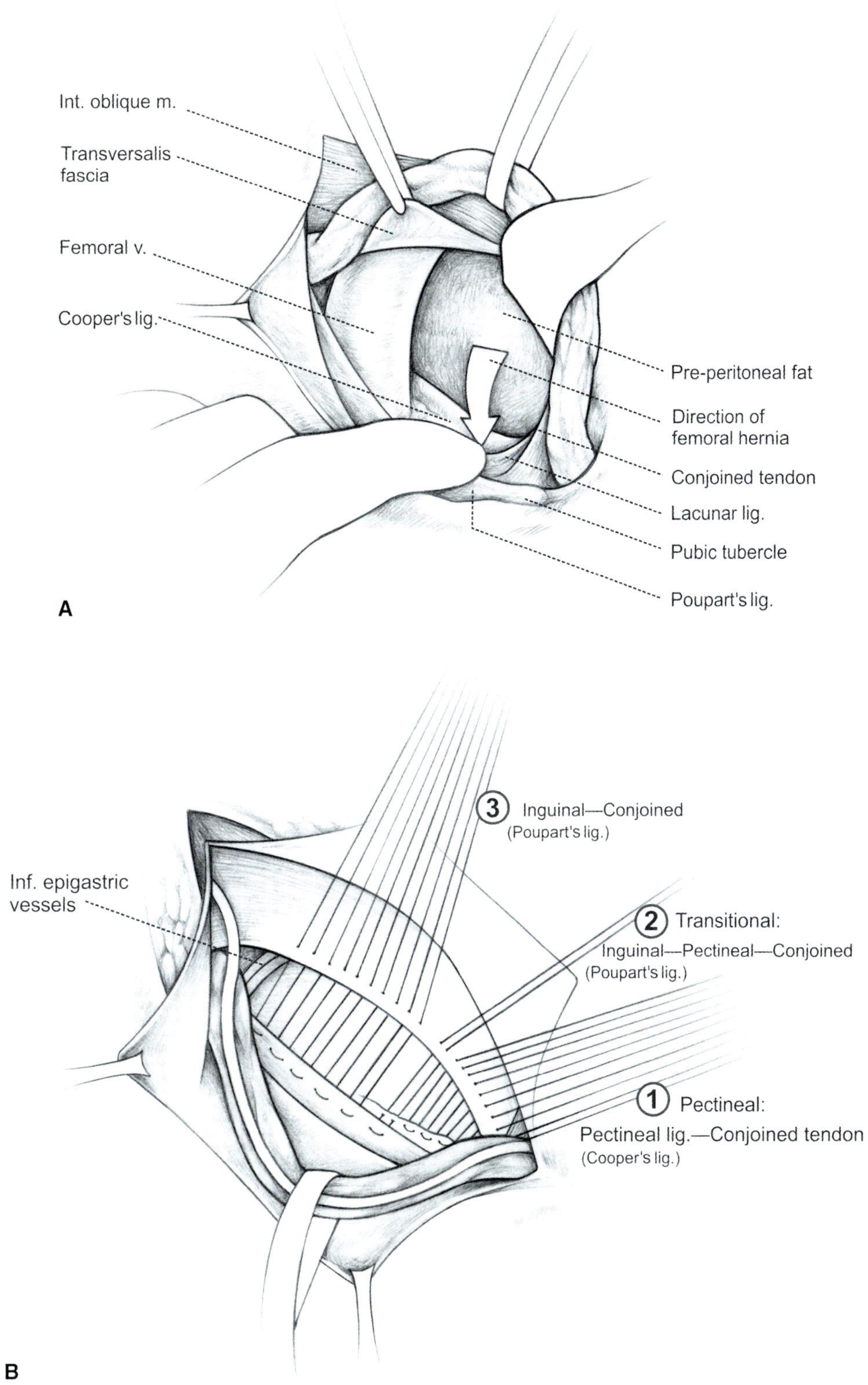

Figure 4.5 **A:** Femoral hernia location. **B:** McVay hernia repair.

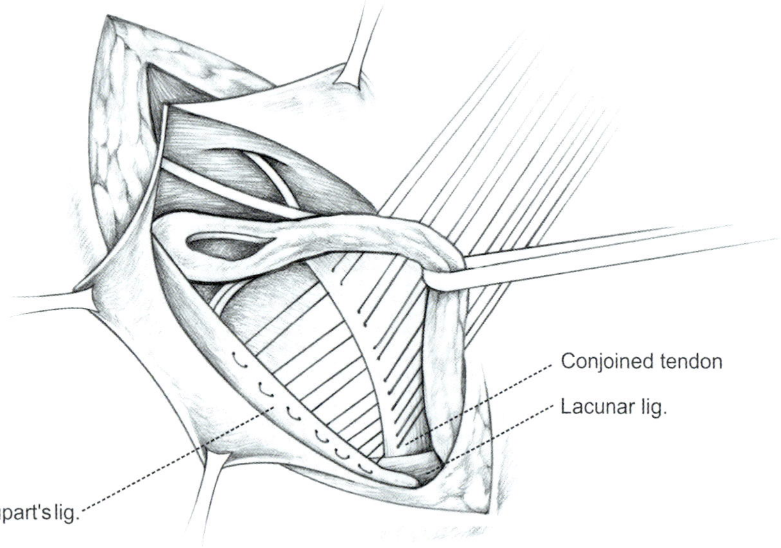

Figure 4.6 Bassini hernia repair.

FINDING COOPER'S LIGAMENT, BASSINI REPAIR, MCVAY REPAIR, LICHTENSTEIN REPAIR

The pectineal (Cooper's) ligament is the firm nearly vertically oriented ligament laying on the superior pubic ramus. Cooper's ligament joins the lacunar (Gimbernat's) ligament, which then joins the inguinal (Poupart's) ligament near the pubic tubercle. Identify internal oblique and transversus muscle fibers. Medially, they form the "conjoined area." Place an Allis clamp on these conjoined fibers and pull them down to the inguinal ligament. Using interrupted suture to join the shelving edge of Poupart's ligament to the conjoined tendon is called a Bassini repair (Fig. 4.6). Using interrupted suture to join Cooper's ligament to the conjoined tendon (using Poupart's ligament lateral to the femoral vein) is called a McVay repair. Remember to place a transition stitch at the transition from Cooper's ligament up to Poupart's ligament near the femoral vein (Fig. 4.5B). A relaxing incision is commonly used to stress relieve these tissue repairs. The relaxing incision is a vertical incision placed medially through elements of the conjoined tendon (Fig. 4.7). Underlying muscle is the rectus abdominis.

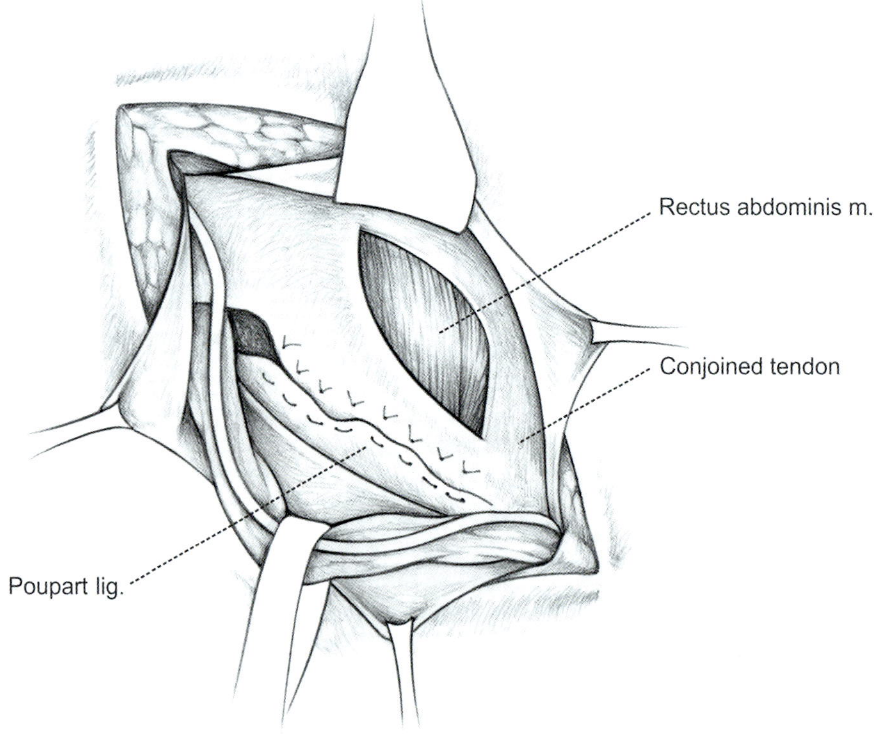

Figure 4.7 Relaxing incision for McVay repair.

After lateral incision of the abdominal wall as directed in Chapter 5, return to this section. When the entire anterior abdominal wall has been mobilized through lateral incisions and the surgeon has returned to this section, elevate the anterior abdominal wall. Use a directional operative light shining at the groin area from the skin side of the abdominal wall. Look from the underside of the abdominal wall. The transversalis fascia in Hesselbach triangle is rather thin and easily transilluminated. This is the area of direct herniation. Lay the anterior abdominal wall back down in its normal position and at the anteromedial portion of the internal ring, place a Kelly clamp through to the intraperitoneal space. Again, elevate the anterior abdominal wall and note the position of the clamp, which is the location of an indirect inguinal hernia.

Tension-free hernia repairs are very popular due to improved postoperative pain and lower recurrence rates. The repair begins the same as the previously performed procedures. A mesh plug is used by some surgeons to control the indirect component anteromedial to the spermatic cord (Fig. 4.8A, B). An overlay mesh

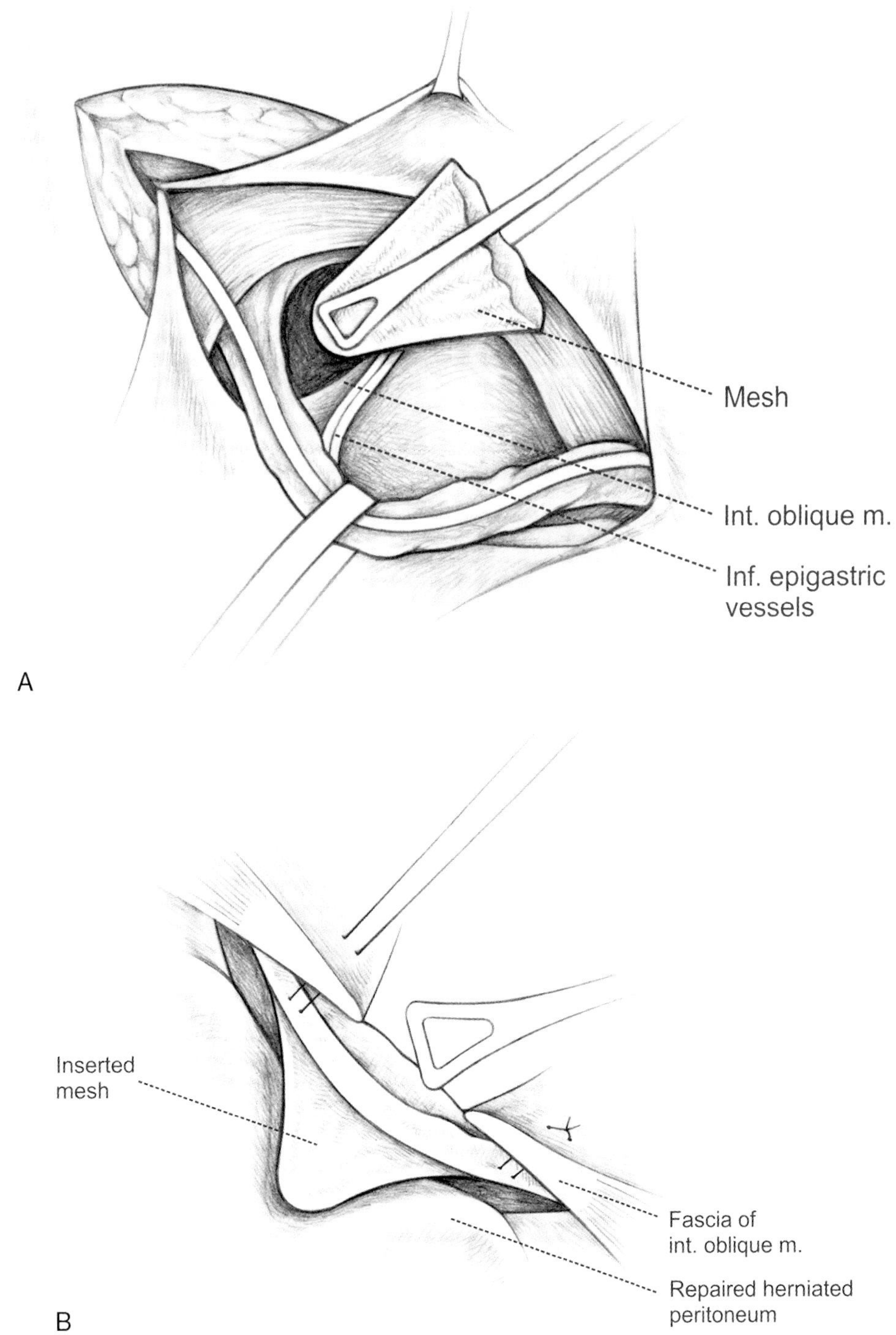

Figure 4.8 **A:** Mesh plug placement for indirect inguinal hernia. **B:** Mesh plug—cross-section view.

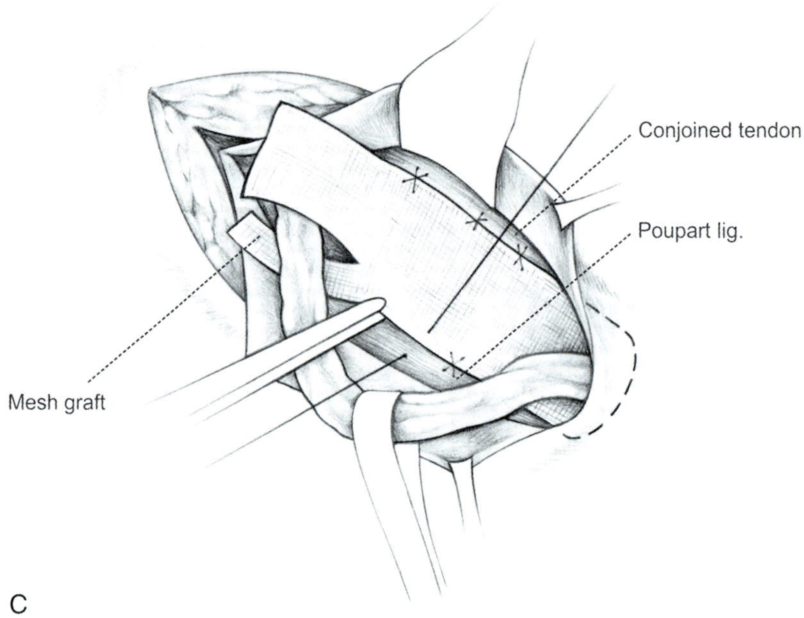

Figure 4.8 (Continued) **C:** Tension-free inguinal hernia repair with mesh.

patch (Lichtenstein repair) is used under the cord. It is affixed to the conjoined tendon and inguinal ligament (Fig. 4.8C). Surgeons use various techniques to affix the mesh: running suture, interrupted suture, suture/no suture at the pubic tubercle, tissue glue, and no suture at all. There is agreement that the mesh must overlay the pubic tubercle by at least 1cm to prevent a medial recurrence. The tails surround the spermatic cord laterally. I never suture the tails; rather they are tucked under the lateral external oblique fascia. By tucking and not suturing, I intend to avoid the problem of impaired venous drainage of the testicle (see Fig. 4.8C).

CLINICAL HIGHLIGHTS

- Indirect inguinal hernias are anteromedial in their relation to the spermatic cord.
- Remember that tight repair at the internal ring will cause constriction leading to impaired venous drainage of the testicle causing atrophy and pain. Some studies report incidence as high as 15%-20%.
- Direct inguinal hernias occur through the transversalis fascia medial to the inferior epipastic vessels in Hesselbach triangle. This is also the most common area of recurrence after herniorrhaphy, the so-called medial recurrence.
- Indirect inguinal hernia is a patent processus vaginalis, which is a congenital condition and may be associated with hydrocele.
- Risk of incarceration and strangulation is highest with femoral hernias and lowest with direct hernias, and indirect hernias are in between.
- Femoral hernias are more common in females than males. However, remember that indirect inguinal hernias are the most common hernias in females.
- Any sign of bowel strangulation requires an abdominal incision and resection through laparotomy.
- The iliohypogastric (L1), inguinal (L1), and genital branches of the genitofemoral nerve (L1, L2) are the sensory nerves of the lower abdomen and groin area. They are variable in their dominance and somewhat variable in their location. The inguinal nerve is typically identified on the surface of the cremaster. However, it is found in its usual location in only 60% of patients (see Fig 4.2). An attempt to identify and document all three nerves should be undertaken.
- The genitofemoral nerve penetrates and lays on the anterior surface of the psoas muscle, continues on the inguinal floor, and descends with the spermatic cord. Although difficult to see, its location can be easily determined by identifying the external spermatic vein or so-called blue line in the OR.
- Remember the structures of the femoral canal in order from lateral to medial: *NAVEL*, femoral *Nerve*, common femoral *Artery*, femoral *Vein*, *Empty* space (femoral hernia, lymphatics), *Lacunar* ligament (see Figs. 4.5A and 10.8).

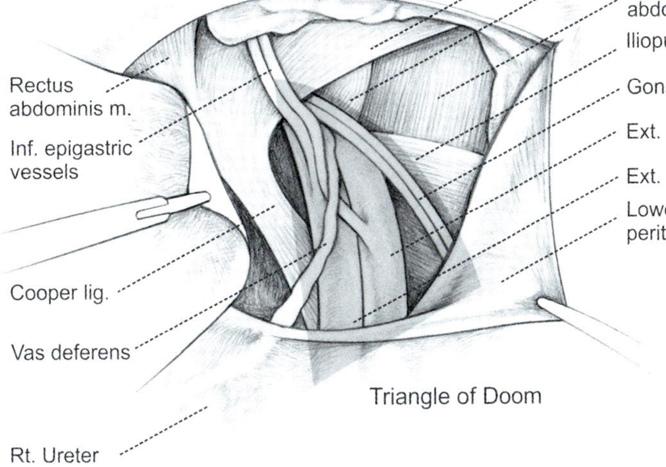

Figure 4.9 Laparoscopic TAPP view after reflection of the peritoneum.

- Triangle of Doom: a term used in laparoscopic hernia repair to emphasize risk of vascular damage if staples are used on Cooper's ligament laterally (Figs. 4.9).
- Circle of Death: a term used to emphasize location of an aberrant obturator artery off the inferior epigastric artery. It is found medial to the vas deferens and descends to the obturator canal (Fig. 4.10). The

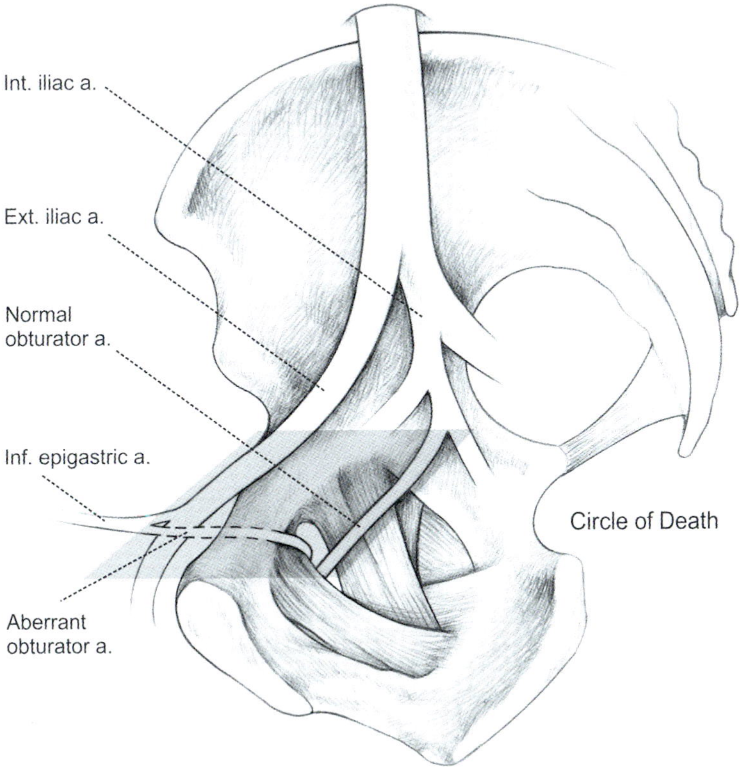

Figure 4.10 Aberrant origin of the obturator artery.

course of this artery will lie on Cooper's ligament and will interfere with a McVay repair or may be injured during hernia repair.

- Laparoscopic hernia repair:
  - Transabdominal preperitoneal repair (TAPP) view after reflecting the peritoneum (see Fig. 4.9).
  - Triangle of doom refers to vascular structures in the area of a TAPP repair.
  - Avoid damage to the lateral femoral cutaneous nerve, which exits the pelvis 1.5 cm medial to the anterior superior iliac spine. Remember when stapling the superior border of the mesh implant not to continue too far laterally into the Triangle of Pain (ToP). ToP is defined inferomedially by the gonadal vessels and superior laterally by the iliopubic tract.

SPIGELIAN HERNIA: LATERAL VENTRAL HERNIA

INCISION

Spigelian hernia occurs at the lateral margin of the rectus sheath inferior to the umbilicus. This defines the intersection of the semicircular line and the lateral rectus border. The semicircular line (of Douglas) is the inferior extent of the posterior rectus sheath.

OPERATIVE PROCEDURE

Make an oblique (or transverse) incision in the area described above extending ~10 cm (Fig. 4.11A). Carry this dissection down to the external oblique and rectus fascia. Place a self-retaining Weitlaner clamp to hold the skin and subcutaneous tissue open. Clear off the rectus and external oblique fascia to gain an appreciation for the so-called spigelian area. Incise in line with the fascial elements the aponeurosis formed by the external oblique muscle with the rectus muscle (Fig. 4.11B). This may entail a single thick aponeurotic layer or three separate and thinner fascial layers. Carry this down to the peritoneum. The surgeon has now created a spigelian hernia (Fig. 4.12).

After mobilization of the anterior abdominal wall, explore the lateral, oblique, and transversus muscles. Note that each is invested (surrounded) by a layer of fascia. Next, examine the rectus sheath and note that the anterior rectus sheath is consistent and thick. It is through this layer that midline incisional (ventral) hernias occur. Flip the abdominal wall inferiorly. Examine the posterior rectus sheath and identify the semicircular line. This line is the point at which the posterior rectus fascia loses its strength and identity. This line can be difficult or impossible to identify in the operating room setting. Return the abdominal wall to its normal position. Repair the defect you created in layers as illustrated (Fig. 4.13).

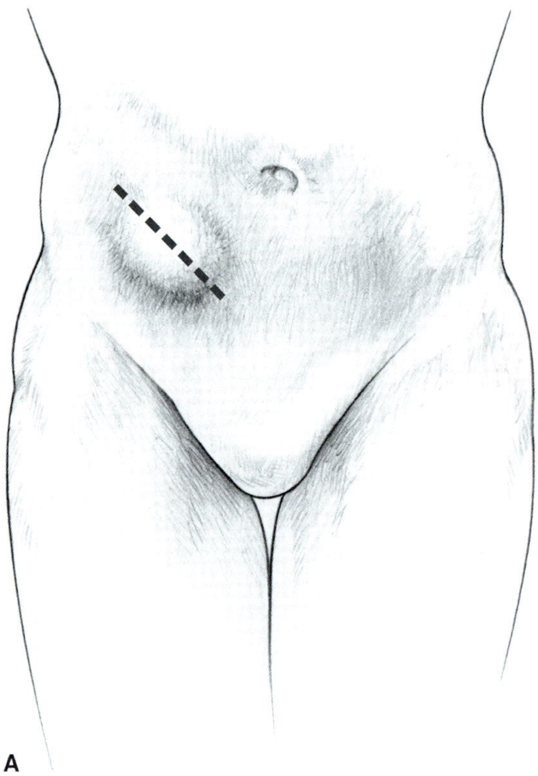

A

Figure 4.11 A: Incision over spigelian hernia.

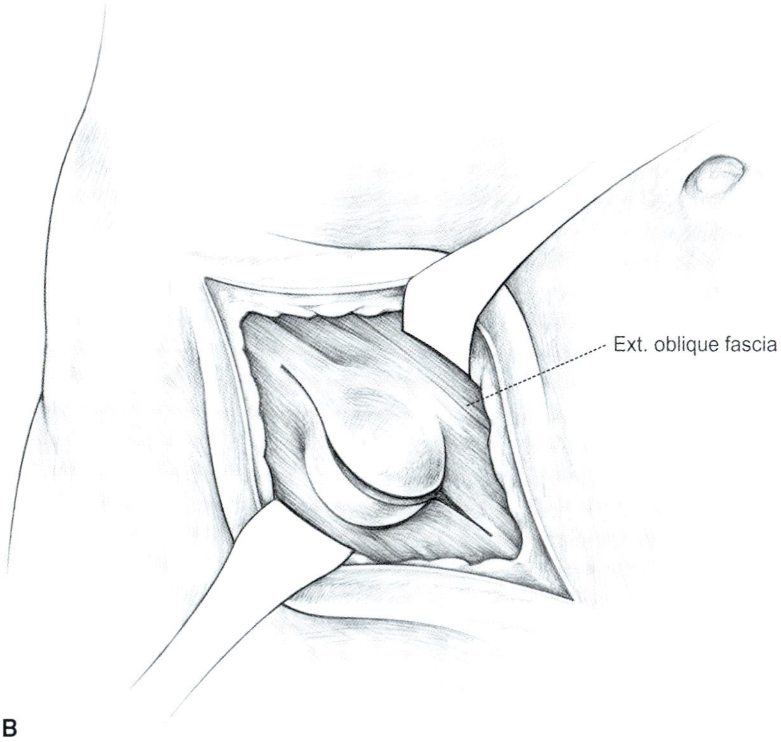

Figure 4.11 (Continued) **B:** Incision of external oblique fascia over spigelian hernia.

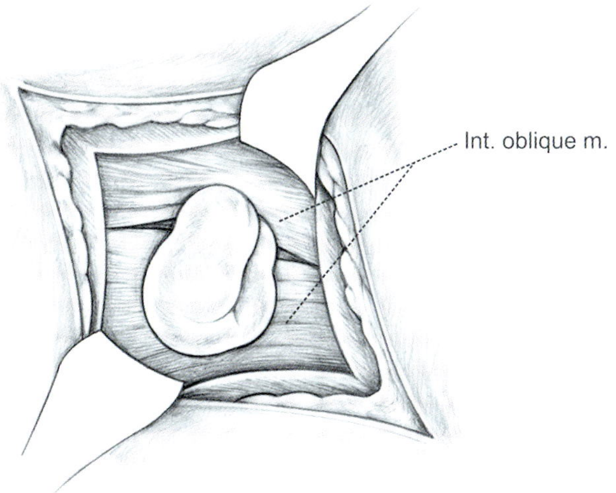

Figure 4.12 Spigelian hernia sac protruding through muscle layers.

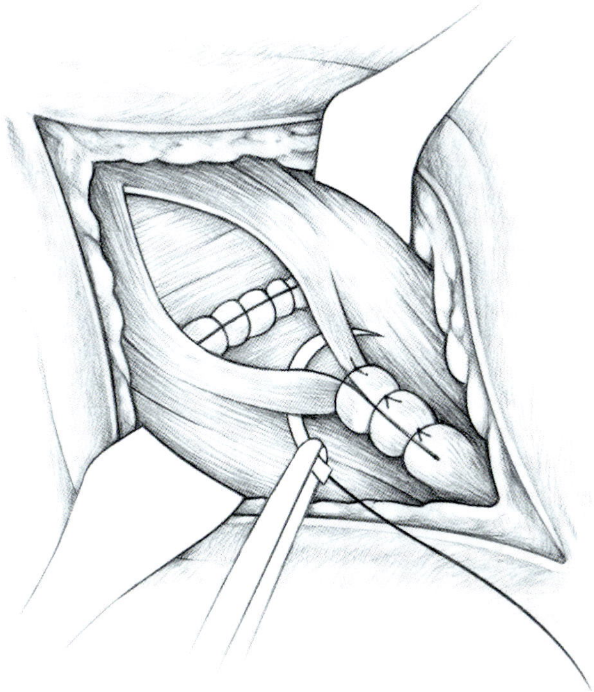

Figure 4.13 Spigelian hernia tissue repair in layers.

CLINICAL HIGHLIGHTS

- Spigelian hernias are rare (1%) and are also known as spontaneous lateral ventral hernias.
- A spigelian hernia may contain preperitoneal fat, a peritoneal sac, or least likely a segment of bowel.
- Spigelian hernia repair technique is much like groin hernias as previously performed. Alternatively, repair may be accomplished like incisional ventral hernias using a laparoscopic tension-free mesh hernia repair.
- These hernias can be very difficult to diagnose, and a high degree of clinical suspicion must be gained from the patient's history and localization of pain.
- If the hernia is incarcerated, incise the medial aspect of the ring toward the rectus sheath.

UMBILICAL HERNIA

INCISION

Create a periumbilical incision (Fig. 4.14). This may be supraumbilical or infraumbilical. Typically, in the operating room, the incision will encompass 50% of the circumference of the umbilicus. Continue the dissection to the attachment of the umbilical skin to the rectus fascia. Divide this attachment. Use a self-retaining Weitlaner retractor to hold skin and subcutaneous tissue from the fascia.

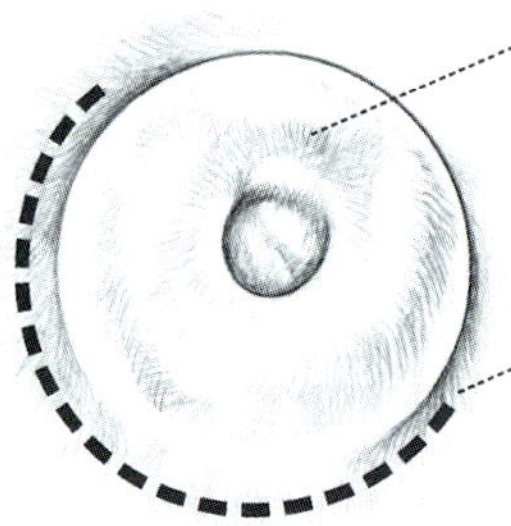

Figure 4.14 Periumbilical incision.

OPERATIVE PROCEDURE

If an umbilical hernia is present, sharply dissect the preperitoneal fat and sac from adjacent fascia (Fig. 4.15). If a hernia is not present, divide the linea alba transversely at the point of umbilical skin attachment. Make the facial incision ~1-2 cm in length. Holding the facial edges open, note that the preperitoneal fat is underneath. Bluntly dissect the preperitoneal fat from the posterior rectus fascia using blunt finger dissection. Observe the space now created. This is the dissection pattern for umbilical herniorrhaphy.

Begin the repair with horizontal mattress sutures (Fig. 4.16A) placed to pull the inferior fascial over the superior fascial edge ("pants-over-vest" repair)

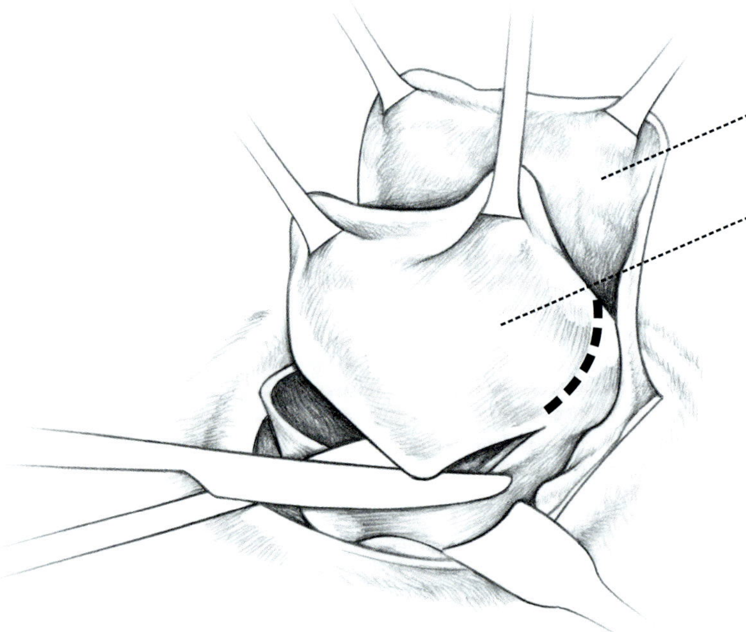

Figure 4.15 Umbilical hernia sac dissection.

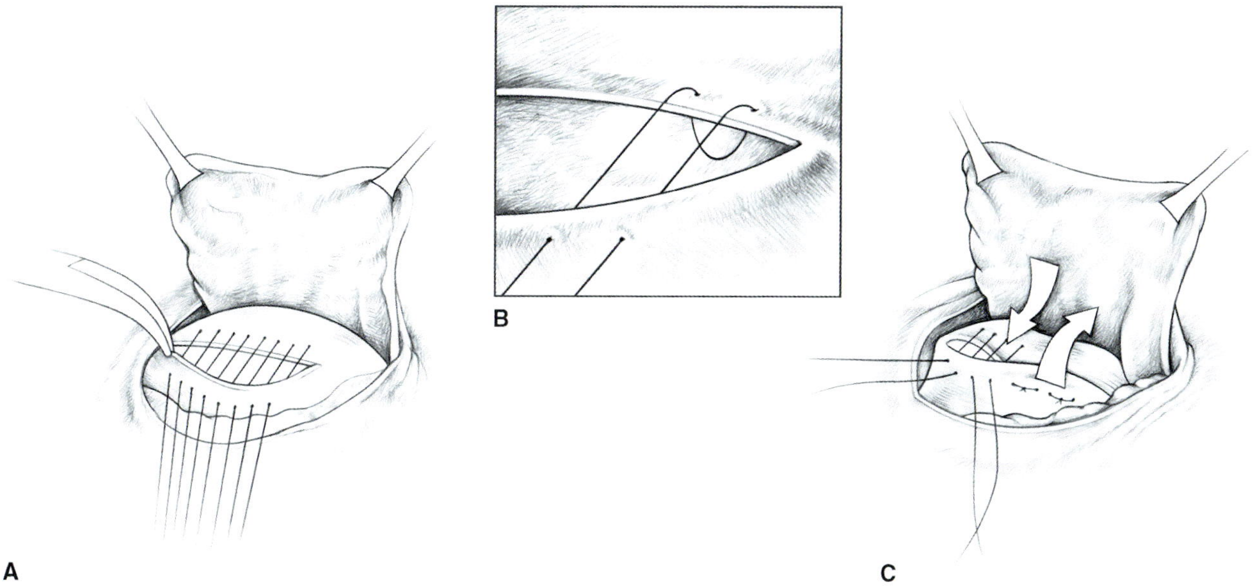

Figure 4.16 **A:** Horizontal mattress suture placement for umbilical hernia repair. **B:** Close-up view of suture placement for umbilical hernia repair. **C:** Fascial closure paths indicated.

(Fig. 4.16B, C). Complete the fascial repair with interrupted suture on the free inferior fascial edge to underlying fascia (Fig. 4.17A). Remember to reconstruct the umbilicus by suturing the central posterior skin to the underlying fascia (Fig. 4.17B). Several other umbilical hernia repair methods may be considered.

CLINICAL HIGHLIGHTS

- A second repair method is simple interrupted suture.
- A third method, which has become more popular particularly in adults to stress relieve the repair, is a tension-free mesh repair. This may be practiced using cheesecloth, which is typically abundant in an anatomy lab and suture. A 2-3 cm overlap of mesh is typically recommended. The mesh is placed below the fascia in the space created by blunt finger dissection. Interrupted suture may be used to anchor the mesh (cheesecloth) in place.
    - Tension-free hernia repair utilizing nonabsorbable mesh has become the dominant method of ventral (incisional) hernia repair in the United States.
- Remember that children <5 years old are not necessarily candidates for umbilical herniorrhaphy unless symptomatic. Many of these umbilical hernias will resolve if 1 cm in size or less.

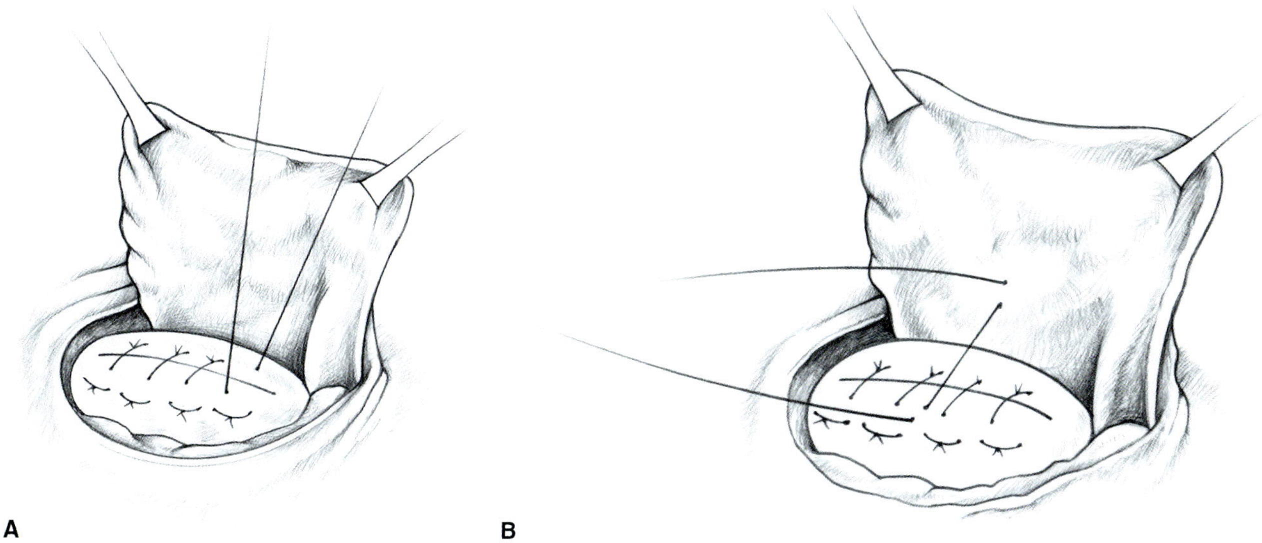

Figure 4.17 **A:** Free fascial edge anchored to underlying fascia. **B:** Umbilical skin sutured to fascial repair.

- With all umbilical hernia repairs, do not forget to suture the umbilical skin down to the underlying fascia. This is done to recreate a more normal-looking umbilicus (see Fig. 4.17B).

OBTURATOR HERNIA

INCISION

Open the abdominal flap as described previously. Displace the pelvic organs (if still remaining) to the left side of the pelvis. Timing of this section is best done after rectal, uterine, and/or bladder procedures have been done. The obturator canal lies on the anterior portion of the pelvic sidewall.

OPERATIVE PROCEDURE

Strip the pelvic peritoneum away from the anterior and lateral pelvis. Identify the obturator canal by identifying the obturator artery, vein, and nerve entering it. Note that the obturator artery and vein typically lie below the obturator nerve. The obturator nerve splits into the anterior and posterior division, and an obturator hernia may follow either of these obturator nerve branches (Fig. 4.18A). The obturator hernia sac will exit the pelvis below the pectineus muscle (see Fig. 10.8).

Clearly identify the contents of the obturator canal (nerve, artery, and vein). Avoiding damage to these structures as they enter the obturator foramen, enlarge the foramen at its inferior wall. This maneuver may be necessary during hernia repair if incarcerated abdominal contents are within the obturator hernia sac. After reduction of hernia contents from the obturator foramen, tissue closure of obturator internus to the pectineal (Cooper's) ligament may be performed (Fig. 4.19). Alternatively, a tension-free repair may be performed with mesh (cheesecloth in the anatomy lab) and suture as illustrated (Fig. 4.18B).

Be aware of possible aberrant obturator vein and artery presence. Instead of coursing from the internal iliac

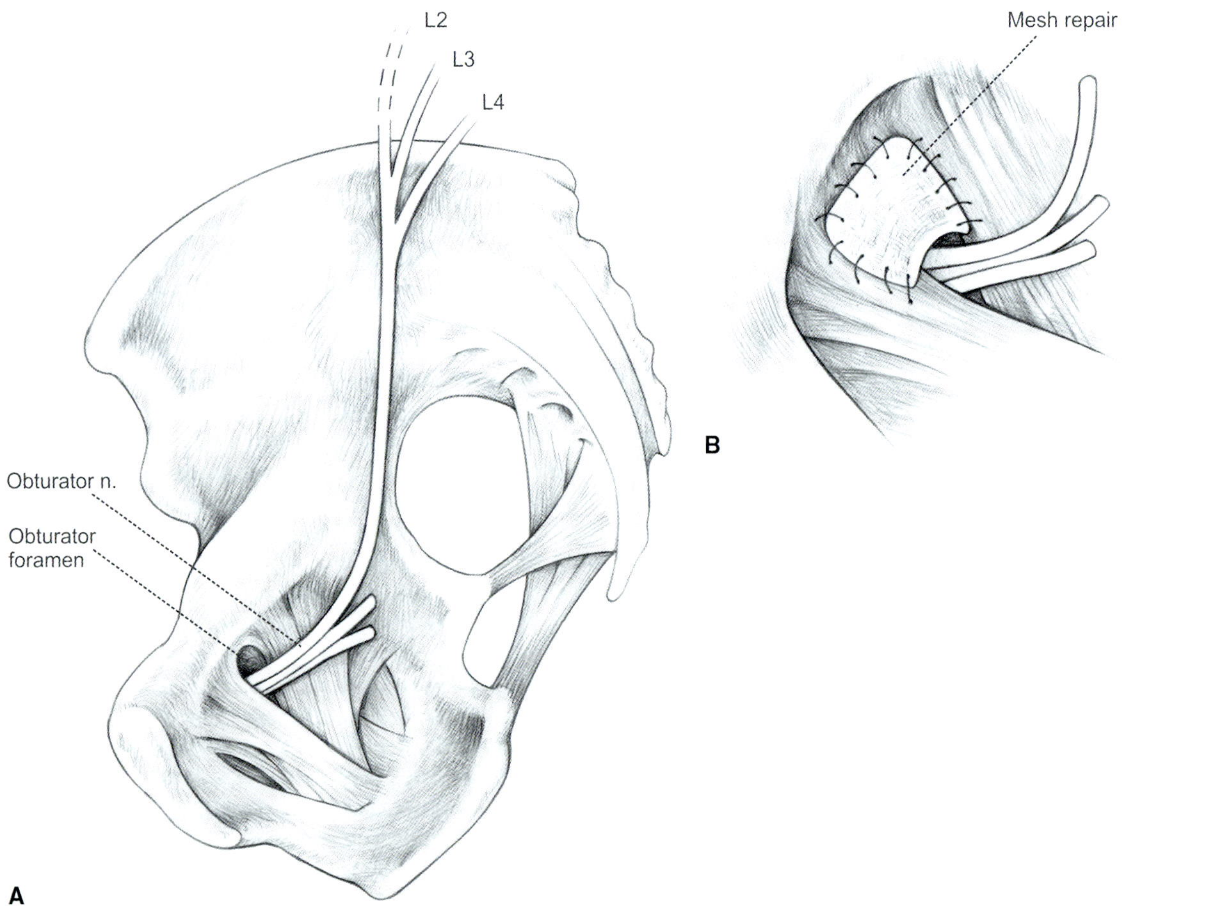

Figure 4.18 A: Normal course of the obturator artery, vein, and nerve to the obturator foramen of the pelvis. **B:** Obturator hernia mesh repair.

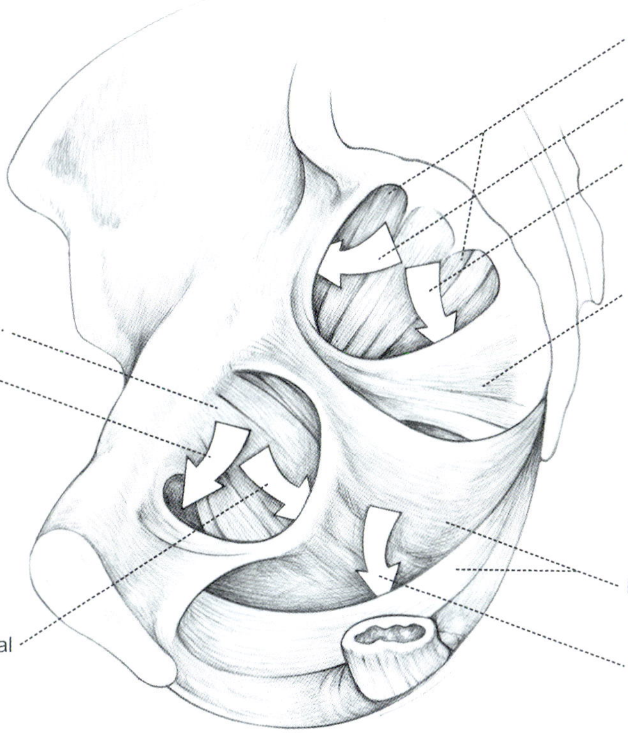

Figure 4.19 Pelvic hernia tracks illustrated by *arrows*.

vessels, these aberrant vessels have a superior course to anastomose with the inferior epigastric artery and external iliac vein, respectively. They interfere with tissue repair in this area. These vessels are dangerous if injured because a damaged segment of vessel may retract through the obturator canal and become very difficult to identify. Historically, aberrant obturator artery and vein are referred to as the Circle of Death (corona mortis) (see Fig. 4.10).

CLINICAL HIGHLIGHTS

- The peritoneal sac is typically medial to the obturator nerve, artery, and vein.
- Compression of the obturator nerve:
  - Produces pain in the medial thigh.
  - Produces weakness of the adductor longus, adductor brevis, adductor magnus, and obturator externus muscles.
  - Symptoms may be reproduced during lateral pelvic wall palpation during rectal or vaginal exam.
  - Internal rotation of the abducted thigh may also reproduce the pain (Howship-Romberg sign).
- Other internal pelvic hernias may also occur (Fig. 4.19).

UNCOMMON HERNIAS

These uncommon hernias are presented here to illustrate the fact that clinical awareness of hernia symptoms should not be dismissed simply because the location is not in a typical anatomical area.

PARADUODENAL HERNIAS

Do this section after opening the abdomen. These hernias occur on the outer edge of the duodenal C-loop. A portion of bowel can invaginate and reduce or remain and incarcerate in the areas illustrated (Fig. 4.20). Repair is dictated by the specific circumstances encountered.

LUMBAR HERNIAS

Do this section along with rectal and buttock dissection.

Type 1 (Grynfelt) superior

Type 2 (Petit) inferior

Two locations have been described. The inferior position occurs at the superior margin of the iliac

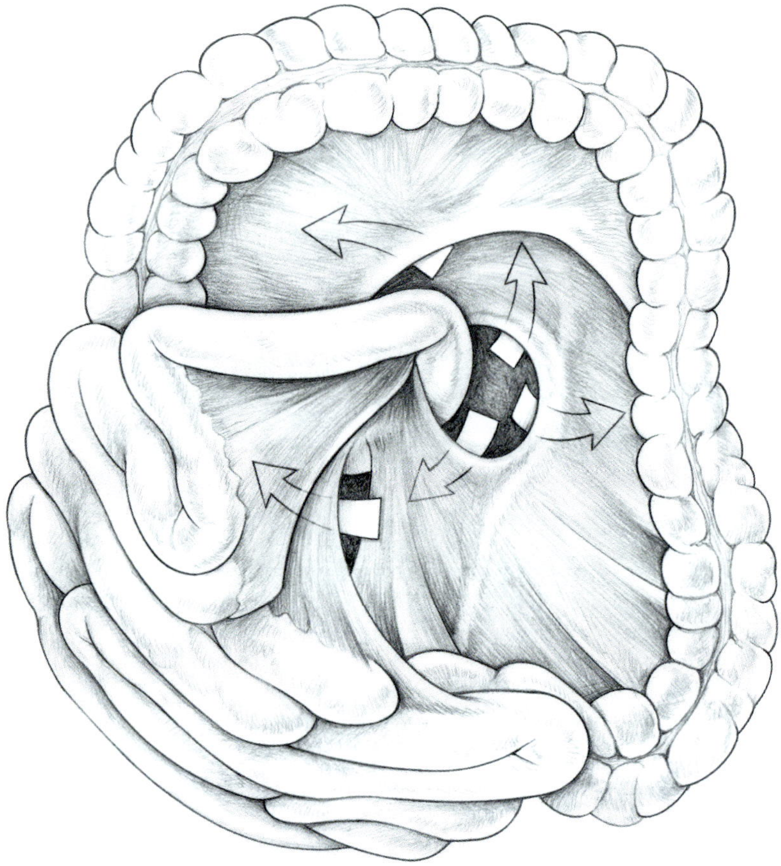

Figure 4.20 Paraduodenal hernia tracks illustrated by *arrows*.

crest and is bordered by the external oblique and latissimus dorsi muscle forming Petit's triangle. The superior position occurs below the 12th rib and is bordered by the external oblique, internal oblique, and erector spinae muscles forming Grynfelt's triangle (Fig. 4.21).

PELVIC HERNIAS

Pelvic hernias occur between muscle layers. Several routes of travel for intermuscular pelvic hernias are illustrated (Fig. 4.19). Other pelvic and perineal hernias are described in the surgical literature.

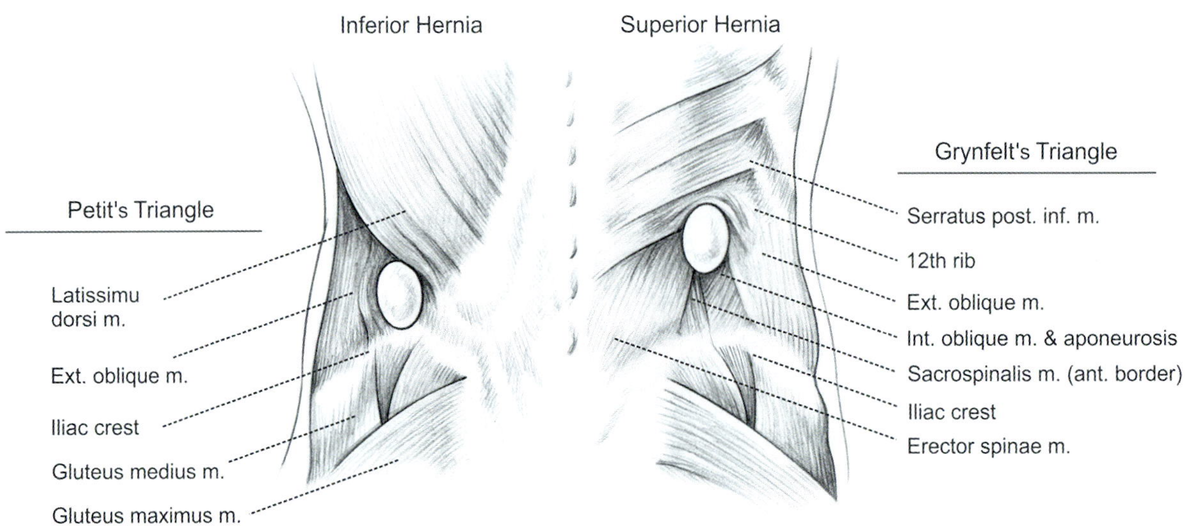

Figure 4.21 Locations of lumbar hernias.

5

Small and Large Intestine

Outline

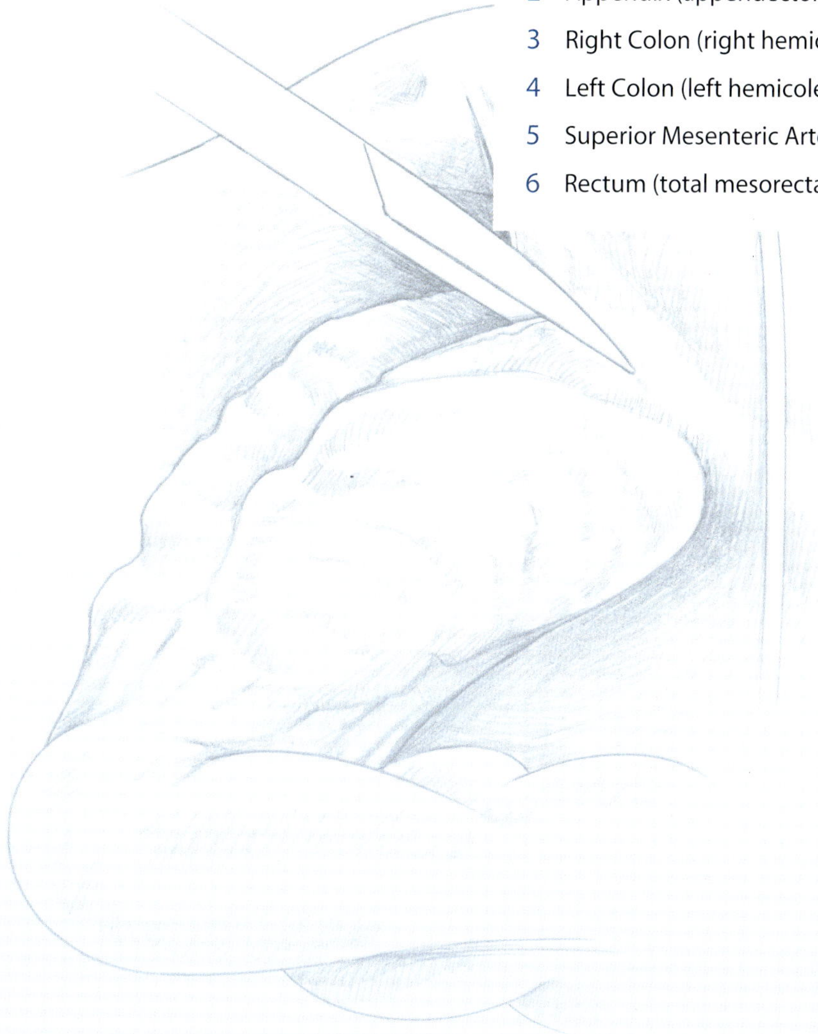

OMENTUM (OMENTAL MOBILIZATION, OMENTECTOMY)

INCISION

Beginning at the xiphoid process, make a full-thickness bilateral subcostal incision through all abdominal wall layers. Use the mid- or posterior axillary line to continue incising the abdominal wall down to the pelvis. Carry this incision anteriorly over the anterior iliac crest. Do not disrupt the inguinal regions. Now lift the anterior abdominal wall. Note that the falciform and round ligaments will have to be divided. The entire anterior abdominal wall can then be flipped caudally onto the thighs to reveal the abdominal contents (see Chapter 3, "Thorax" for chest wall mobilization to facilitate the abdominal sections.) (Fig. 5.1).

OPERATIVE PROCEDURE

Identify the blood supply to the left side of the omentum coming from the left gastroepiploic artery, which is a branch of the splenic artery. Lift the omentum to find the avascular plane between the transverse colon and omentum. Completely take the omentum away from the colon and note the attachments of the splenic flexure to the omentum and to the spleen. During transverse or left colectomy, splenic damage may occur due to traction on these attachments.

Identify the blood supply to the omentum on the right side. Notice that the blood supply comes from the right gastroepiploic artery (Fig. 5.2). (This critical artery will be addressed when dealing with the stomach.) Dissect the greater omentum free of the right gastroepiploic artery. Preserve the right gastroepiploic artery and mobilize the omentum toward the left side of the abdomen. The free pedicle of omentum may be used to wrap an anastomosis to prevent a leak.

CLINICAL HIGHLIGHTS

- The blood supply to the omentum comes from both the right and left gastroepiploic arteries. The omentum can be mobilized from the left side of the abdomen or from the right side of the abdomen.

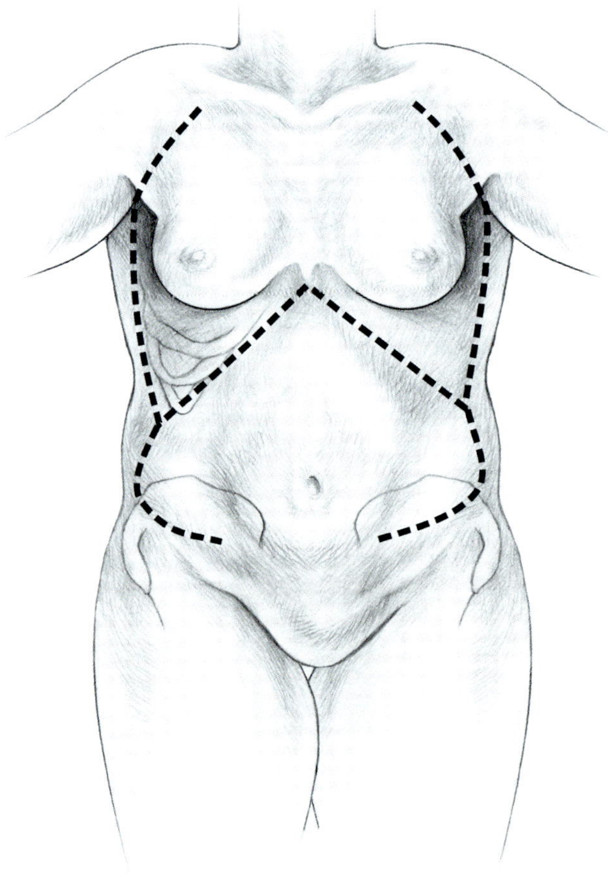

Figure 5.1 Pattern of incisions on the thorax and abdomen.

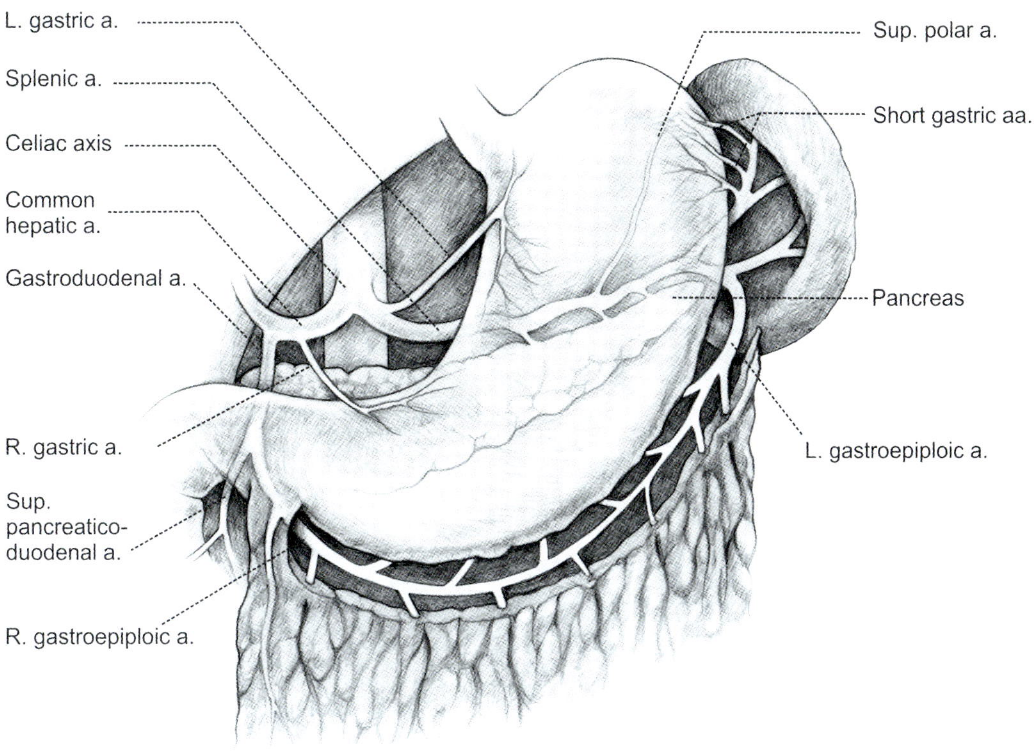

Figure 5.2 Blood supply to the stomach and omentum.

- The anastomosis between right and left gastroepiploic arteries is attenuated, commonly absent (see Fig. 5.2). However, a rich crossover vascular network between the right and left is present in the body of the omentum.
- Preserving gastroepiploic blood supply to the omentum allows it to be used as a wrap. Omental reinforcement of anastomoses has been described in many areas of the body. However, the length necessary to reach these areas is attained by understanding how to mobilize a viable omental pedicle as you have done in this dissection.
- The left gastroepiploic artery arises from the splenic artery and travels in the presplenic fold making it difficult to identify (see Fig. 5.2).

APPENDIX (APPENDECTOMY)

INCISION

An often used incision for appendectomy is a transverse Rocky-Davis incision situated at the lateral border of the rectus muscle transversely between the iliac crest and the umbilicus (Fig. 5.3). The internal oblique and

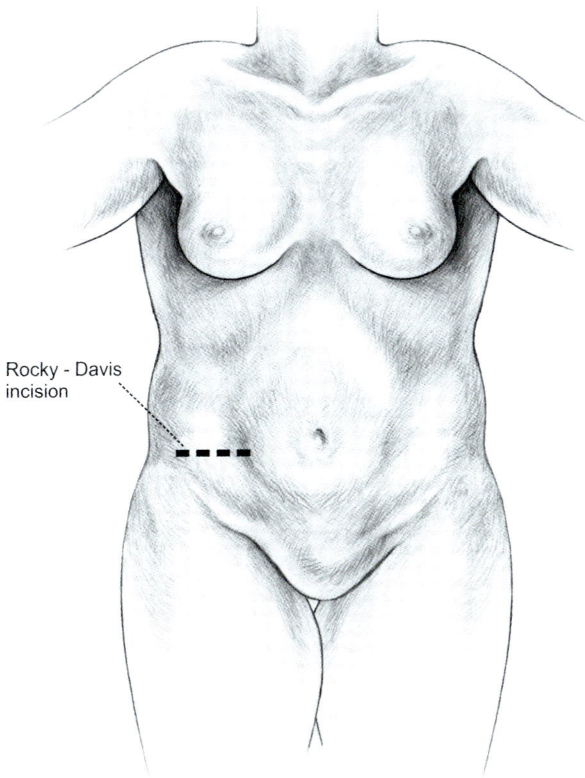

Figure 5.3 Appendectomy, Rocky-Davis incision.

transversus muscles are split in line with their fibers. The peritoneum is opened and retracted to gain access to the inflamed appendix.

OPERATIVE PROCEDURE

The lateral peritoneal attachments of the cecum to the lateral abdominal wall are sharply and bluntly dissected free. In the operating room, a Babcock clamp can be used to assist delivery of the appendix into the wound. Use the ileal fat pad (Fig. 5.4) to guide you to the base of the appendix. This is a very reliable anatomic marker and is found on the antimesenteric border of the terminal ileum. Identify the mesentery of the appendix and ligate it with 2-0 or 3-0 silk (Fig. 5.5). With one click on a Kocher clamp, crush the base of the appendix. Then move the clamp distally on the appendix. At the crushed impression, tie off the base of the appendix with an 0-chromic ligature (Fig. 5.5). Divide the appendix along the Kocher clamp. Remove the appendix and clamp. Some surgeons use cautery and invert the base of the appendix.

CLINICAL HIGHLIGHTS

- The ileal fat pad is a consistent anatomic marker, which is very useful in the inflamed right lower quadrant. Use it to orient yourself and to find the appendix. It will reliably lead you down to the appendix even if other landmarks are obscured by inflammation or abscess formation.
- Inversion of the base of the appendix is optional. If you do, a "Z" stitch may be easier to use than a purse string seromuscular stitch.

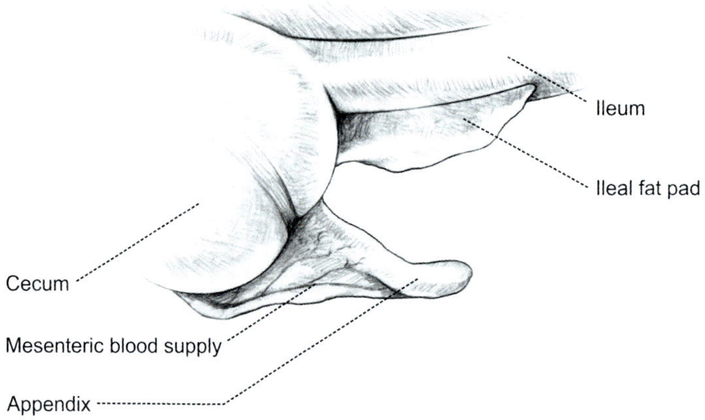

Figure 5.4 Blood supply to the appendix and the ileal fat pad as a guide to find the appendix.

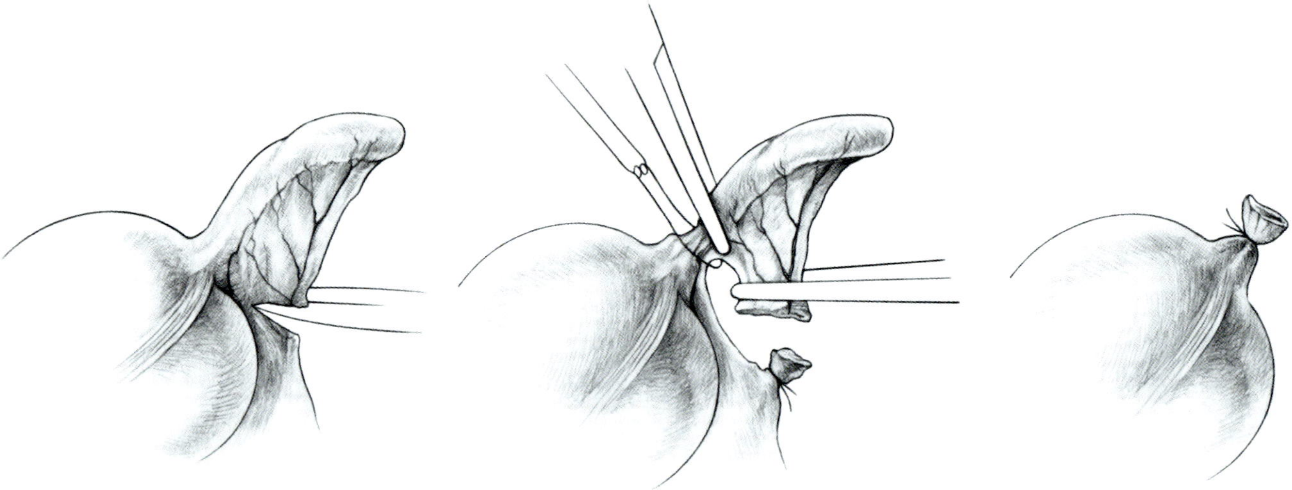

Figure 5.5 Steps to complete an appendectomy.

RIGHT COLON (RIGHT HEMICOLECTOMY)

INCISION

Refer to Omentum procedure.

OPERATIVE PROCEDURE

Mobilize the lateral border of the right colon from its peritoneal wall attachments. Take down the hepatic flexure by dividing the omental and peritoneal attachments (Fig. 5.6). Identify the blood supply to the right colon by tracing the middle colic artery and ileocolic artery branches, which supply the right colon from the superior mesenteric artery (SMA) (Fig. 5.7).

Mobilize the entire right colon medially in a Cattell maneuver taking down all peritoneal attachments (Figs. 5.6 and 5.8). While doing this, notice that the entire posterior attachment of the right colon consists of filmy areolar bands. Dissect from laterally to medially on the posterior abdominal wall staying anterior to the kidney. At the hepatic flexure of the colon, the surgeon will first encounter the ureter and then the gonadal vessels (Fig. 5.8). At the pelvic brim, the surgeon will encounter first the gonadal vessels and then the ureter. Watch for this consistent relationship in the operating room (Fig. 5.9).

Incise the posterior areolar attachments of the hepatic flexure. These are the renocolic and duodenocolic ligaments. The right colon is now mobilized and prepared for devascularization and resection (Fig. 5.10).

Divide the terminal ileum, the ileocolic artery, the right colic artery at the SMA, and the right branch of the middle colic artery. Note that the middle colic artery may be preserved depending on the location of the pathology in the colon. Once these vessels have been divided and the omentum cleared from the colon, the right colon and hepatic flexure can be divided from the transverse colon and removed.

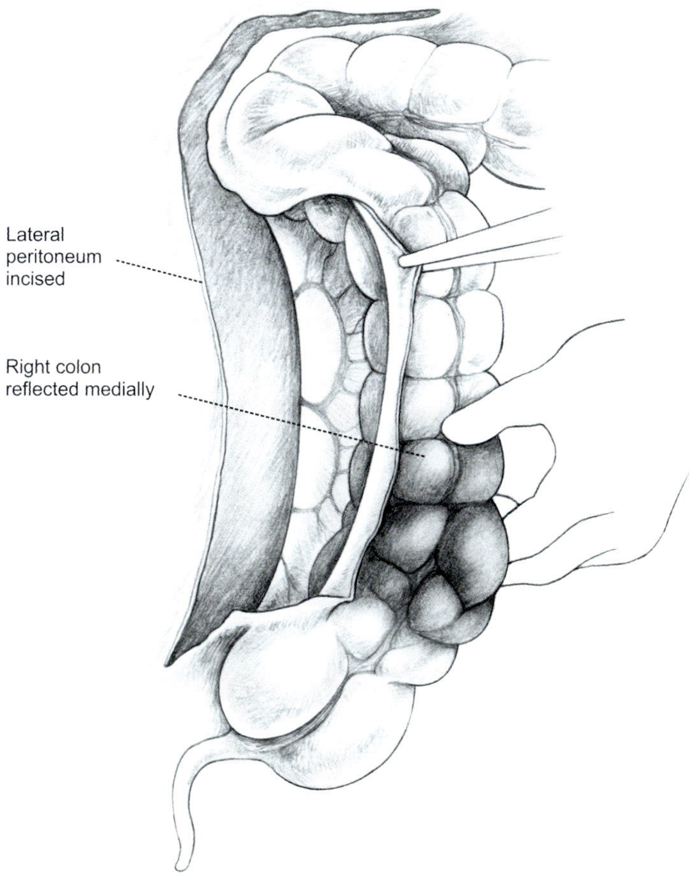

Figure 5.6 Mobilization of the right colon, initial part of a Cattell maneuver.

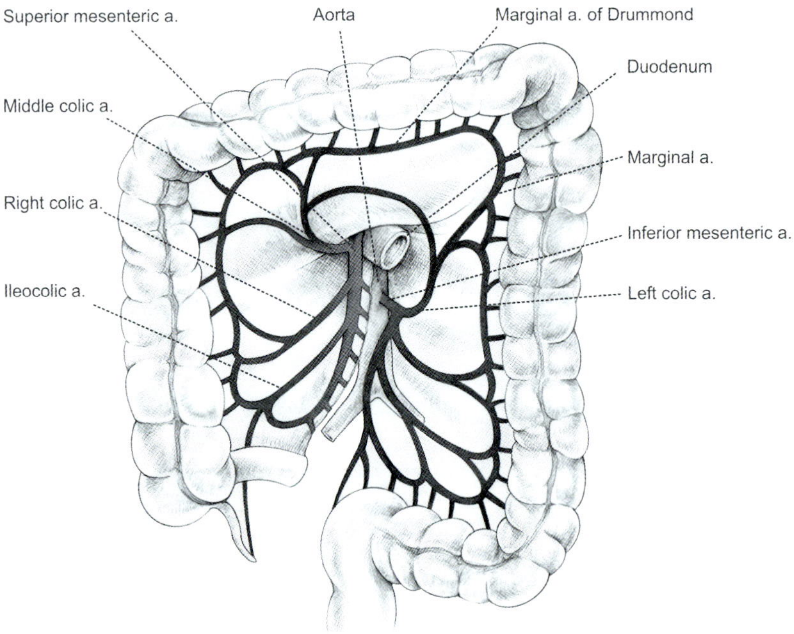

Figure 5.7 Blood supply of the colon: find the ileocolic, right, and middle colic arteries supplying the right colon.

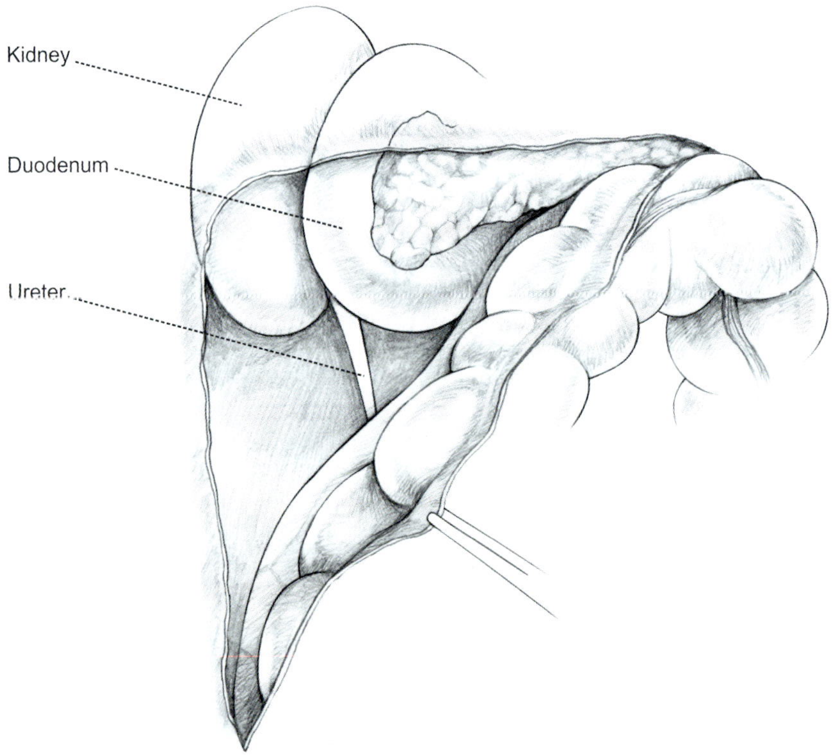

Figure 5.8 Mobilization of the hepatic flexure of the right colon.

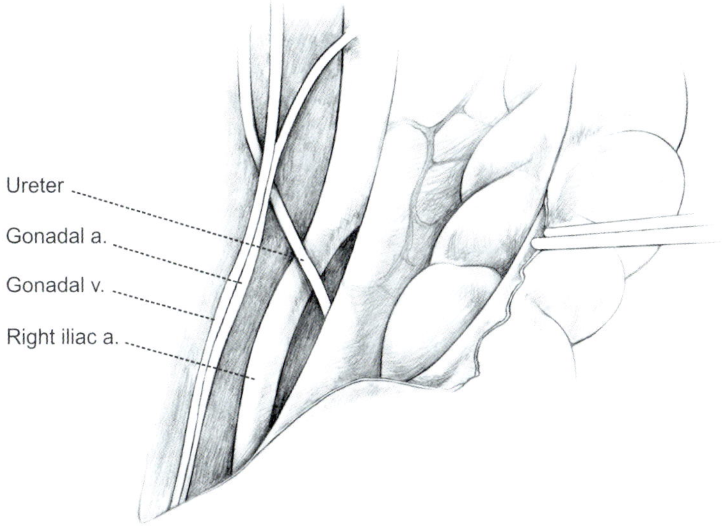

Figure 5.9 Right ureter and gonadal vessels change relative positions as they descend down to the pelvis.

CLINICAL HIGHLIGHTS

- Note the common blood supply to the terminal ileum and the cecum. To perform a right hemicolectomy, be sure to trace the colonic branches back to the main trunk of the ileocolic artery coming from the SMA. Do not ligate the SMA (see Fig. 5.10).

- While mobilizing the posterior areolar attachments of the colon, note the proximity to the right kidney and duodenum. A tumor of the ascending colon may have local extension into the duodenum or right kidney. Oncologic resection will entail resection of these structures en bloc if invaded by tumor (see Fig. 5.8).

- The gonadal vessels cross over the ureter en route down to the pelvis. Cephalad, the gonadal vessels are medial. *At the pelvis, the ureter is medial* (see Fig. 5.9).

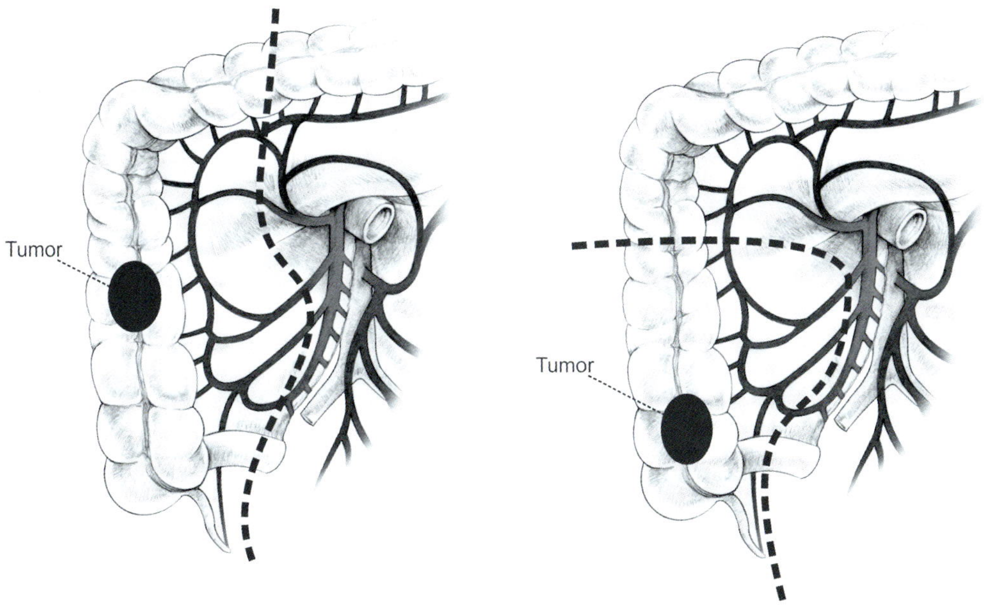

Figure 5.10 Right hemicolectomy resection margins depend on tumor position and blood supply.

LEFT COLON (LEFT HEMICOLECTOMY)

INCISION

Refer to Omentum procedure.

OPERATIVE PROCEDURE

Fully divide the remaining avascular plane between the colon and the omentum by elevating the omentum. Note the attachment of the splenic flexure to the spleen, the splenocolic ligament (Fig. 5.11). Identify the pancreaticocolic ligament. Divide these ligaments. Continue lateral mobilization of the left colon in the paracolic gutter along the line of Toldt. Extend this peritoneal division to the brim of the pelvis (Fig. 5.12).

Mobilize the left colon medially taking down the areolar attachments to the kidney and posterior abdominal wall. Identify the left gonadal vessels and left ureter at the pelvic brim. Trace these structures cephalad and remember these important relationships (Fig. 5.13).

Shine a light through the mesentery and identify the left inferior mesenteric vein (IMV) and left inferior mesenteric artery (IMA). Note that these two vessels do not run side by side. The IMV can be found just lateral to the small bowel at the ligament of Treitz. The IMV then joins the splenic vein. These join the superior mesenteric vein (SMV) to form the portal vein. This junction is located behind the pancreas. The IMA is a direct branch from the aorta, and this junction is in the midabdomen. Now trace the IMV back to the splenic vein and the IMA back to the aorta. Divide the IMV at the ligament of Treitz and IMA at its origin (Fig. 5.14). Divide the sigmoid colon above the sacral promontory (to preserve anatomy relevant to the upcoming TME, Section 6), and remove the left colon and sigmoid colon (Fig. 5.15).

CLINICAL HIGHLIGHTS

- The splenocolic ligament is typically avascular. However, this ligament is clinically troublesome due to its attachment to the delicate capsule of the spleen. Traction on this ligament during colectomy can cause a tear of the spleen necessitating splenectomy in 7% of cases.
- The lienorenal (splenorenal) ligament may rarely contain aberrant splenic vessels from the inferior pole. The left gastroepiploic artery may also be injured during splenic flexure mobilization of the colon.
- Remember that ligation of the blood supply for resection of a segment of colon is dependent on the location of the colonic pathology. By remembering the vascular anatomy of the colon, well-perfused anastomotic ends under no tension will assure the best opportunity for healing without clinical leak (see Fig. 5.7).

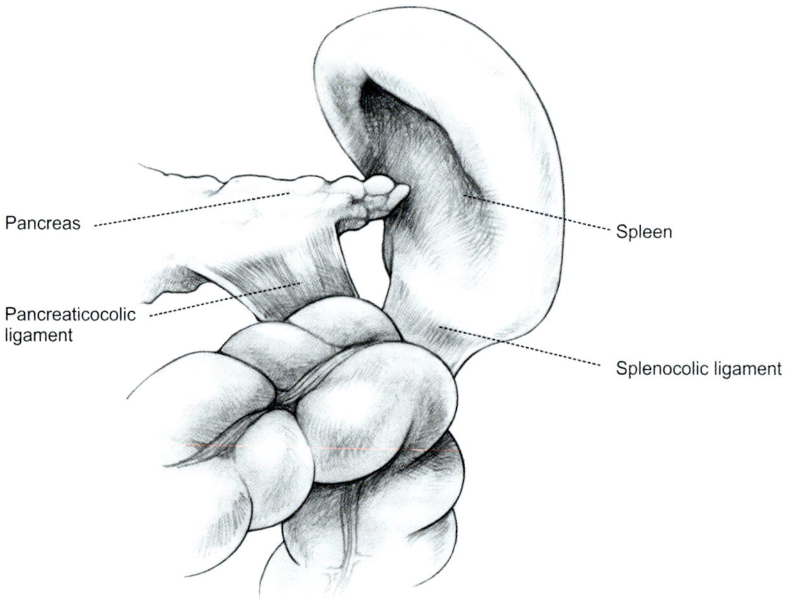

Figure 5.11 Mobilization of the splenic flexure of the left colon.

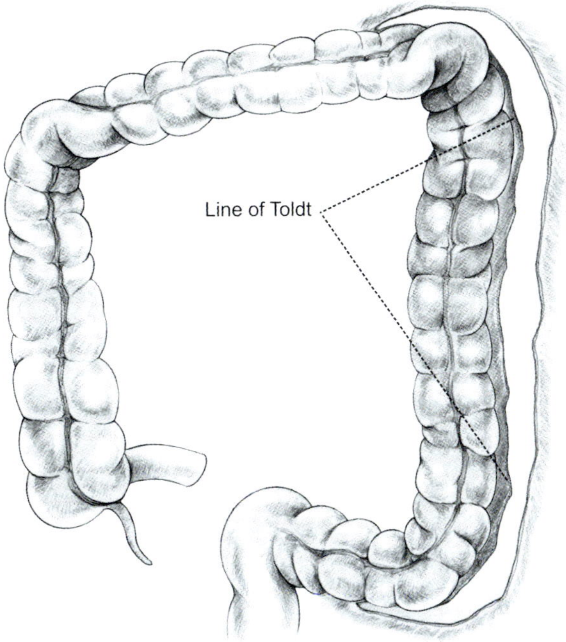

Figure 5.12 Peritoneal division along the line of Toldt.

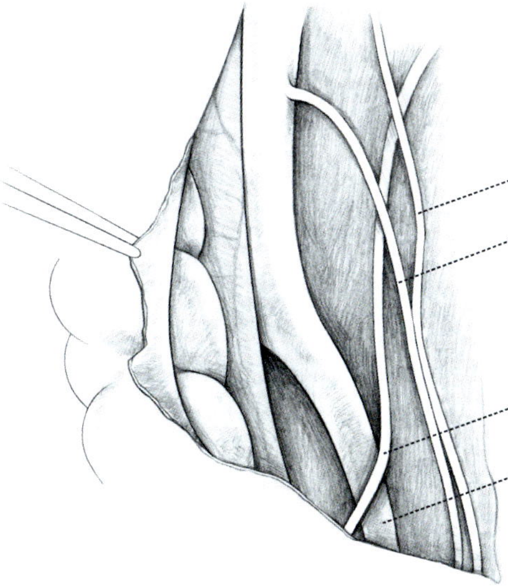

Figure 5.13 Left ureter and gonadal vessels change relative positions as they descend down to the pelvis.

- The arc of Riolan is an inconsistent anastomotic artery between middle and left colic arteries. Present in 61%, minimal in 32%, and absent in 7% (Fig. 5.16).
- Note the space of Riolan below the marginal artery of Drummond. This avascular space may be used for retrocolic placement of Billroth II gastroenterostomy or passage of the Roux limb for gastric bypass (Fig. 5.17).
- Transverse colectomy encompasses colon in the middle colic artery distribution (Fig. 5.18).
- The gonadal vessels cross over the ureter en route down to the pelvis. Cephalad, the gonadal vessels are medial. *At the pelvis, the ureter is medial* (see Fig. 5.13).

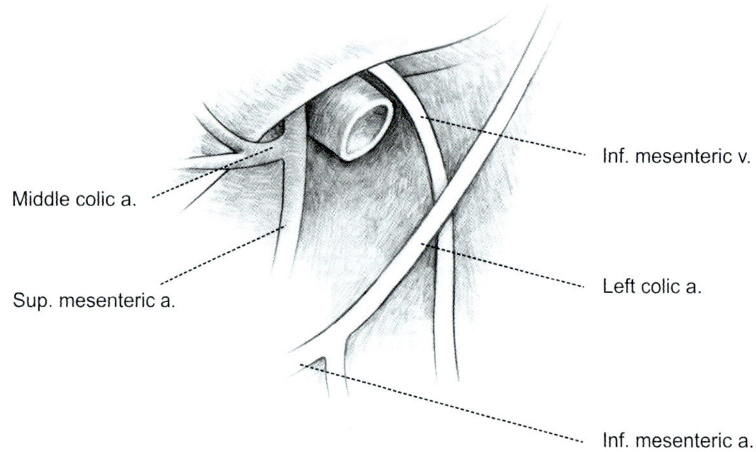

Figure 5.14 Left colic artery and vein diverge at the proximal jejunum near the ligament of Treitz.

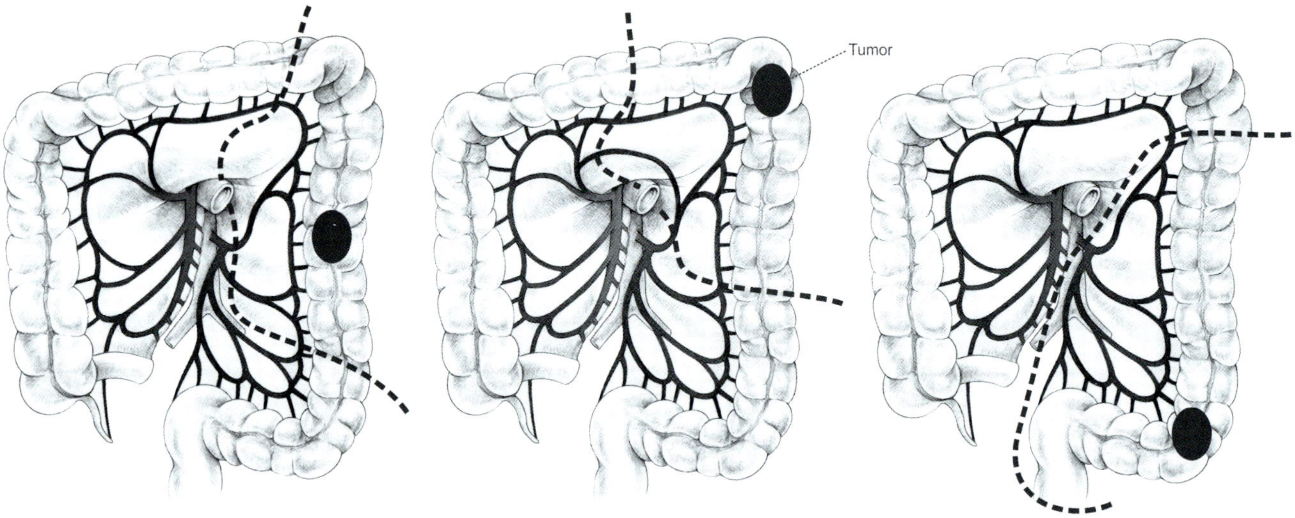

Figure 5.15 Left hemicolectomy and sigmoid colectomy resection margins depend on tumor position and blood supply.

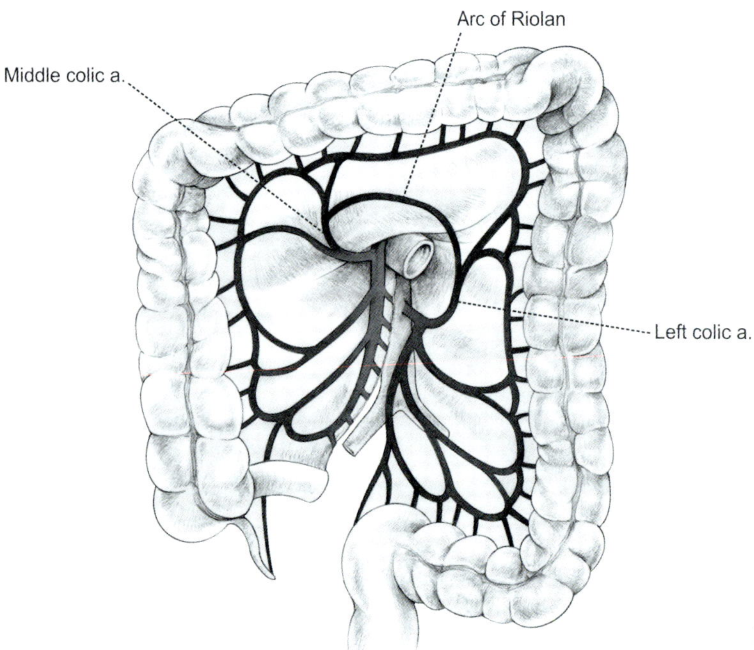

Figure 5.16 Arc of Riolan.

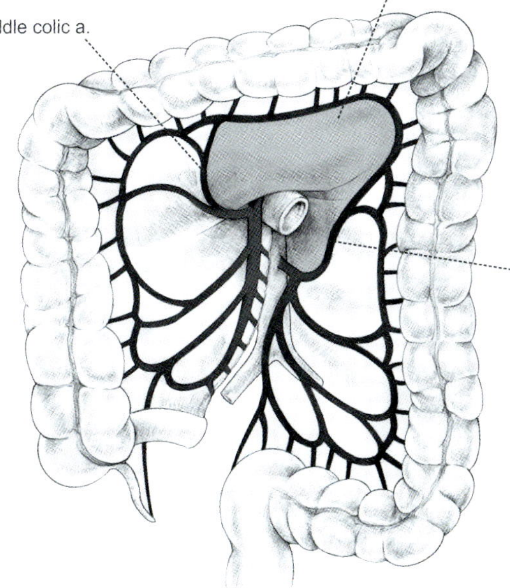

Figure 5.17 Space of Riolan used for gastroenteric connections.

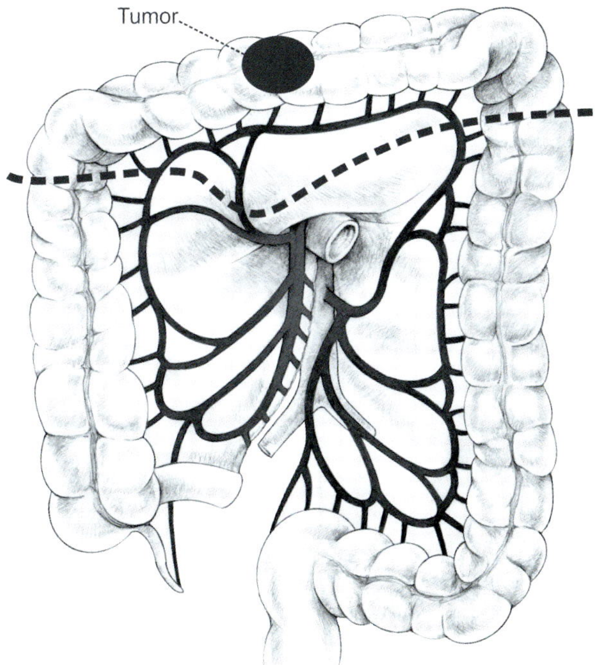

Figure 5.18 Transverse colectomy in middle colic artery distribution.

SUPERIOR MESENTERIC ARTERY (ROUX-EN-Y PROCEDURE)

INCISION

Refer to Omentum Procedure.

OPERATIVE PROCEDURE

Find the ligament of Treitz by elevating the transverse colon. Be sure to keep the ligament of Treitz intact for future reference. Incise the peritoneum over the SMA to identify the middle colic artery, the main body of the SMA, and its branches (Fig. 5.19). Note that

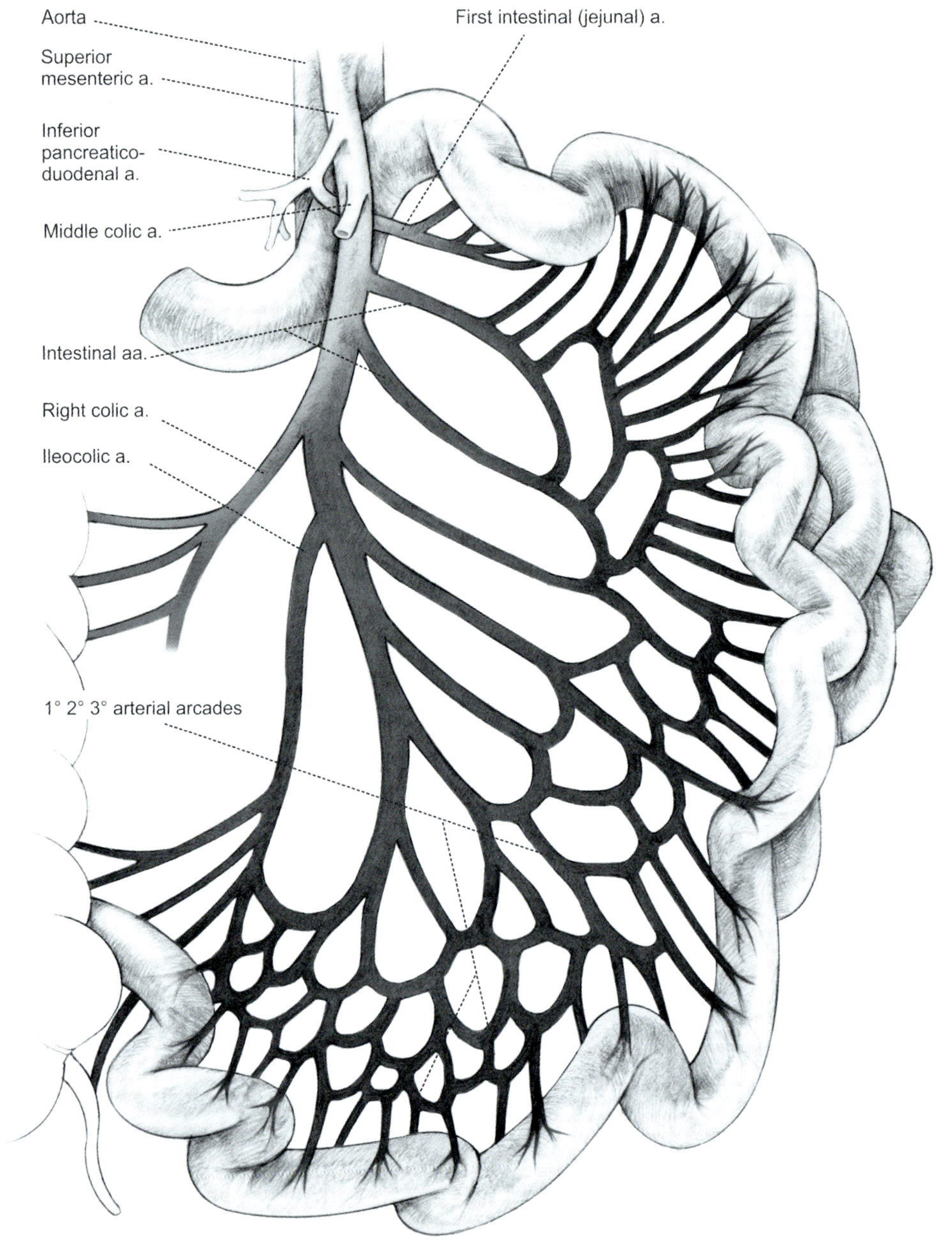

Figure 5.19 Superior mesenteric artery distribution and intestinal artery collateral pattern.

the main body of the SMA heads down to the ileocecal junction. The branches of the SMA come off to the left and inferiorly to supply the small bowel. Trace one of the branches of the SMA all the way to the small bowel. Also, trace the SMA to its terminal ileal and cecal branches (Fig. 5.19).

Note the 1°, 2°, 3° arterial arcades (first degree, second degree, third degree). In some literature, the 1° arcades are referred to as the "intestinal" arteries. The 1° arcades become shorter moving distally from jejunum to ileum (see Fig. 5.19).

Next, expose the SMA as it traverses over the third portion of the duodenum. Dissect it back to its origin from the aorta under the pancreas. Identify the inferior pancreaticoduodenal artery coming off the first jejunal branch of the SMA or the SMA itself (Figs. 5.20 and 5.21). Note that the angle formed by the SMA and aorta is normally >25°. The fat that you are dissecting around the SMA prevents the SMA syndrome. An aortic-SMA angle <15° predisposes duodenal obstruction (Fig. 5.22).

Examine the blood supply from the SMA to the proposed biliary limb of the Roux-en-Y connected to the fourth portion of the duodenum. It is critical to note that the blood supply to the biliary limb must be maintained by preserving the first branch artery off the SMA to the small bowel (see Figs. 5.20 and 5.21). Also, note

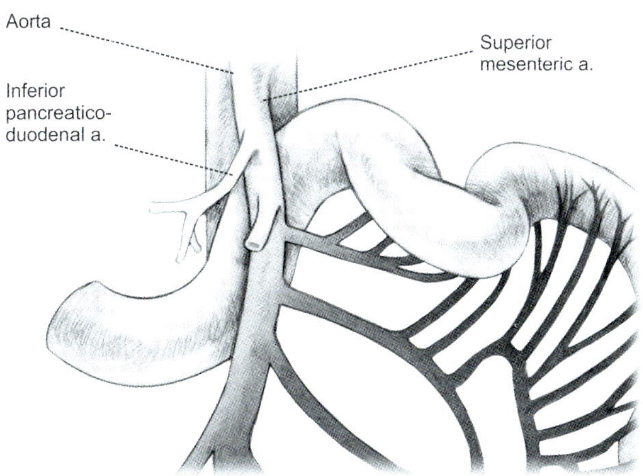

Figure 5.20 First intestinal artery bifurcation from SMA.

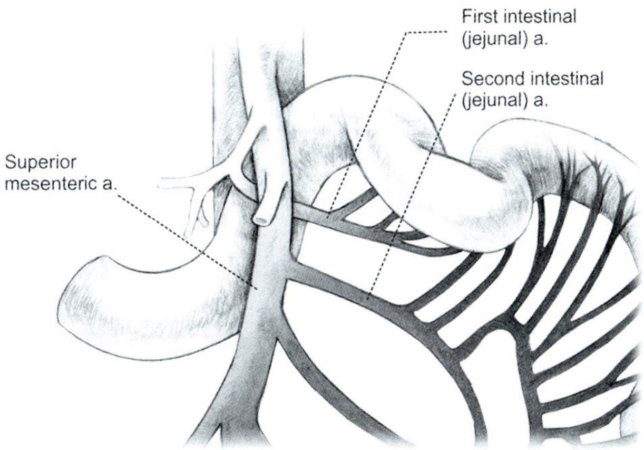

Figure 5.21 First intestinal artery bifurcation from inferior pancreaticoduodenal artery.

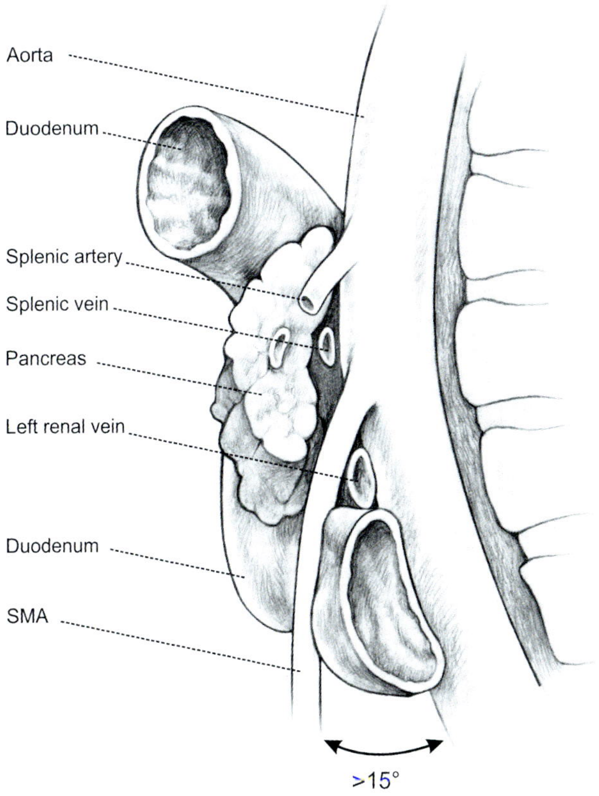

Figure 5.22 Etiology of the SMA syndrome.

that the rich anastomotic network stemming from the second-degree arcade is not present for that first branch artery off the SMA. Disruption of this first branch artery could result in necrosis of the biliary limb.

To construct the Roux limb, remember that the classic (nonbariatric) construction begins 15-20 cm downstream from the ligament of Treitz (Fig. 5.23). Divide the small bowel at this point. On the Roux limb mesentery, divide one 1° arcade ("intestinal" artery) of the SMA at the first bifurcation level. Measure a 40 cm Roux limb that will be the defunctionalized Roux limb. Place the biliary limb on the antimesenteric side of the Roux limb. For the anastomosis, use 3-0 or 2-0 suture in a single layer. (If the cadaver bowel wall is very thin, you may need to suture full thickness instead of the seromuscular stitch you will use in the OR.) Alternatively, a two-layer closure consisting of outer seromuscular and inner mucosal layers may be performed. Be certain to include mucosa in every inner-layer stitch on both bowel elements in the anastomosis.

CLINICAL HIGHLIGHTS

- Note that the SMA branches generally have three degrees (or levels) of arcade (see Fig. 5.19). The first level arteries are main arterial branches, the second level has medium-sized vessels, and the third level contains the terminal branches to the small bowel. More complex patterns of branching may be seen (see Fig. 5.19).
- Lateral arcades are formed between the 1° and 2° arterial arcades as well as the 2° and 3° arcades. No collateral circulation exists at the level of the vasa recta or at the mesenteric bowel surface (Fig. 5.24).
- The first bifurcation level vessels (referred to as "intestinal arteries" in Hollinshead) are optimal to take down for mobilizing a defunctionalized Roux-en-Y limb. At the 1° level vessel, a maximum of two 1° vessels may be taken down for Roux limb mobilization.
- Remember the critical nature of the first jejunal branch artery off the SMA. It may be an end artery without collateral supply found in all other branches off the SMA. Disruption of this artery may result in necrosis of the biliary limb (see Figs. 5.20 and 5.21).
- You can now see a classic (nonbariatric) Roux-en-Y before you. In your "mind's eye," see that cysts, abscesses, pseudocysts, obstructed stomachs, obstructed biliary systems, and many other clinical conditions could be addressed with Roux-en-Y internal drainage.
- Remember Roux-en-Y construction: biliary limb 20 cm and Roux limb 40 cm (see Fig. 5.23).

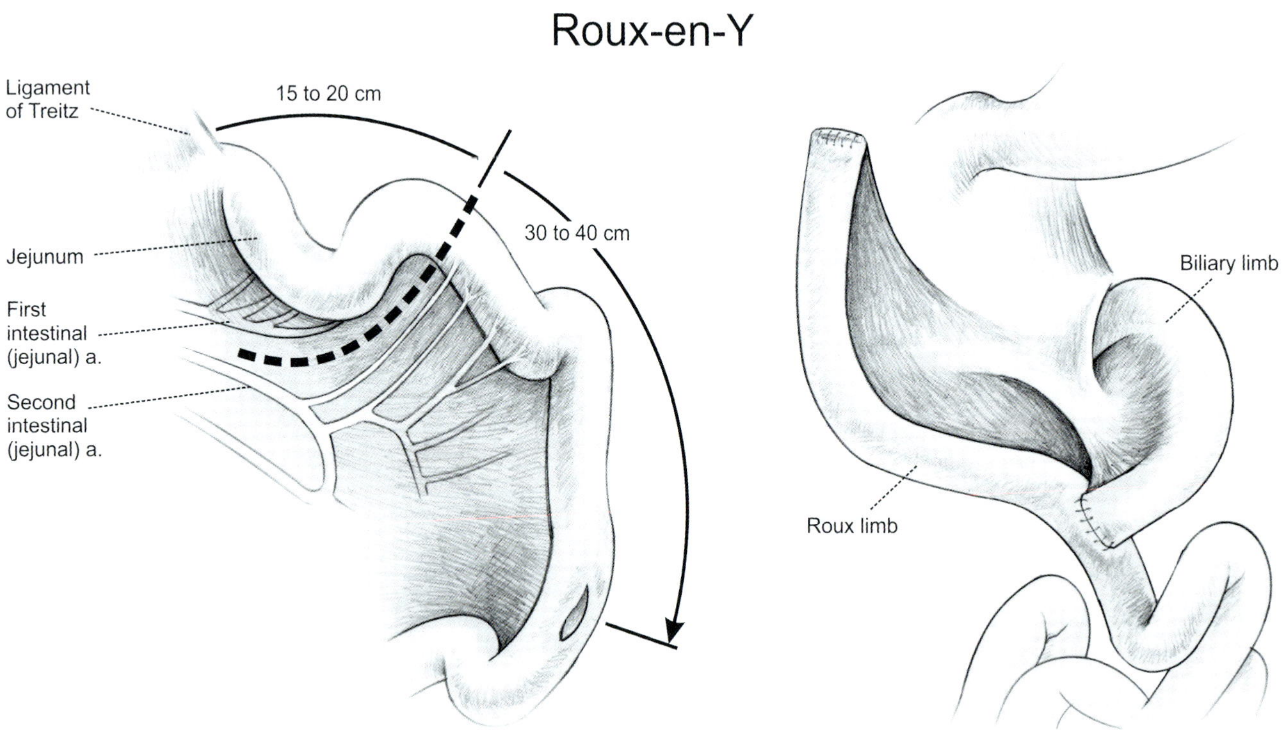

Figure 5.23 Roux-en-Y design used for gastric, biliary, abscess, or pseudocyst drainage.

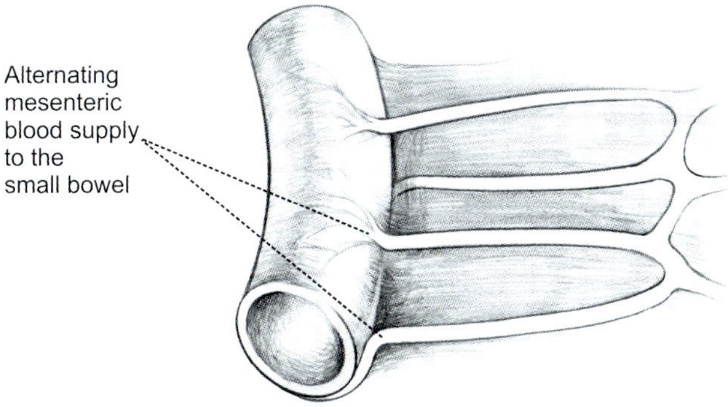

Figure 5.24 Alternating pattern of terminal arteries supplying the small bowel.

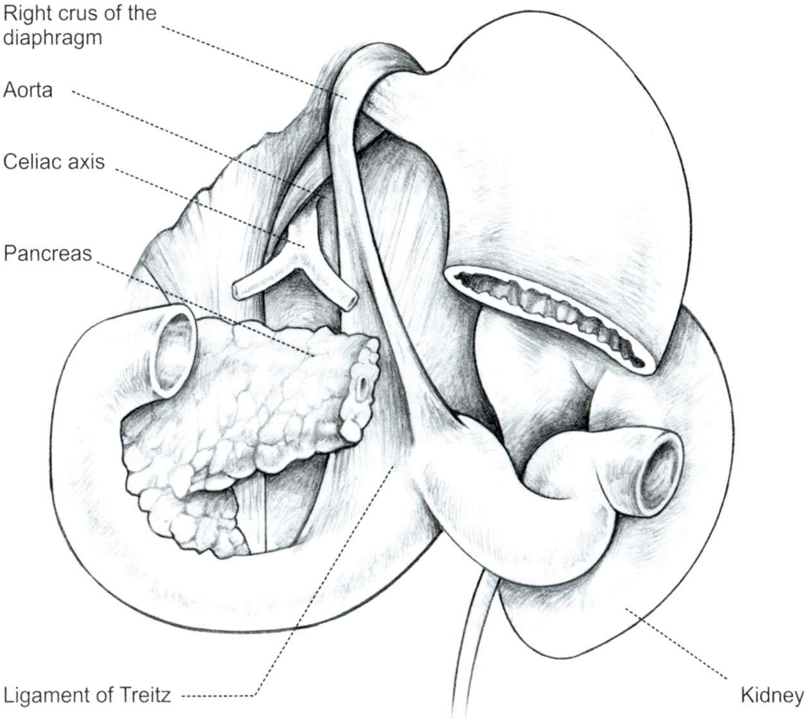

Figure 5.25 Ligament of Treitz location and suspension from the right diaphragmatic crus.

- The SMA syndrome occurs when the artery pinches the third portion of the duodenum. Thin, young females tend to be the risk group for this uncommon condition.

- Note that the ligament of Treitz descends from the *right* crus of the diaphragm. Also note that the left crus originates with the right crus. The ligament of Treitz passes posterior to the pancreas and anterior to the left kidney. Dissect it. Now is your chance to see the entire ligament of Treitz (Fig. 5.25).

- If conservative management fails for the SMA syndrome, division of the ligament of Treitz (see Chapter 6, "Stomach and Duodenum") or duodenojejunostomy may be employed for relief of the condition.

- Terminal arterial branches to the small bowel have an alternating pattern of blood supply to the bowel wall. Consider this when looking to divide or anastomose the bowel (Fig. 5.24).

RECTUM (TOTAL MESORECTAL EXCISION, TME)

INCISION

Refer to Omentum Procedure.

OPERATIVE PROCEDURE

Mesorectal dissection begins at the level of the sacral promontory. An avascular cleavage plane exists between the visceral layer of the mesorectum and the presacral fascia (parietal layer). During this dissection, identify the hypogastric nerve plexus in the presacral fascia (Fig. 5.26). The sympathetic left and right hypogastric nerves are identified and preserved in the presacral fascia. Lateral margins of dissection are defined by the lateral pelvic walls. Halfway between the pelvic brim and the tip of the coccyx, the surgeon will encounter the rectosacral ligament (Fig. 5.27), which must be divided sharply (Fig. 5.28). This step, division of the rectosacral ligament, is the key feature of TME allowing the node bearing tissue to be excised en bloc. The assistant continues to help by elevating the rectum out of the pelvis (Fig. 5.29).

Anterior dissection begins in the rectovesical pouch in the male. Proceed with sharp anterior dissection inferiorly down to the rectoprostatic (Denonvilliers) fascia and by retracting the rectum posteriorly away from the GU tract. Denonvilliers' fascia will join the prostate deep in the pelvis (see Fig. 5.27).

In the female, or the rectouterine pouch (pouch of Douglas), dissection begins sharply in the plane between the rectum and the posterior vaginal wall. Proceed down below the cervix.

In both sexes, mobilize the posterior aspect of the rectum staying on the visceral mesorectum (see Fig. 5.28). Separate the mesorectum from the presacral fascia using Metzenbaum scissors allowing access to the muscular pelvic floor. Fibroelastic fibers below the mesorectum consist of the longitudinal muscle coat of the rectum and its fascia that blend with the levator ani and its fascia.

Lateral stalks, or ligaments, at this level also include sacral parasympathetic fibers. The medial edge of the pelvic autonomic nerve plexuses (PANPs) are divided at the lateral margin of the mesorectum to preserve all pelvic autonomic nerves except those directly entering the rectum. The "lateral stalk" dissection is completed on the contralateral side sharply under direct vision (see Fig. 5.29). Tumor invasion is the only reason to violate the planes. Levator ani muscles are now visible. The rectal dissection can now be completed circumferentially to the pelvic floor. Be sure you can palpate below the tip of the coccyx to simulate a very low dissection (Fig. 5.30).

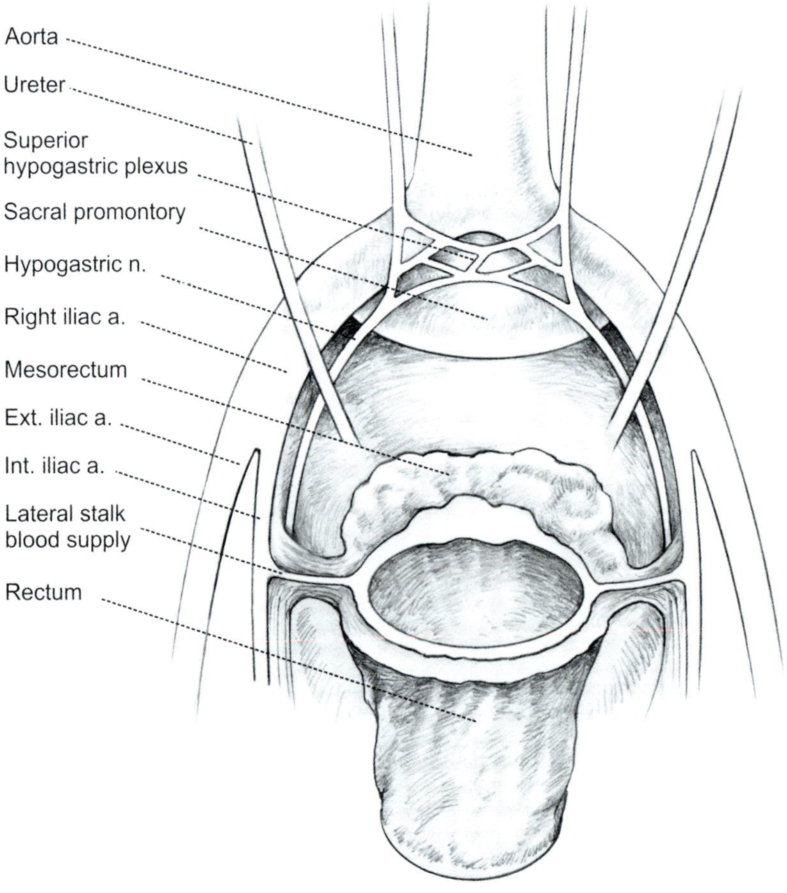

Figure 5.26 Hypogastric nerves guide beginning of total mesorectal excision and location of the "lateral stalks."

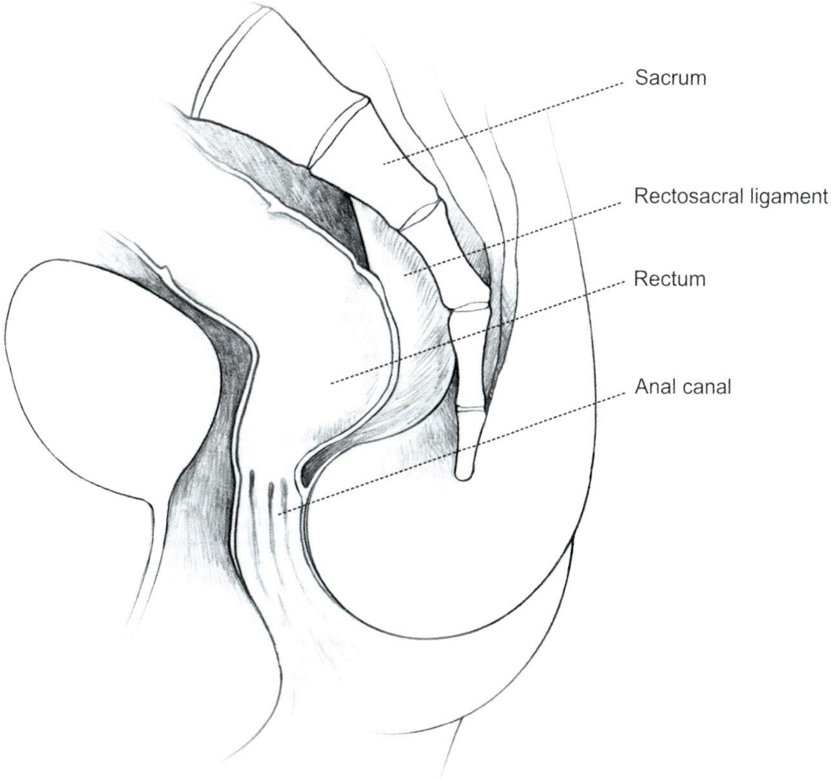

Figure 5.27 Rectosacral ligament (side view).

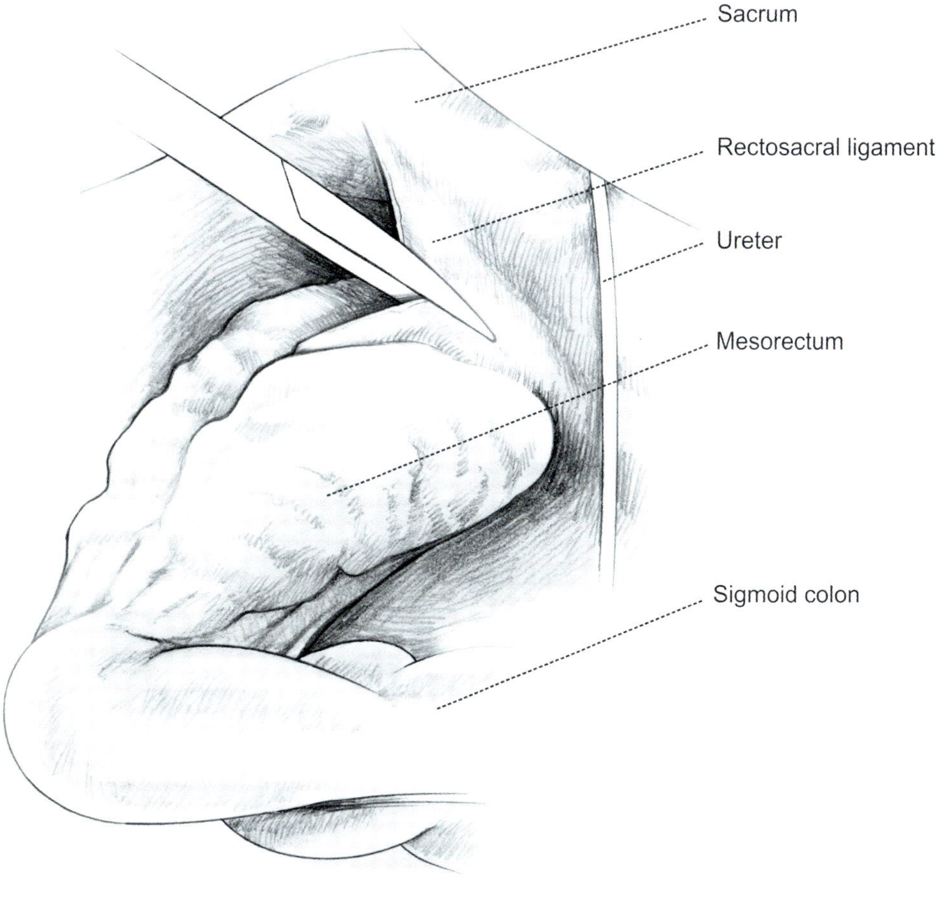

Figure 5.28 Division of the rectosacral ligament for TME.

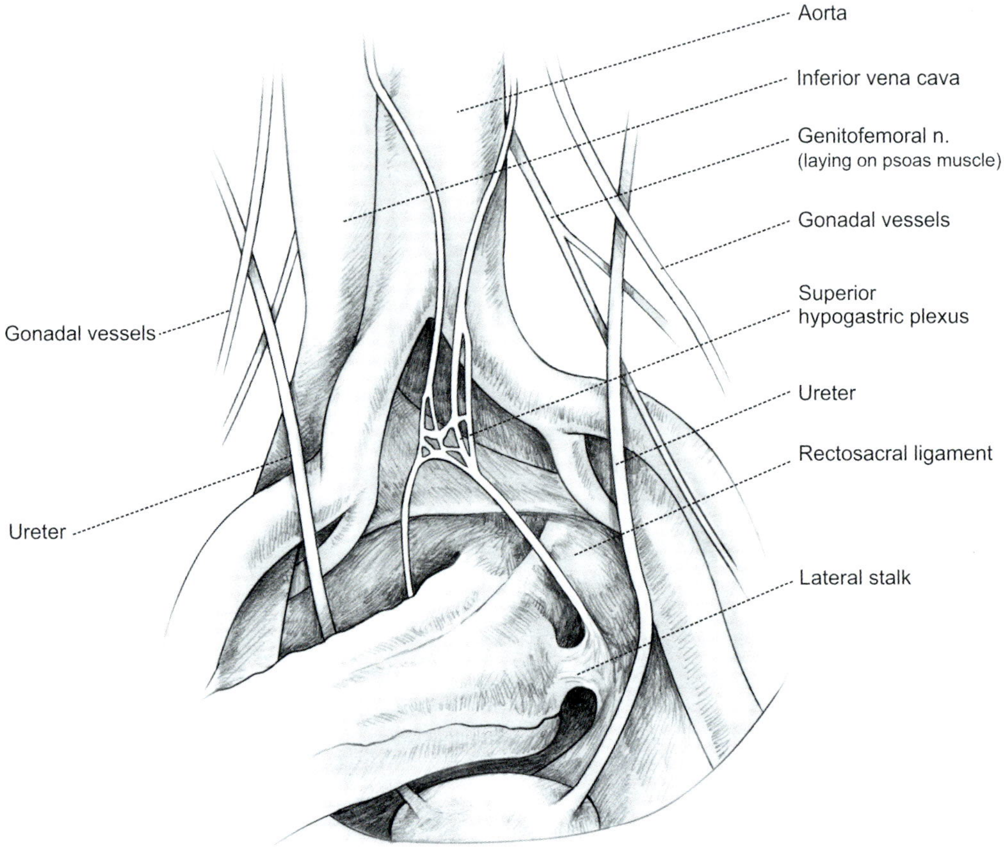

Figure 5.29 Oblique view of lateral stalk and ureter.

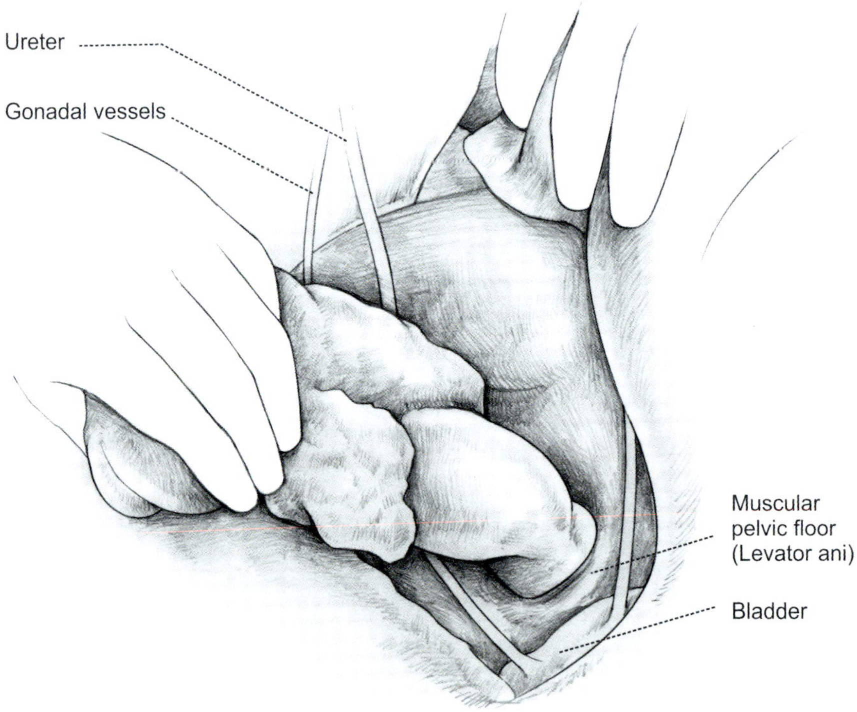

Figure 5.30 TME dissection to the pelvic floor (levator ani).

CLINICAL HIGHLIGHTS

- Be sure to identify the left and right ureters at the pelvic brim and follow them into the pelvis to insure their preservation during mesorectal dissection.

- On examination of the rectal specimen, notice a posterior bilobed encapsulated appearance of the specimen. It merely looks like a posterior lipoma. This identifies a good mesorectal dissection.

- If a tumor is located anteriorly in the rectum, excision of Denonvilliers' fascia, along with a portion of the GU tract or the posterior vaginal wall, may be necessary for good margins.

- Note that the superior hypogastric plexus in the presacral fascia is an extension of the aortic plexus (sympathetic). Identifiable nerve trunks become apparent as the hypogastric plexus courses into the right and left hypogastric nerve bundles.

- The sigmoid arteries form an arcade (like the small bowel). The rectal arteries, however, directly enter the bowel wall. 2% of cases will have rectal arterial arcades, which may cause intraoperative bleeding.

- The inferior rectal artery consistently arises from the internal pudendal arteries and supplies the terminal 10 cm of rectum and anus (Fig. 5.31).

- The venous drainage of the terminal rectum and anus is to the internal iliac vein. This is systemic drainage; not portal as is the rest of the intestine. Remember that this is an area of portacaval anastomosis. The superior rectal vein (portal drainage) abundantly interconnects with the middle and inferior rectal veins (caval drainage) (Fig. 5.31).

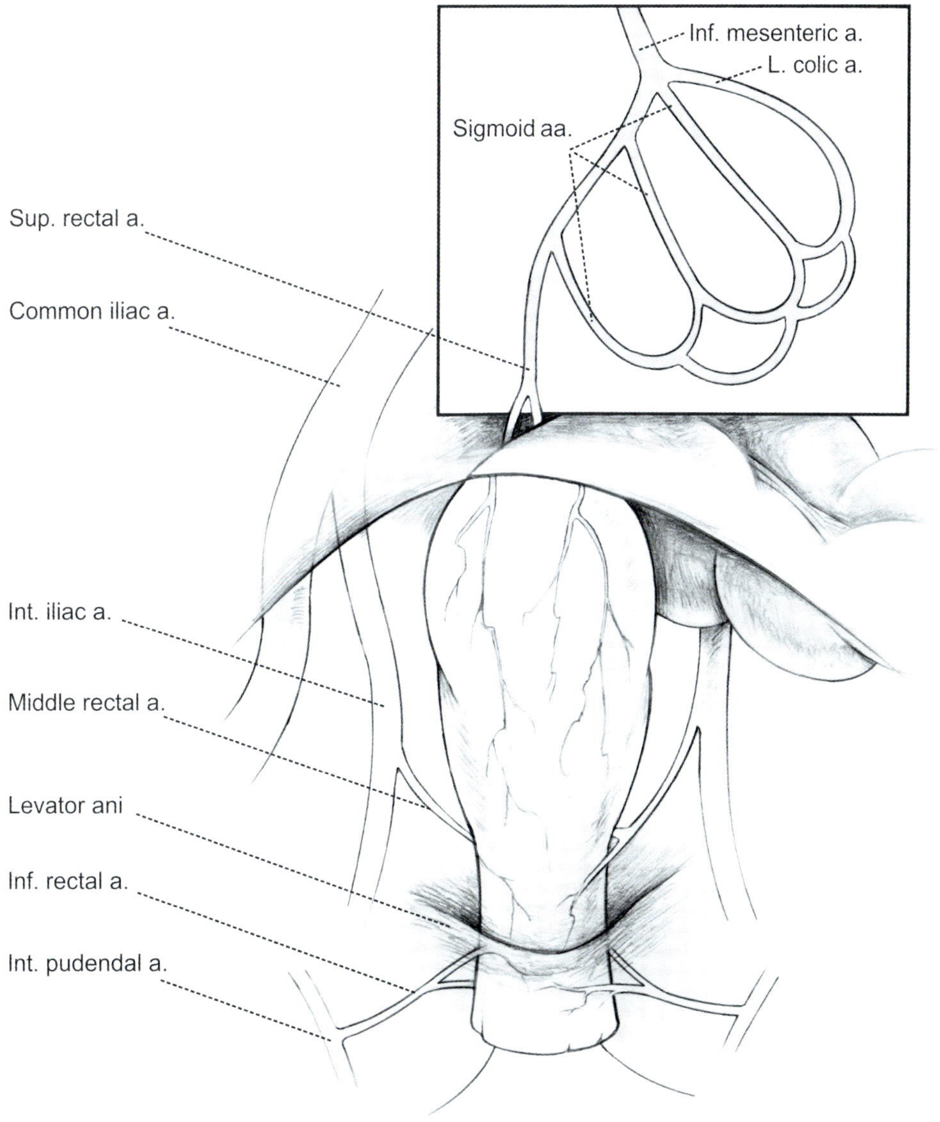

Figure 5.31 Pattern of blood supply to terminal 10 cm of the rectum and anus.

6

Stomach, Duodenum, and Esophagus

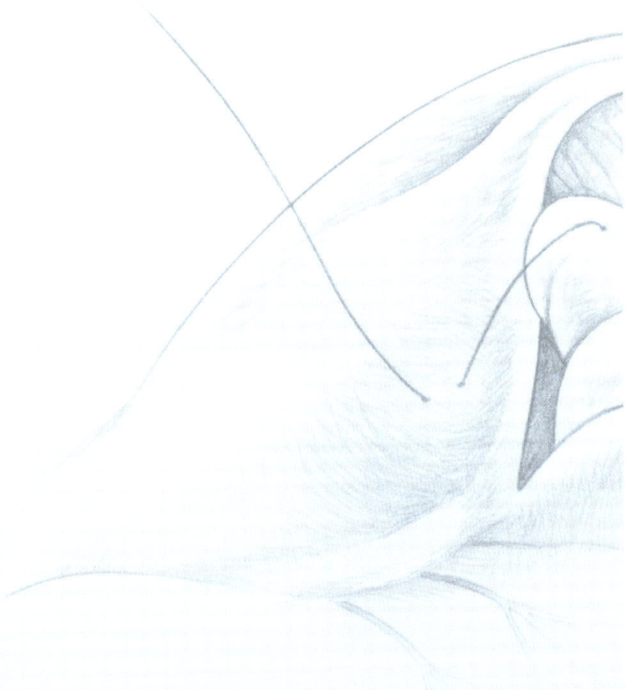

Outline

1 Vagus Nerves (truncal and highly selective vagotomy)
2 Esophageal Hiatus (nissen fundoplication)
3 Greater Curvature of the Stomach and Short Gastric Vessels (mobilization for esophagectomy or fundoplication)
4 Distal Esophagus (Heller myotomy)
5 Pylorus (pyloromyotomy, pyloroplasty, bleeding ulcer, Ladd bands)
6 Duodenal Segments (ligament of Treitz/SMA syndrome)
7 Gastrotomies and Gastrostomies
8 Esophagus (transhiatal esophagectomy)
9 Gastric Resection (antrectomy, hemi-, sub-total, complete)

VAGUS NERVES (TRUNCAL AND HIGHLY SELECTIVE VAGOTOMY)

INCISION

Access to the chest and abdomen will facilitate this procedure (Figs. 3.2 and 5.1). If the chest has been dissected, remove the lungs and identify the thoracic esophagus.

OPERATIVE PROCEDURE

Through the thoracic exposure, identify the vagus nerve plexus on the thoracic esophagus. Notice that it is a true plexus but most commonly goes through the esophageal hiatus as an anterior and posterior dominant trunk. This pattern occurs in 88% of patients. Four vagal structures are identified in 7% and more than four vagal structures are identified in 5% of patients (Fig. 3.31A, B). In the author's practice, it has been observed over many years and many patients that the anterior trunk often has more than one branch at the esophageal hiatus and the posterior trunk is a large dominant trunk at that level.

Perform a truncal vagotomy through an abdominal approach. Incise the peritoneal covering over the abdominal esophagus to expose the esophagus and surrounding attachments (Figs. 6.1 and 6.2). Mobilize the esophagus with blunt dissection and place a Penrose drain around it. Dissect the anterior surface of the esophagus and search for the anterior trunk of the vagus nerve and any branches. The vagus is identified by palpation of a firm, "bowstring" structure on the anterior surface of the esophagus. Using blunt finger dissection, elevate the esophagus away from the aorta. A similar bowstring identifies the posterior trunk by your index finger behind the esophagus. The "criminal nerve of Grassi" (Fig. 3.31A, B) is a branch to the proximal stomach, which splits off the posterior trunk in the chest and descends through the esophageal hiatus separate from the main posterior vagal trunk. Look for the criminal nerve. Secure the posterior and anterior trunks with a tonsil clamp. Transect and remove a 1-cm segment of each nerve, which in the OR is sent to pathology for confirmation. The two ends of each trunk are ligated with hemoclips.

To perform a highly selective vagotomy (HSV), identify and preserve the "crow's foot" of the anterior division of the nerve of Latarjet to the distal antrum and pylorus (Fig. 6.3) 6 cm proximal to the pylorus. Select the next proximal branch of the nerve of Latarjet to the stomach and transect it. In the OR, you will use metal clips. Continue the dissection proximally cutting these fine branches to the lesser curve of the stomach from the anterior nerve of Latarjet. Notice that small vessels often accompany these fine branches off the nerve of Latarjet. Continue the dissection proximally to the

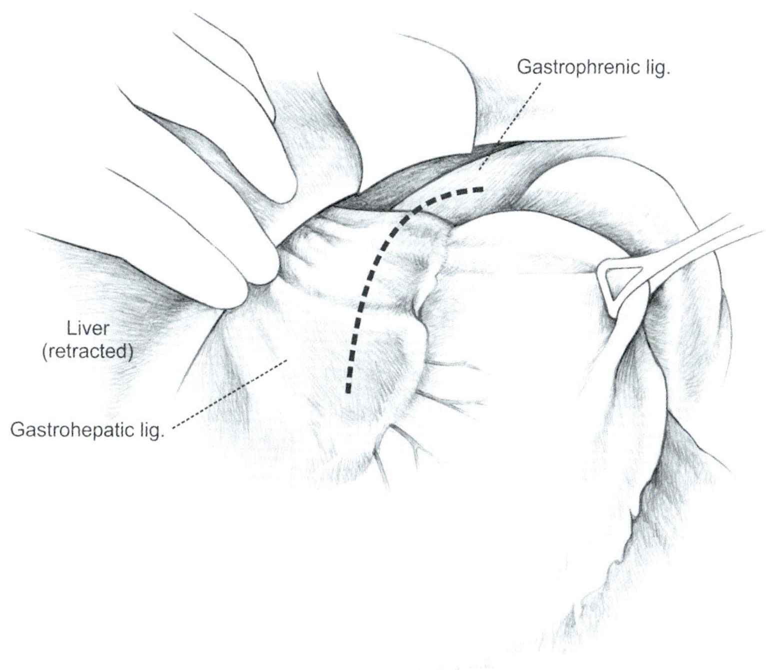

Figure 6.1 Incision of the gastrohepatic ligament.

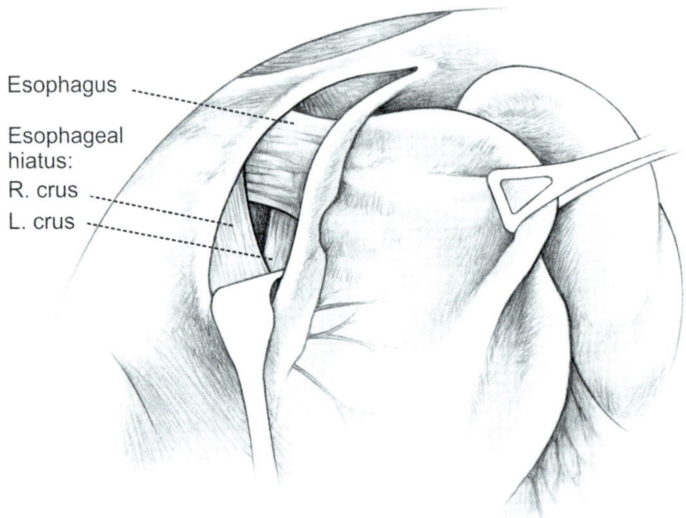

Figure 6.2 View of the right and left crus of the esophageal hiatus.

esophageal hiatus and maintain the continuity of the anterior nerve of Latarjet and the anterior vagus nerve. Some surgeons recommend a limited takedown of the short gastric vessels to facilitate the 6-cm esophageal clearing step of HSV (Fig. 6.4).

Flip the stomach superiorly and look at the posterior surface of the stomach. This can also be performed from an anterior approach if the greater curve of the stomach has not yet been mobilized. Find the "crow's foot" to be preserved. Select the next proximal branch

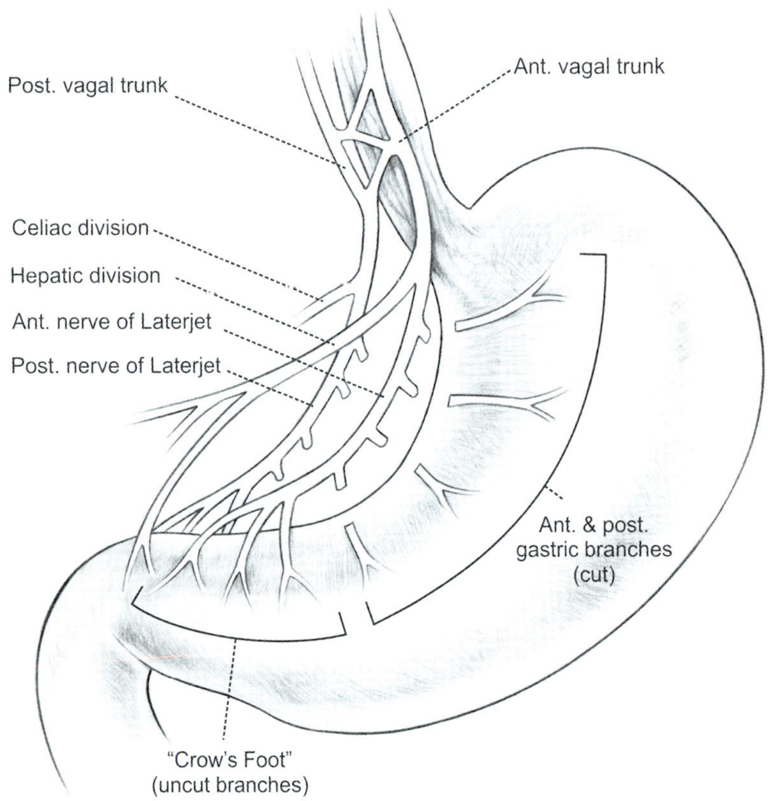

Figure 6.3 Highly selective vagotomy and identification of "crow's foot."

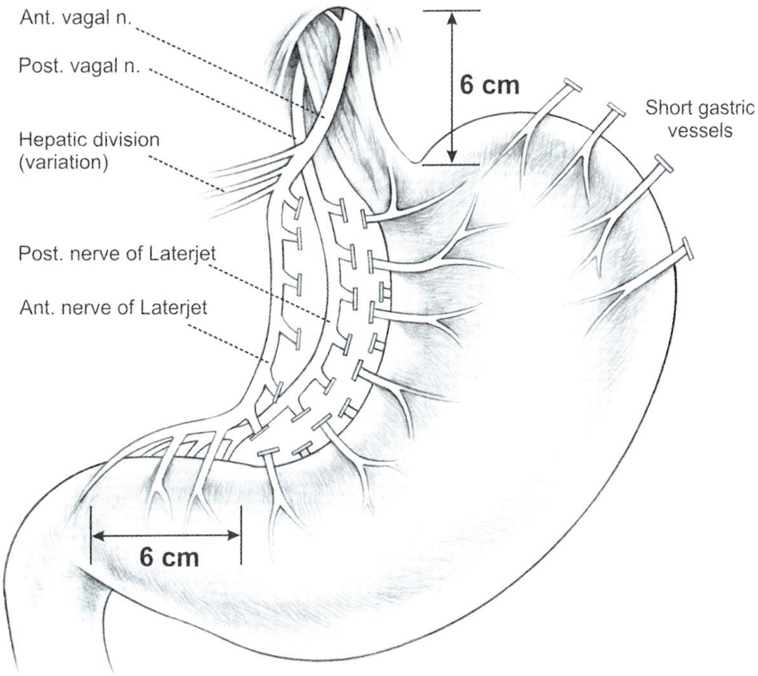

Figure 6.4 Dimensions of HSV. View of HSV from anterior approach with clips applied.

of the posterior nerve of Latarjet and transect it. In a similar fashion to the anterior side, continue the dissection staying very close to the lesser curve of the stomach, which will preserve the posterior nerve of Latarjet. Continue this dissection proximally transecting each branch from the nerve of Latarjet. When this is completed, flip the stomach down into its normal position. Now, clean off all small branches and areolar attachments from the anterior and posterior nerves of Latarjet to the lesser curve of the stomach.

Observe the lesser curve of the stomach and note that it has a 2- or 3-cm. bare area exposing the outer muscle fibers. Some surgeons recommend reperitonealizing this bare area to avoid the occasional necrosis caused by stripping the lesser curve of the stomach. Interrupted silk suture may be used to oppose the anterior and posterior borders of stomach serosa over the bare area created by HSV.

Vagus nerve trunks are apparent after HSV and can easily be divided to simulate truncal vagotomy.

CLINICAL HIGHLIGHTS

- Transthoracic vagotomy is done in the distal thoracic esophagus near the esophageal hiatus. This procedure is utilized to salvage prior truncal vagotomy that was clinically incomplete (see Chapter 3).
- The "criminal nerve of Grassi" is simply the most superior branch of the posterior vagus. This particular branch divides from the right vagus in the chest and descends separately from the posterior trunk into the abdomen. Be aware that small branches of the vagus may descend through the esophageal hiatus separate from the main trunks. It is particularly important to remember to transect the criminal nerve during vagotomy to treat peptic ulcer disease (Fig. 3.31A, B).
- Highly selective vagotomy, superselective vagotomy, proximal gastric vagotomy, and parietal cell vagotomy all refer to the same operation.
- Damage to the crow's foot during HSV may inhibit relaxation of the pylorus.
- Nerves of Latarjet can have highly variable structure but will always have gastric branches that are the target of HSV.
- Truncal vagotomy is always accompanied by a drainage procedure such as pyloroplasty. Otherwise, 40% of patients with truncal vagotomy will have pyloric spasm and will require reoperation for gastric drainage.
- Remember that truncal vagotomy may involve more than a single vagal nerve trunk in either the anterior or posterior position.

- Although transection of the nerve in truncal vagotomy should be sufficient, many surgeons remove a 1-cm segment of each nerve and send it to pathology for confirmation.

- Clips placed on the vagal ends will show on future chest x-rays to show a vagotomy was performed.

- To ensure complete transection of the vagal trunks, it is recommended to clear the distal esophagus superiorly for 5-6 cm of all investing tissues.

ESOPHAGEAL HIATUS (NISSEN FUNDOPLICATION)

INCISION

Access to the chest and abdomen will facilitate this procedure (Figs. 3.2 and 5.1). Reflect the anterior chest wall superiorly and the anterior abdominal wall inferiorly.

OPERATIVE PROCEDURE

This portion will require two participant surgeons. The assistant surgeon will vigorously elevate under the liver displacing the left lateral lobe superiorly. The assistant surgeon may choose to stand at the head of the table or on the cadaver's left side. The primary surgeon should stand on the right side of the cadaver. Dissection commences in the gastrohepatic ligament (Fig. 6.1). This ligament is typically a thin, translucent veil of tissue. However, be aware that aberrant left hepatic vessels (replaced or accessory) may be identified in this location (Fig. 6.5). A "replaced" vessel replaces a major named vessel. "Accessory" means the normal vessel is present in addition to the accessory vessel.

Continue dissection to the right crus of the esophageal hiatus. Clear the edge of the right crus and separate the wall of the esophagus from it. Divide the peritoneum over the esophagus anteriorly and find the areolar plane between the esophagus and the right crus (Fig. 6.2). Once this areolar plane is identified, continue dissection of the right crus posteriorly. Next, develop this areolar plane anteriorly and separate the esophagus also from the left crus of the esophageal hiatus by dividing the gastrophrenic ligament (Fig. 6.6).

Continue dissection of the left crus and the esophageal hiatus posteriorly. Clear all peritoneal attachments from this edge. Further develop the plane between the esophageal hiatus and the esophagus to mobilize the distal 4 cm. of the esophagus into the abdomen. Develop the plane between the esophagus and the diaphragm posteriorly with the right index finger coming from the left crus, behind the esophagus. During this dissection, identify a tight band of tissue, which causes a "bowstring" effect. The bowstring is the posterior vagus nerve. Include the posterior vagus with the esophagus as the esophagus is mobilized.

Examine the entire esophageal hiatus of the diaphragm and note that the right and left crus of the esophageal hiatus arise from the right crus of the diaphragm (Fig. 6.7). The gastrohepatic and gastrophrenic ligaments (Fig. 6.1), posterior esophageal areolar attachments, and peritoneal attachments of the cardia of the stomach must all now be mobilized and free. Place the right index finger behind

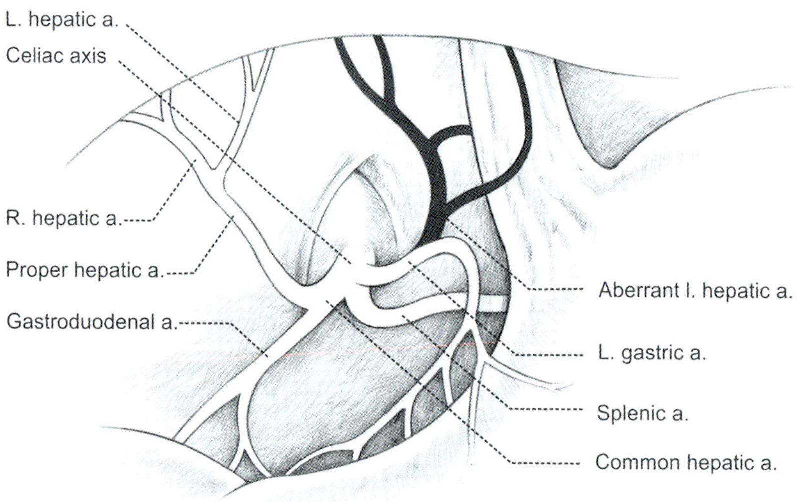

Figure 6.5 Aberrant hepatic vessels in the gastrohepatic ligament.

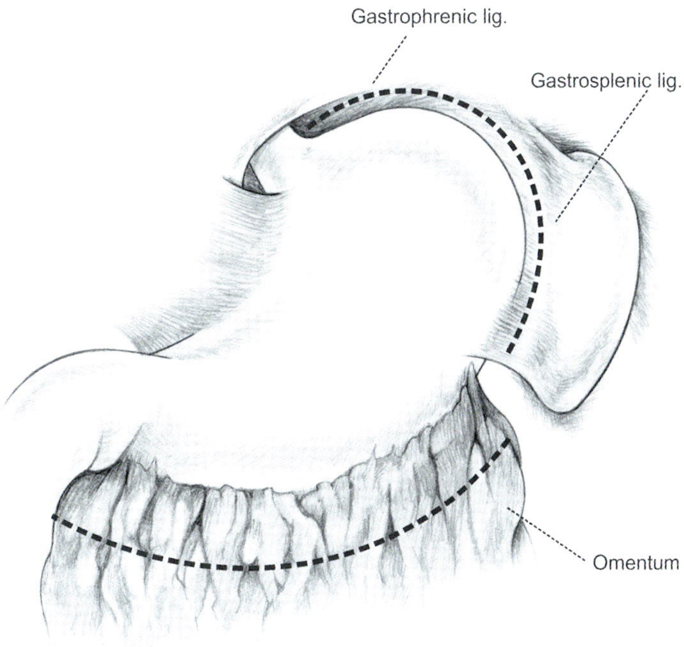

Figure 6.6 Mobilization of the stomach.

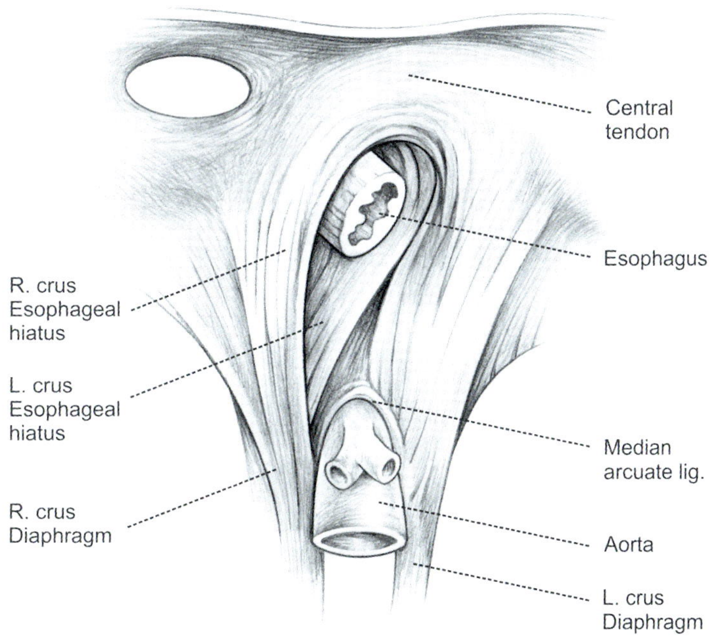

Figure 6.7 Right and left crus of the esophageal hiatus originating from the right crus of the diaphragm.

the esophagus through the space now created. Repair the hiatal hernia posteriorly by suturing the right and left crus together (Fig. 6.8). The cardia of the stomach can be alternately pushed and pulled through this created space and a Nissen fundoplication can be performed (Fig. 6.9). Due to the stiff nature of cadaveric tissue, the fundoplication wrap may need to be formed and held with two hands or two Babcock clamps prior to suture (Fig. 6.10). Using 2-0 silk, place deep submucosal sutures beginning on the left side of the stomach, then picking up anterior esophageal muscle fibers and finally ending with a good deep submucosal bite of the stomach on the right side. Place three of these sutures causing a 3- to 3.5-cm wrap (Fig. 6.11). At this point, note how good mobilization of the greater curvature of the stomach aids in effective

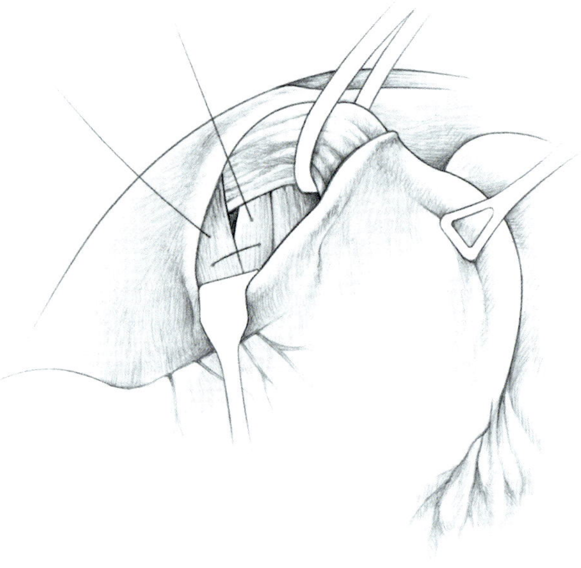

Figure 6.8 Repair of esophageal hiatus for hiatal hernia.

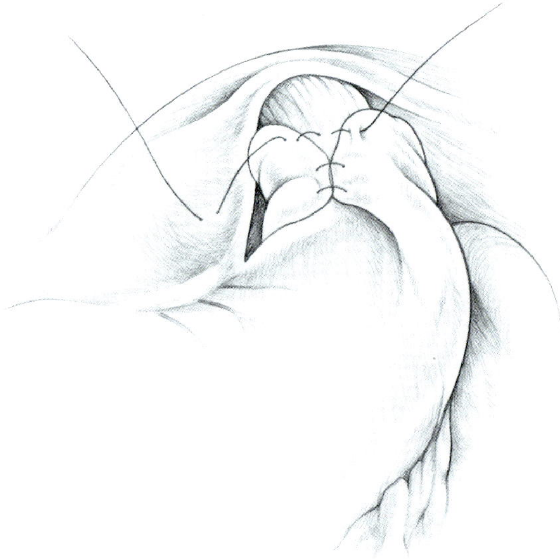

Figure 6.11 Sutures for Nissen fundoplication.

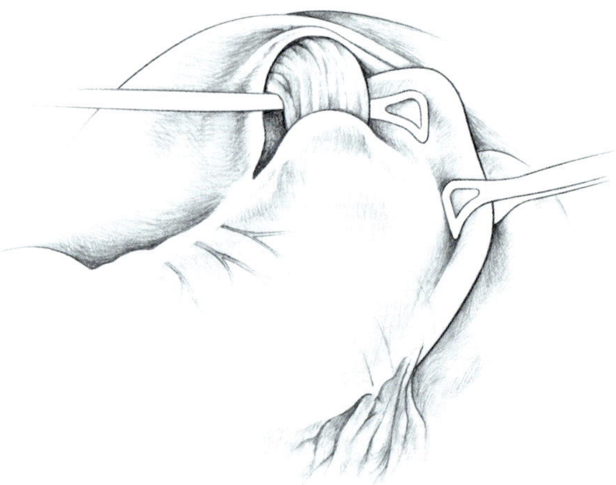

Figure 6.9 Traction of cardia for Nissen fundoplication.

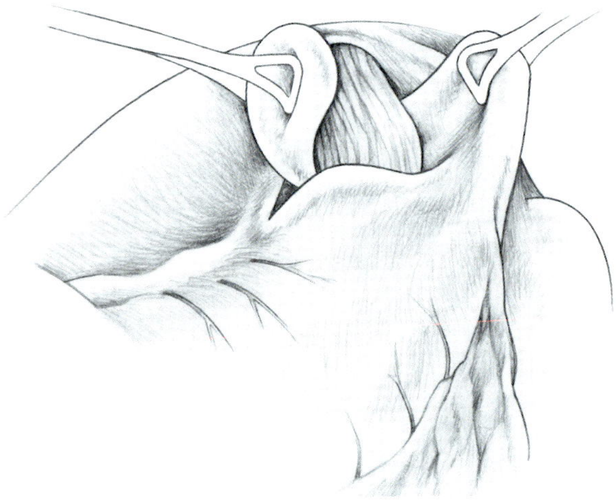

Figure 6.10 Formation of gastric walls for Nissen fundoplication.

fundoplication. The author places the superior fundoplication stitch to incorporate the right crus. This is done to prevent rupture of the fundoplication into the chest through the repaired hiatus (Fig. 6.11).

CLINICAL HIGHLIGHTS

- The gastrohepatic ligament will contain aberrant hepatic arteries in ~11% of patients (Fig. 6.5). Preservation should be attempted. The vessels can be divided if necessary without causing harmful ischemia to the left lobe of the liver.
- Right and left crus of the esophageal hiatus both arise from the right crus of the diaphragm in most cases (Fig. 6.7).
- Right and left pleural reflections are close to the mediastinal dissection of the esophagus and may be torn during surgery. After Nissen fundoplication, whether open or laparoscopic, remember to obtain a postop chest x-ray to rule out pneumothorax.
- With the right index finger placed posterior to the esophagus to create the space for a Nissen wrap, note the transverse angle of the finger. Now, think about port placement for a laparoscopic Nissen. It is easy to see that the right medial subcostal port needs to be very close to the rib margin in order to have a Babcock clamp follow this transverse trajectory (Fig. 6.9).

- Remember during Nissen fundoplication, use esophageal dilators of 50-60 French with or without the presence of a nasogastric tube when suturing the wrap to avoid postoperative dysphagia.

- Esophageal perforation during Nissen fundoplication is most commonly associated with severe esophagitis. Preoperative endoscopy should be performed by the surgeon for anatomical considerations and for evaluation of the status of the esophagus. Esophagitis must be medically controlled preoperatively to avoid perforation at surgery.

- The Nissen fundoplication re-establishes the angle of His. An acute angle of His must be established for the collar of Helvetius (oblique muscle band) to act as an active part of the lower esophageal sphincter in concert with the inner circular muscle layer of the esophagus. Place an instrument in the newly established angle of His after your Nissen to demonstrate (Fig. 6.12A, B).

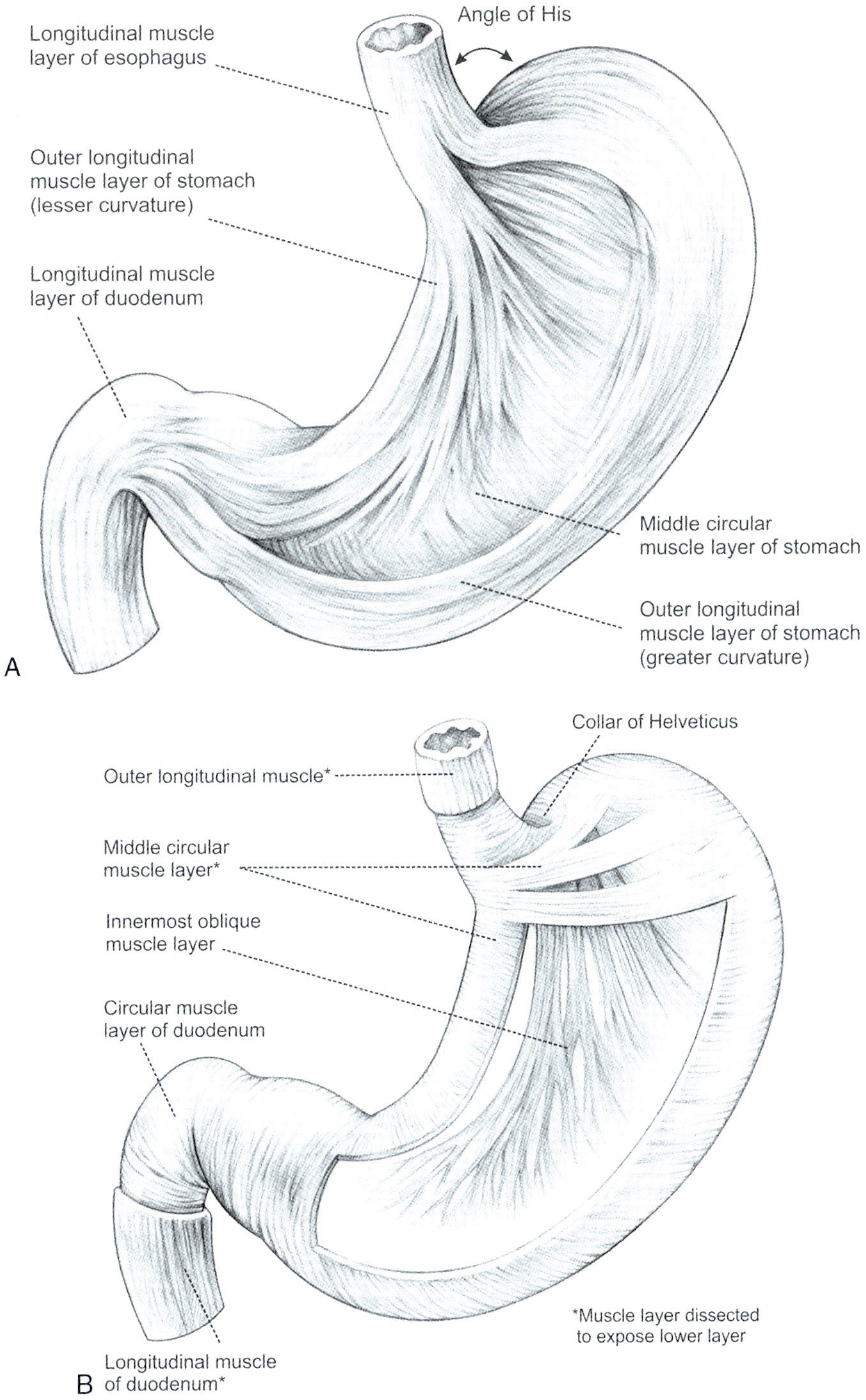

Figure 6.12 A, B: Muscular layers of the stomach emphasizing the circular lower esophageal sphincter, angle of His, and collar of Helvetius.

GREATER CURVATURE OF THE STOMACH AND SHORT GASTRIC VESSELS (MOBILIZATION FOR ESOPHAGECTOMY OR FUNDOPLICATION)

INCISION

Access to the chest and abdomen will facilitate this procedure (Figs. 3.2 and 5.1). Reflect the abdominal wall inferiorly and the anterior chest wall superiorly. Identify the stomach. Identify the right gastroepiploic artery on the greater curvature of the stomach and be certain to preserve it.

OPERATIVE PROCEDURE

Elevate the stomach and identify the right gastroepiploic artery. Trace this artery back under the pylorus as it then branches off the gastroduodenal artery. Note the close proximate relationship between the gastroduodenal artery and the posterior duodenal bulb (Fig. 6.13). A posterior duodenal ulcer, which erodes through the wall of the duodenum, may erode into the gastroduodenal artery. If the penetrating ulcer causes profuse hemorrhage, it will require emergent surgery.

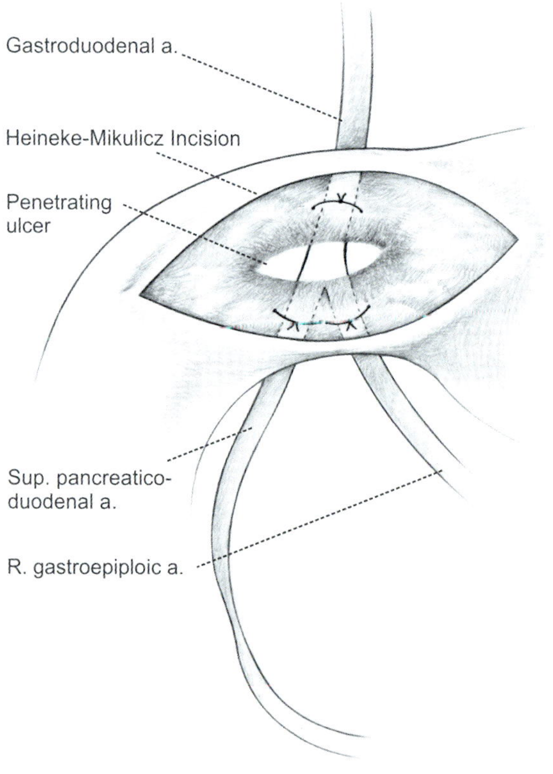

Figure 6.13 Exposure of gastroduodenal, right gastroepiploic, and superior pancreaticoduodenal arteries to demonstrate their locations for hemorrhage due to penetrating peptic ulcer.

Enter the lesser sac. See how you gain access to the posterior stomach, pancreas, or pancreatic pathology such as pancreatic pseudocyst. Note that the transverse mesocolon may be at risk (Fig. 6.14). Inflammatory or malignant adhesions may cause the greater omentum and the transverse mesocolon to fuse. This may put the blood supply to the transverse mesocolon at risk when entering the lesser sac. So begin by entering the lesser sac laterally and mobilize the greater curve medially (Fig. 6.14).

Trace the right gastroepiploic artery distally in its course along the greater curvature of the stomach (Fig. 5.2). This artery must be preserved during transhiatal esophagectomy when the stomach is used as the replacement conduit. The majority of the blood supply to the stomach is supplied by the right gastroepiploic artery after mobilization of the stomach. Free the remainder of the greater curvature of the stomach from the greater omentum (Fig. 6.6). Moving to the left, attempt to identify the left gastroepiploic artery. This artery arises from the splenic artery and has a tortuous, variable course along the left gastric border (Fig. 5.2). Continue dissection along the left greater curvature. Dissection in this area is through the gastrosplenic ligament. This may be a single sheet of tissue or two sheets of tissue (Fig. 6.15).

Continue sharp transection of the gastrosplenic ligament close to the stomach superiorly. Note the short gastric vessels encountered, which are transected and ligated individually. Carry this sharp dissection up to the diaphragm through the gastrophrenic ligament (Fig. 6.6). This will mobilize the entire greater curvature and cardia of the stomach. Elevate the stomach and look into the lesser sac. Identify the anterior wall of the pancreas. Inferior to the pancreas, identify the transverse mesocolon.

CLINICAL HIGHLIGHTS

- The gastrosplenic ligament may have one or two well-defined folds. During mobilization of the greater curvature of the stomach, if present, the presplenic fold may be confusing to the surgeon who is unaware of this anatomic variable (Fig. 6.15).
- Remember that short gastric vessels may come off the superior-most aspect of the spleen. These so-called supreme short gastric vessels may be identified only at the uppermost portion of the cardia of the stomach in the gastrophrenic ligament. Be aware that inferior phrenic vessels may also be found in this area arising from branches of the celiac axis and traversing posterior to the esophagus (Fig. 5.2).

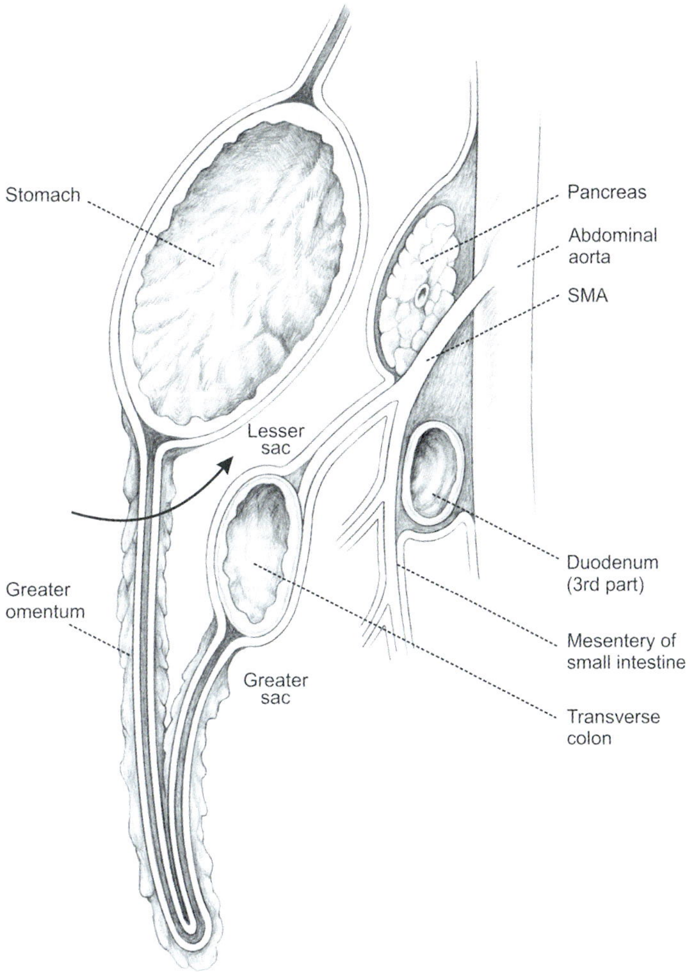

Figure 6.14 Entrance into the lesser sac; relationship of the greater omentum and transverse mesocolon.

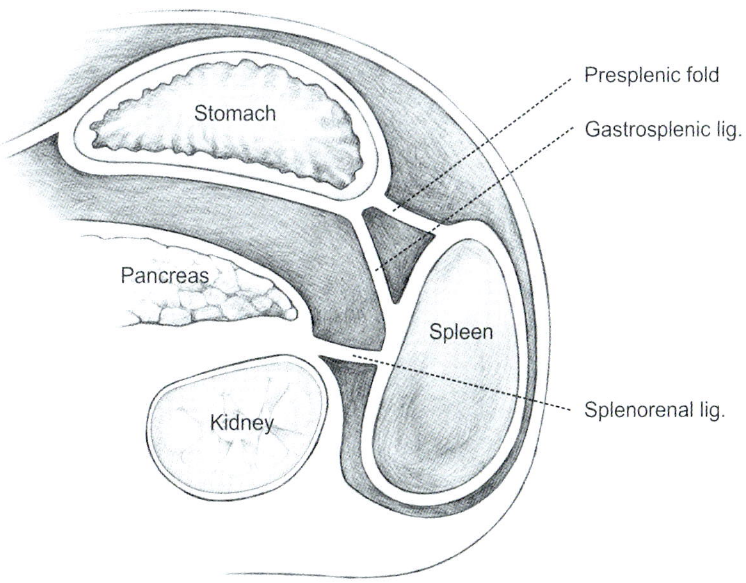

Figure 6.15 Presplenic fold; gastrosplenic ligament with short gastric arteries.

- In the operating room, the short gastric vessels can be divided and ligated individually or transected with a vessel coagulation device. Depending on the body habitus, the supreme short gastric vessels can be difficult to identify and control. Appropriate retraction, visualization, and careful dissection in this area are key to prevent massive hemorrhage.

- When taking down the short gastrics, the surgeon must exercise care in retraction of the stomach, as overzealous retraction can lead to avulsion of the short gastrics from the splenic hilum, necessitating splenectomy.

- Note that a pancreatic pseudocyst in the lesser sac will likely be adjacent to the posterior wall of the stomach. On the body, re-evaluate this relationship. The surgeon can now see why a posterior wall gastrocystostomy is a convenient and anatomically sensible method of pseudocyst drainage for lesser sac pseudocysts.

- Remember the critical nature of the right gastroepiploic artery when performing transhiatal esophagectomy. The right gastroepiploic artery, the right gastric artery, and the extremely rich interconnected nature of the gastric submucosal plexus allow complete mobilization of a viable stomach for an esophageal replacement conduit.

DISTAL ESOPHAGUS (HELLER MYOTOMY)

INCISION

Access to the chest and abdomen will facilitate this procedure (Figs. 3.2 and 5.1). Open the chest and abdominal cavities as previously performed.

OPERATIVE PROCEDURE

From the abdominal side, identify the distal 6 cm of esophagus (Fig. 6.4). Split the outer longitudinal and inner circular fibers down to the mucosa using a curved hemostat for dissection and forceps for retraction (Fig. 6.16A, B). Continue this dissection superiorly and inferiorly along the esophagus. Carry the myotomy onto the anterior surface of the stomach for ~2 cm (Fig. 6.17). Be careful of the outer mucosal layer because in the preserved specimen, this is a rather thin layer. When performing this procedure for achalasia, the muscle is thicker and the mucosal layer is more clearly defined.

Separate the muscular layer from the mucosa laterally on each side for ~1 cm. Suture the cardia of the stomach to the left esophageal lateral muscular wall (Fig. 6.17). Wrapping the gastric wall posteriorly and suturing the

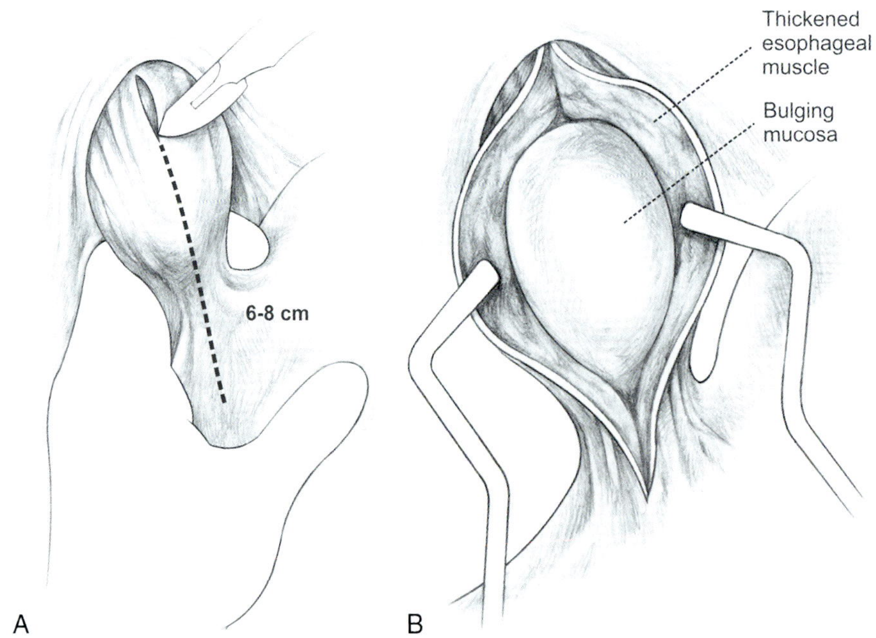

Figure 6.16 **A:** Dimensions and incision for Heller myotomy of the esophagus. **B:** Heller myotomy with bulging mucosa.

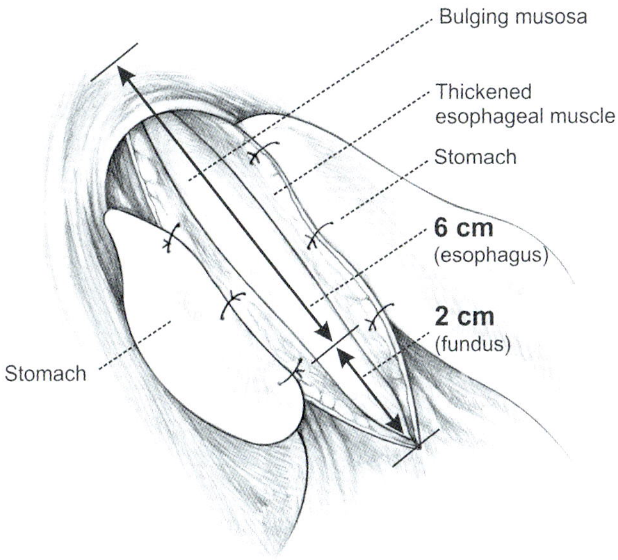

Figure 6.17 Dimensions and formation of Toupet fundoplication after Heller myotomy.

gastric wall to the right lateral myotomy for 3-4 cm will achieve a Toupet fundoplication (Fig. 6.17).

The author's preference is the Dor fundoplication, shown in Figure 6.18A–C. The left edge of the myotomy is sutured to the stomach to create the angle of His (Fig. 6.18A). The right edge of the myotomy and right crus of the diaphragm are sutured to the stomach to create an anterior fundoplication as illustrated (Fig. 6.18B, C).

CLINICAL HIGHLIGHTS

- Remember that esophagoscopy and barium esophagram may both be necessary to define the anatomy of achalasia.
- Heller myotomy is performed for achalasia and remains the gold standard for prevention of

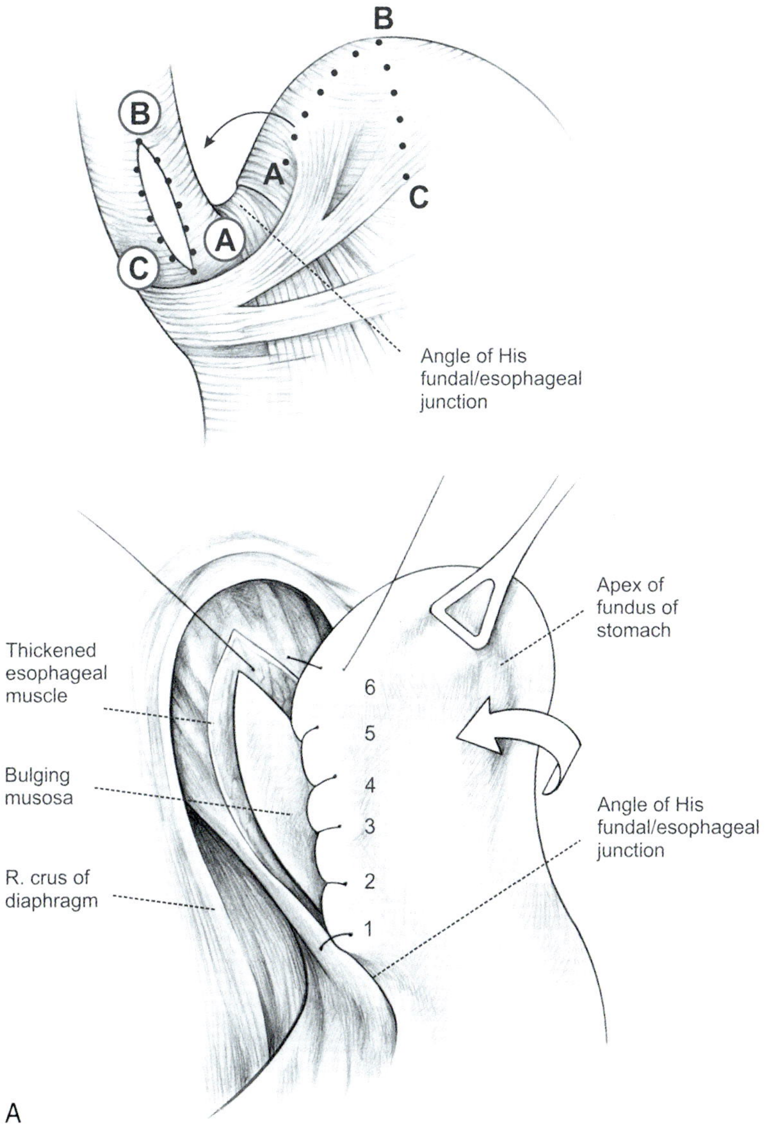

Figure 6.18 **A:** Dimensions, design, and beginning of Dor fundoplication after Heller myotomy.

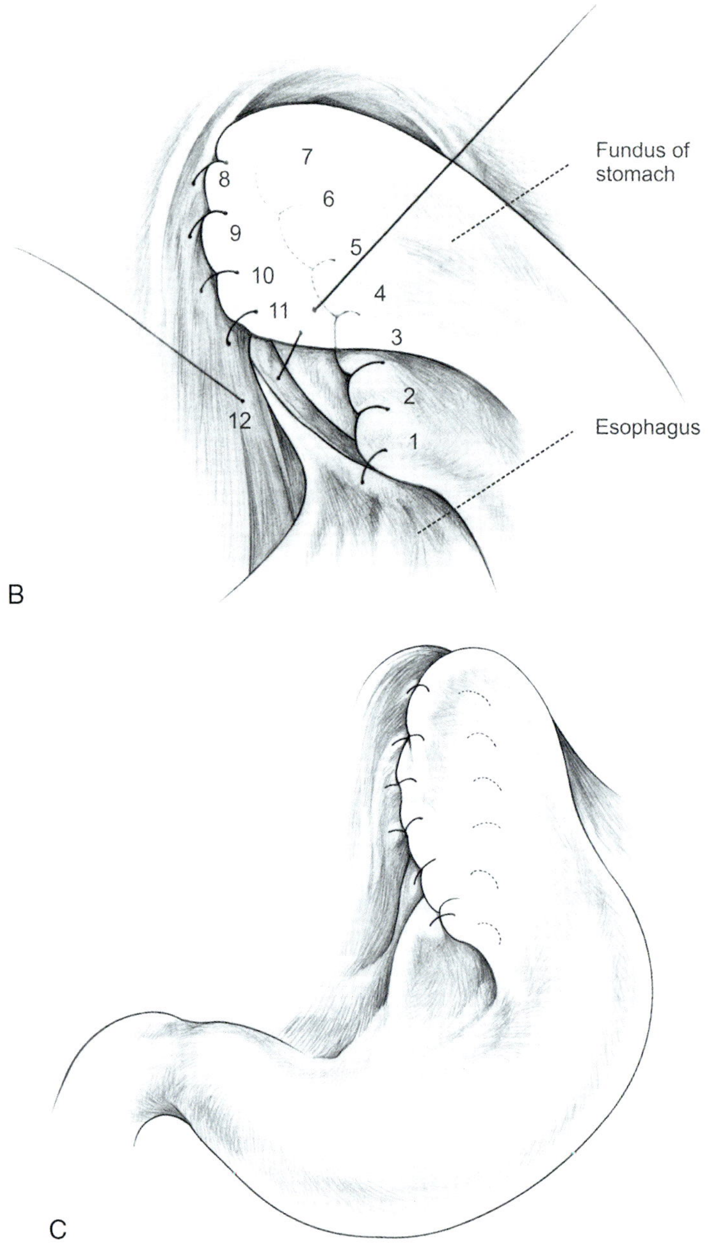

Figure 6.18 (Continued) **B:** Suture placement for Dor fundoplication after Heller myotomy. **C:** Completed Dor fundoplication.

recurrent dysphagia and optimal long-term outcomes. Preoperative evaluation with endoscopy and esophageal manometry is required.

- Gastroesophageal reflux disease (GERD) commonly accompanies Heller myotomy. Consideration should be given to partial fundoplication to improve GERD symptoms postoperatively.
- Dor or Toupet are preferred over Nissen fundoplication after Heller myotomy to avoid obstructive symptoms in the proximal megaesophagus.
- Remember to perform completion esophagoscopy confirming adequacy of the myotomy proximally and distally.

PYLORUS (PYLOROMYOTOMY, PYLOROPLASTY, BLEEDING ULCER, LADD BANDS)

INCISION

Full access to the abdomen will facilitate these procedures (Fig. 5.1). Reflect the abdominal wall onto the thighs.

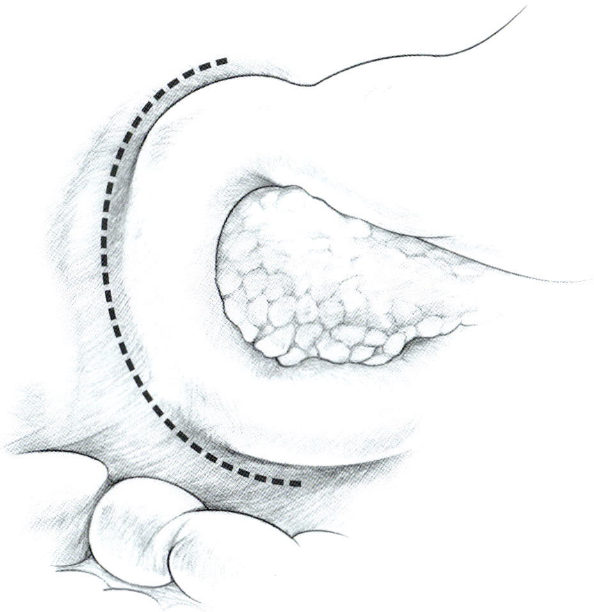

Figure 6.19 Peritoneal incision to begin Kocher maneuver.

OPERATIVE PROCEDURE

The first step toward mobilizing the duodenum is a Kocher maneuver. Begin the Kocher maneuver by incising the lateral peritoneal attachments of the duodenum (Fig. 6.19). Elevate the duodenum and continue the incision of the peritoneum upward toward the common bile duct (CBD) and downward toward the third portion of the duodenum. Notice that peritoneal attachments to the kidney and posterior abdominal wall are easily manipulated with blunt dissection. Continue the medial dissection to the inferior vena cava (IVC) (Fig. 6.20). A complete Kocher maneuver is accomplished by duodenal mobilization from the CBD to the fourth portion of the duodenum (Fig. 6.21) and medially to the IVC.

Once the duodenum has been freed from its posterior attachments, observe the inner sweep of the duodenum and its segmental blood supply. The intimate association of the duodenal blood supply, duodenum, and pancreas is why those structures are often dealt with as a single surgical unit when considering resection (see pancreaticoduodenectomy in Chapter 7, Section 7).

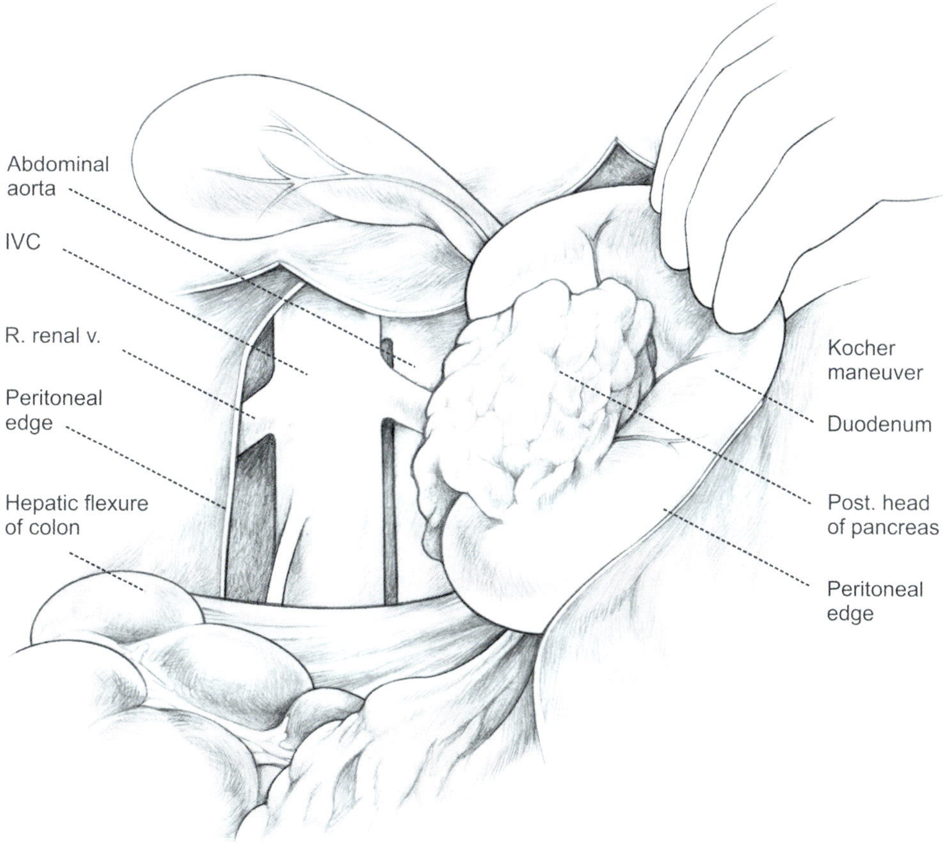

Figure 6.20 Completed Kocher maneuver.

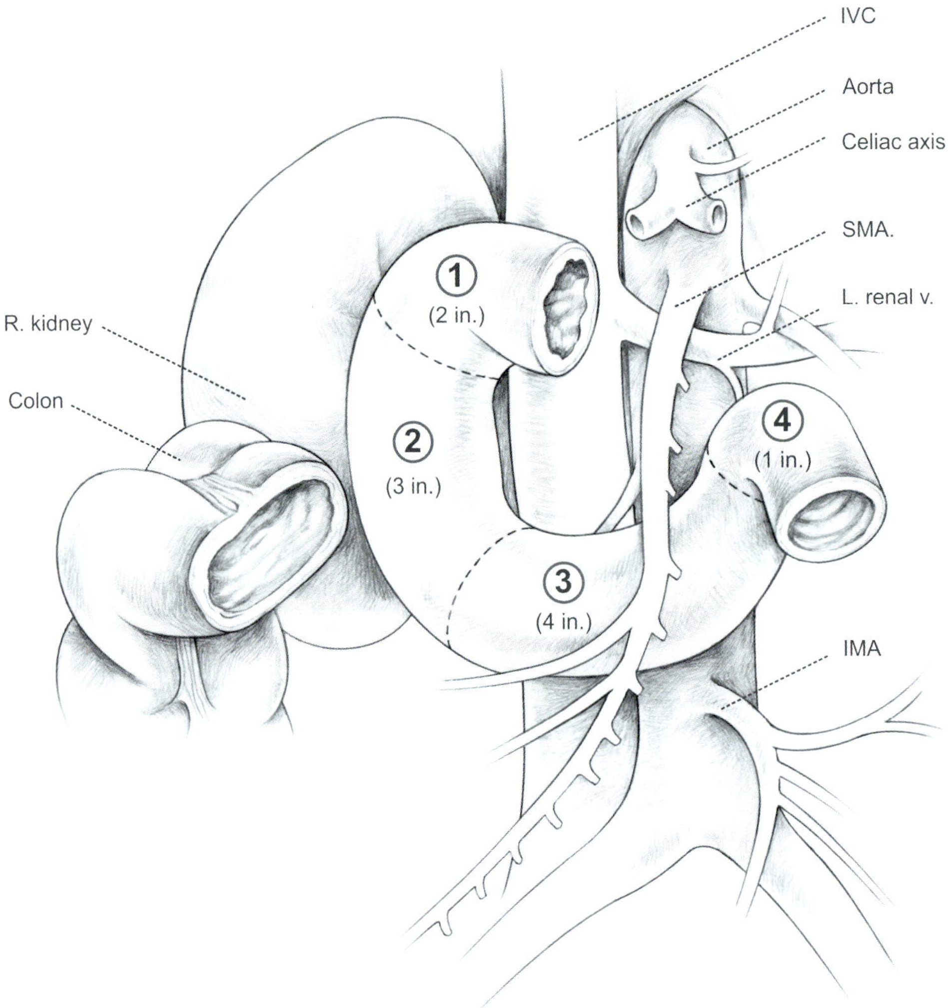

Figure 6.21 Relationship of duodenal segments, superior mesenteric artery, right kidney.

The Ladd procedure for incomplete rotation of the intestine is demonstrated on the accompanying diagram (Fig. 6.22). Complete release of the Ladd bands and derotation of the midgut are important elements to relieve this cause of congenital obstruction. Your duodenal dissection at this point gives a clear example of duodenal release and inspection. Note the descending trajectory of the second and third segments of the duodenum.

Next, identify the pylorus by its bulky nature and by the overlying (prepyloric) vein of Mayo (Fig. 6.23). A pyloromyotomy is performed by dividing the serosa (Fig. 6.24) over the pylorus and carefully spreading the circular pyloric muscle (Fig. 6.25). Be aware that the duodenal mucosa of the duodenal bulb flares out immediately after the pyloric muscle is divided and may even lay retrograde onto the circular pyloric muscle (Fig. 6.26). Divide all these fibers down to the underlying mucosa. Be careful in the area of the fornix. The mucosa of the fornix is subject to perforation.

If you entered the duodenal mucosa during your pyloromyotomy or if you set out to do a pyloroplasty, incise through the mucosa. Carry the longitudinal pyloroplasty incision onto the stomach and down the duodenum just sufficient to divide all pyloric fibers (Fig. 6.27). Stay sutures or retraction may be used to facilitate exposure, anatomy, and technique. The goal is transverse closure of the pylorus (Fig. 6.28). Transverse closure is performed with an inner layer of 3-0 running absorbable suture followed by an outer layer of 3-0 or 2-0 silk Lembert suture. If peptic ulcer disease has caused a great deal of scarring, a single interrupted layer of

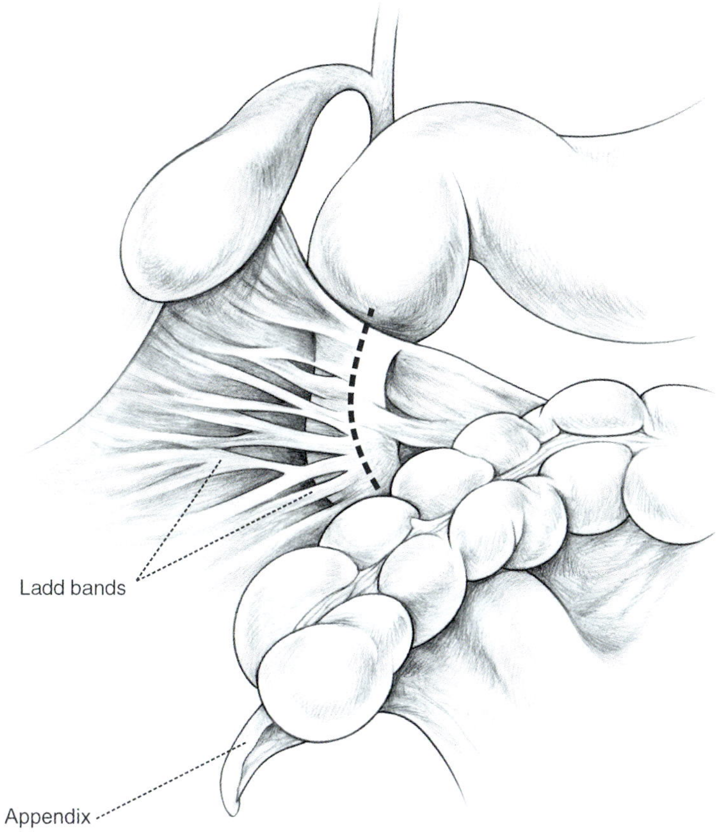

Figure 6.22 The Ladd bands found in cases of congenital malrotation of the bowel.

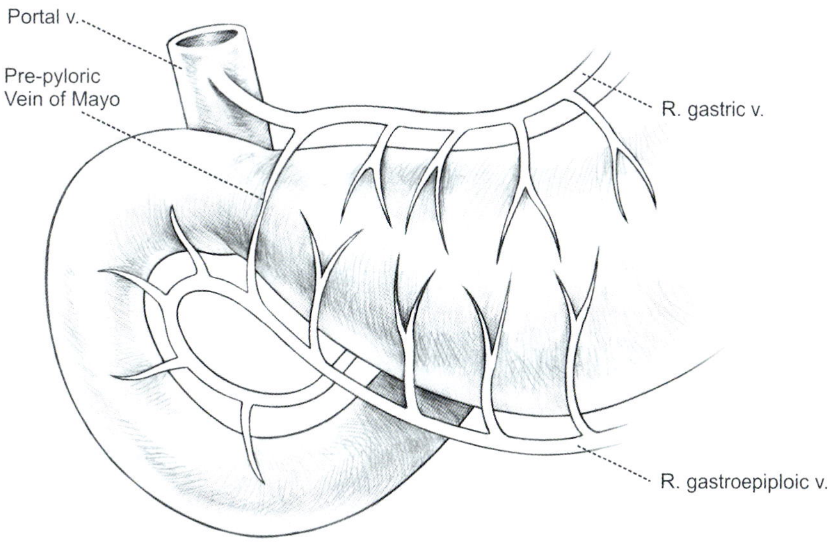

Figure 6.23 Pylorus identified by vein of Mayo.

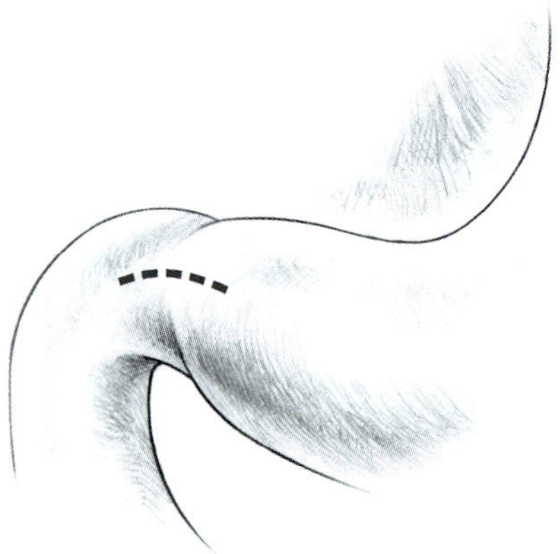

Figure 6.24 Pyloromyotomy incision.

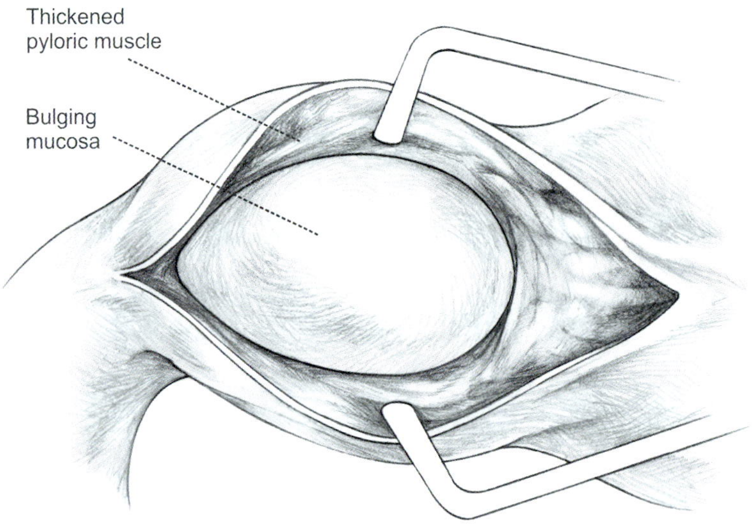

Figure 6.25 Pyloric muscle split and subsequent bulging mucosa to treat congenital pyloric stenosis.

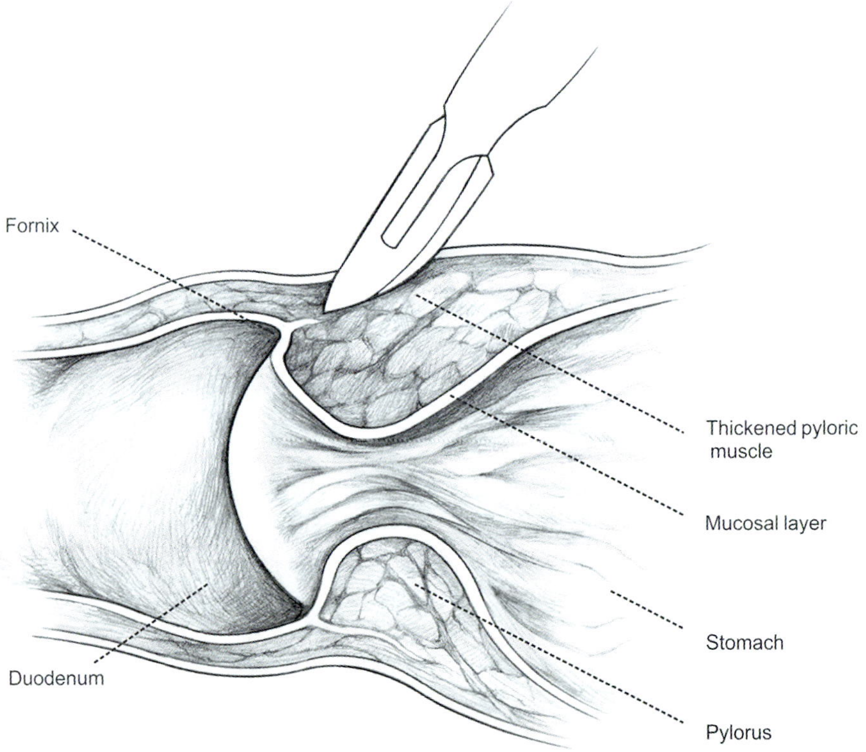

Figure 6.26 Cross-section of duodenal bulb emphasizing location of fornix.

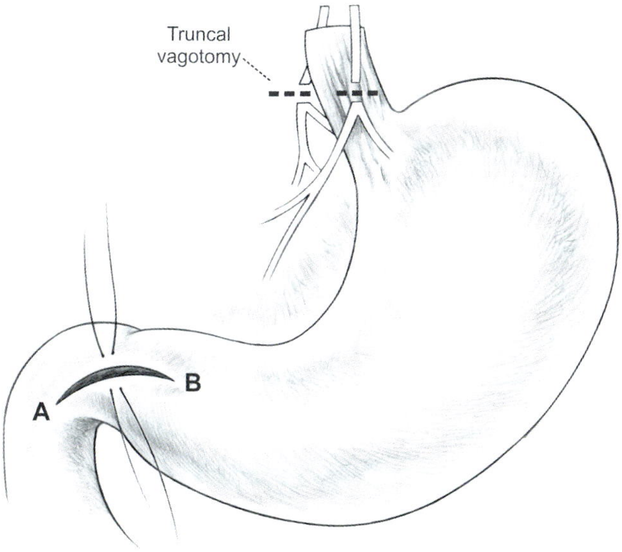

Figure 6.27 Truncal vagotomy, incision, and stay sutures for H-M pyloroplasty.

suture may be all that is possible. This is a Heineke-Mikulicz (H-M) pyloroplasty (Fig. 6.29).

A Finney pyloroplasty is simply an extended pyloroplasty that can be used for severe fibrosis and inflammation from long-standing peptic ulcer disease (Fig. 6.30). To perform this pyloroplasty, take down the sutures placed for the previous pyloroplasty. Extend the duodenal portion of the pyloroplasty another 3 cm. Place a back row of interrupted 3-0 or 2-0 silk suture to approximate the distal stomach and proximal duodenum (Fig. 6.31). This is followed by an inner layer of 3-0 running absorbable and an anterior outer layer of interrupted 2-0 silk is used to complete the Finney pyloroplasty (Fig. 6.32). In the OR, if there is significant inflammation, a single layer may be necessary. A submucosal suture method for single-layer closure is demonstrated (Fig. 6.33). To help understand the Finney pyloroplasty, a cross-section of the closure is shown (Fig. 6.34).

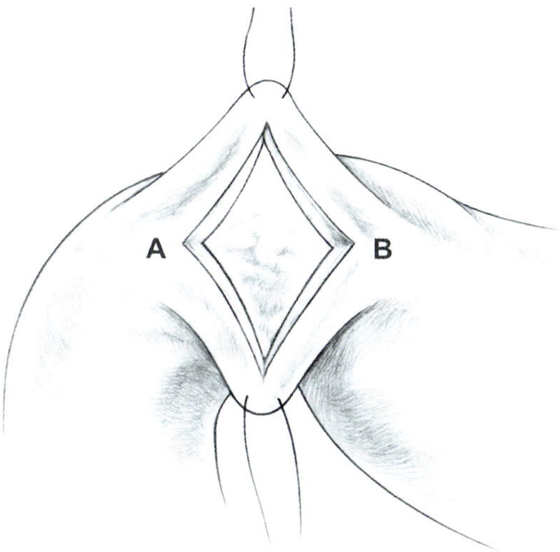

Figure 6.28 Transverse alignment for H-M pyloroplasty.

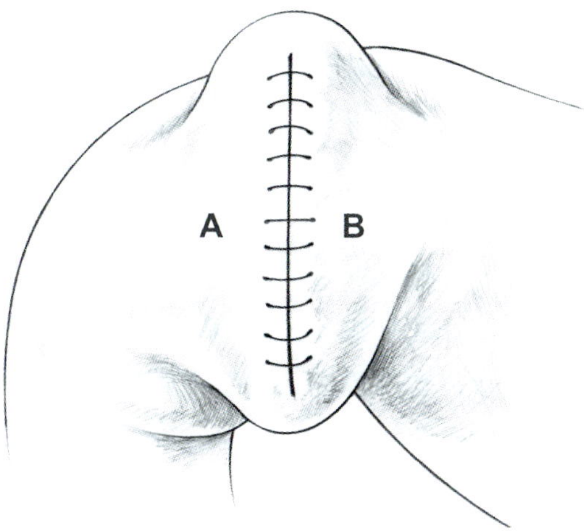

Figure 6.29 Completed H-M pyloroplasty.

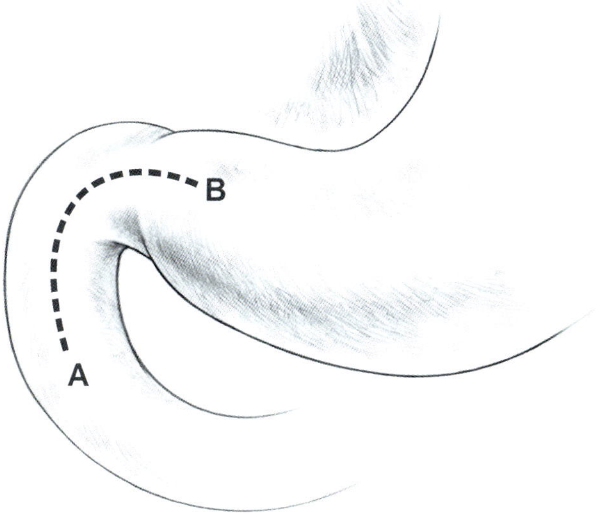

Figure 6.30 Incision for Finney pyloroplasty.

The Jaboulay pyloroplasty is another useful drainage procedure used in the setting of a scarred or inflamed duodenal bulb. A distal gastrotomy and proximal

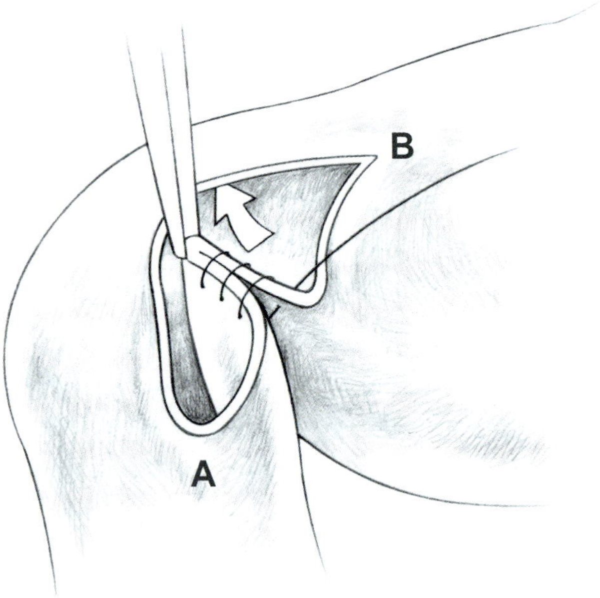

Figure 6.31 Back wall construction for Finney pyloroplasty.

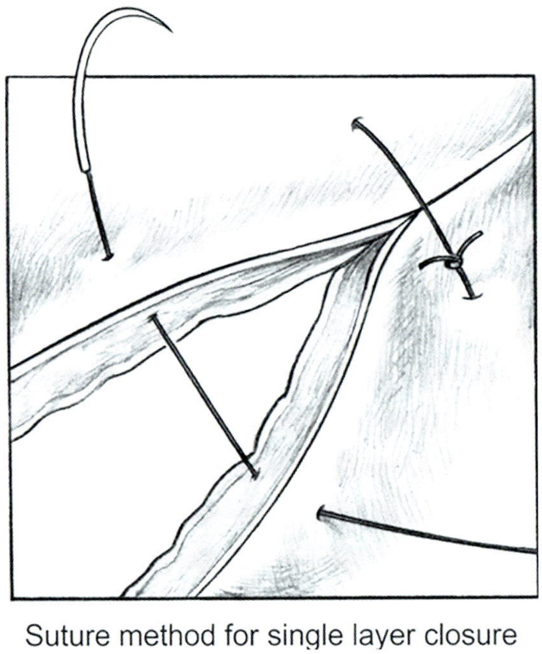

Figure 6.33 Submucosal suture track for single-layer closure.

duodenotomy are closed in layers as demonstrated in the diagram. Practice for this type of pyloroplasty may be performed with gastric and duodenal incisions placed rather distant from each other and repaired in side-to-side fashion in two layers (Fig. 6.35A–C).

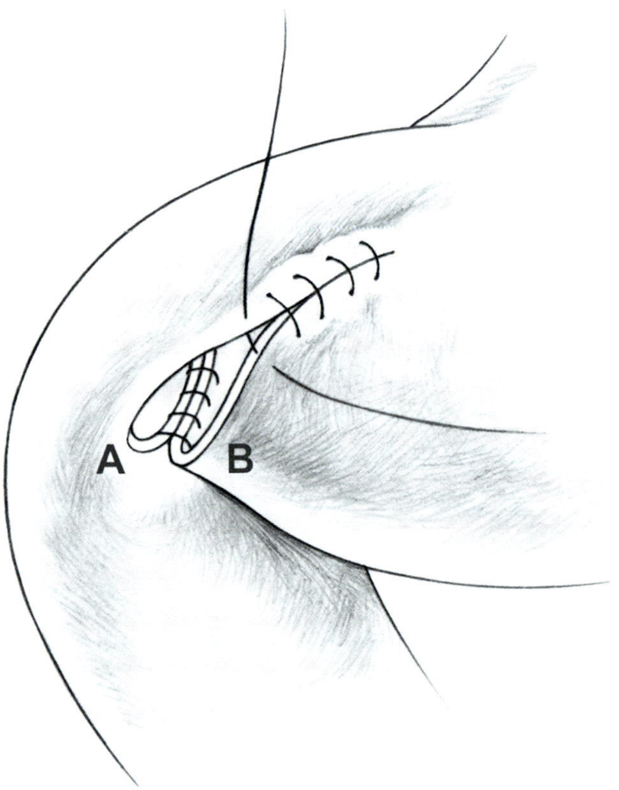

Figure 6.32 Front wall construction for Finney pyloroplasty.

When massive bleeding is encountered from a posterior penetrating ulcer in the duodenal bulb, it is necessary to perform a triple ligation of the gastroduodenal artery and it branches. This is done through a pyloroplasty-type incision to examine the posterior duodenal wall. Take down any previously sutured pyloroplasty in preparation to examine the source of bleeding from a penetrating duodenal ulcer. Identify the gastroduodenal artery as it lays in direct contact below the duodenal bulb (Fig. 5.2). Open the posterior wall of the duodenal bulb to expose the gastroduodenal artery. Remember this association when confronted with a

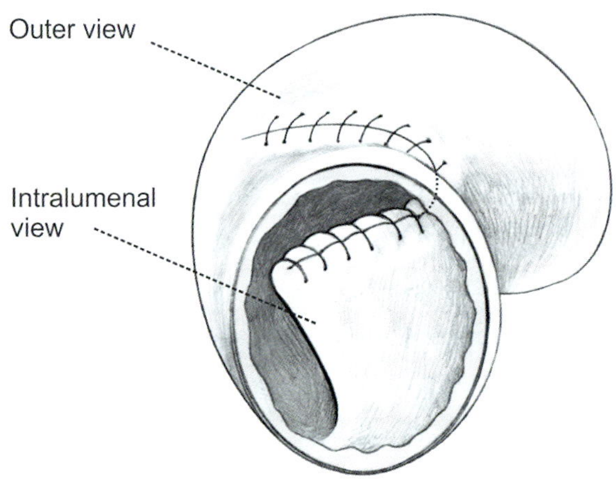

Figure 6.34 Cross-section of Finney pyloroplasty showing inner and outer views.

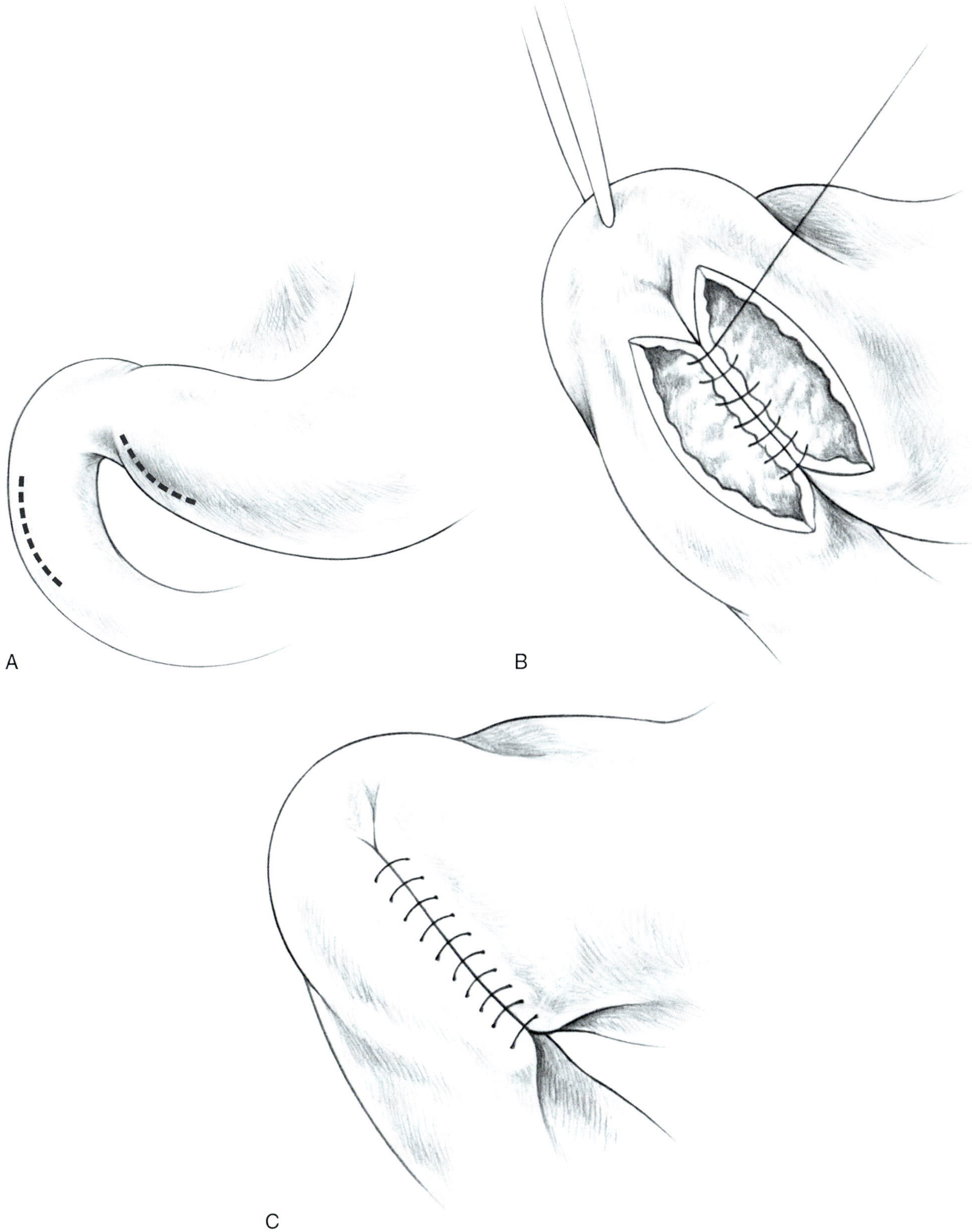

Figure 6.35 A–C: Incisions, back wall construction, and completion of Jaboulay pyloroplasty.

bleeding ulcer. Note the vertical orientation of the gastroduodenal artery and branches sent laterally under the distal duodenum. The dissecting surgeon can now see the rationale for placing 2-0 silk suture at the superior edge of the penetrating ulcer, the inferior edge of the ulcer, and the right lateral edge of the ulcer to control bleeding (Fig. 6.13). It now becomes obvious that repair of access to the bleeding ulcer is simply repaired in the Heineke-Mikulicz (H-M) fashion previously described.

Follow the gastroduodenal artery as it appears at the inferior border of the duodenal bulb. Note that it then becomes the right gastroepiploic artery. Continue to trace the gastroepiploic artery along the greater curve of the stomach. Remember that this is the dominant blood supply of the stomach when used as a conduit for transhiatal esophagectomy and can be traced by palpation or if necessary intraoperative Doppler in the OR.

CLINICAL HIGHLIGHTS

- Pyloromyotomy in an infant is done for a hypertrophic pylorus described as an epigastric "olive" on physical examination of the child.
- The hypertrophic pyloric muscle separates easily when meticulous care is used to avoid damage to the duodenal mucosa. Mucosal closure with absorbable suture is employed if mucosa is entered.
- Heineke-Mikulicz, Finney, and Jaboulay are the most commonly performed pyloroplasties.
- The type of pyloroplasty performed depends on the condition of the pylorus and duodenum and the disease for which it is being performed. The Heineke-Mikulicz is the most common. The surgeon should be familiar with all three as it may be necessary to convert to a different type due to the conditions encountered intraoperatively.
- Triple ligation of the gastroduodenal artery and its major branches (right gastroepiploic artery and superior pancreaticoduodenal artery) is necessary for adequate control of massive life-threatening hemorrhage from a penetrating posterior duodenal bulb ulcer that has eroded into the gastroduodenal artery. All three vessels must be ligated due to the extensive collateral circulation around the duodenum and head of the pancreas through the hepatic and superior mesenteric arteries (Fig. 6.13).
- When performing emergency surgery for a bleeding ulcer, access to the duodenal bulb is best obtained by the anterior incision for a Heineke-Mikulicz pyloroplasty. Remember the superior inferior and right lateral sutures used to control the underlying gastroduodenal artery (Fig. 6.13).
- Kocher maneuver may be used to inspect for evidence of obstruction from congenital problems of annular pancreas and the windsock deformity.

DUODENAL SEGMENTS (LIGAMENT OF TREITZ/SMA SYNDROME)

INCISION

Full access to the abdomen will facilitate this procedure (Fig. 5.1). The first part of the duodenum has been dealt with during the discussion of pyloroplasty. The second part of the duodenum descends. The third portion of the duodenum is transversely located and is intimately associated with the uncinate process of the pancreas and the overlying superior mesenteric artery (SMA). The fourth portion of the duodenum contains the suspensory ligament of Treitz (Figs. 6.21 and 6.36A, B).

OPERATIVE PROCEDURE

Fully mobilize the first and second portions of the duodenum as well as the proximal third portion of the duodenum using the Kocher maneuver (Figs. 6.19 and 6.20). This should yield mobilization of the duodenum from the CBD around the entire sweep of the duodenum all the way to the retroperitoneal aorta. Note that the anterior surface of the IVC has been exposed. Also note the intimate relationship and blood supply of the duodenum and pancreas.

Return the first and second portions of the duodenum to their normal position. Identify the fourth portion of the duodenum and the suspensory ligament of Treitz. Dissect this ligament vertically to its normal attachment to the right crus of the diaphragm (Fig. 5.25). During this dissection, identify and preserve the SMA. Note that the SMA lies transversely over the third portion of the duodenum. This may rarely cause duodenal obstruction known as the SMA syndrome (Fig. 5.22).

Trace the ligament of Treitz proximally. Observe its course anterior to the left renal vein and posterior to the pancreas to its origin from the right crus of the diaphragm (Fig. 5.25). While exposing the ligament of Treitz, note that it may take various forms from a broad-based ligament to a much narrower ligament to several leaves of ligaments (Fig. 6.36A, B).

Surgical treatment for the SMA syndrome may include division of the suspensory ligament of Treitz and relocation of the duodenum to a more inferior position. Divide the ligament of Treitz and place the duodenum two fingerbreadths inferiorly. Note that the mobilization of the third and fourth portions of the duodenum is necessary to perform this maneuver adequately (Fig. 6.37). Look for all fibers that may contribute to the suspensory ligament of Treitz.

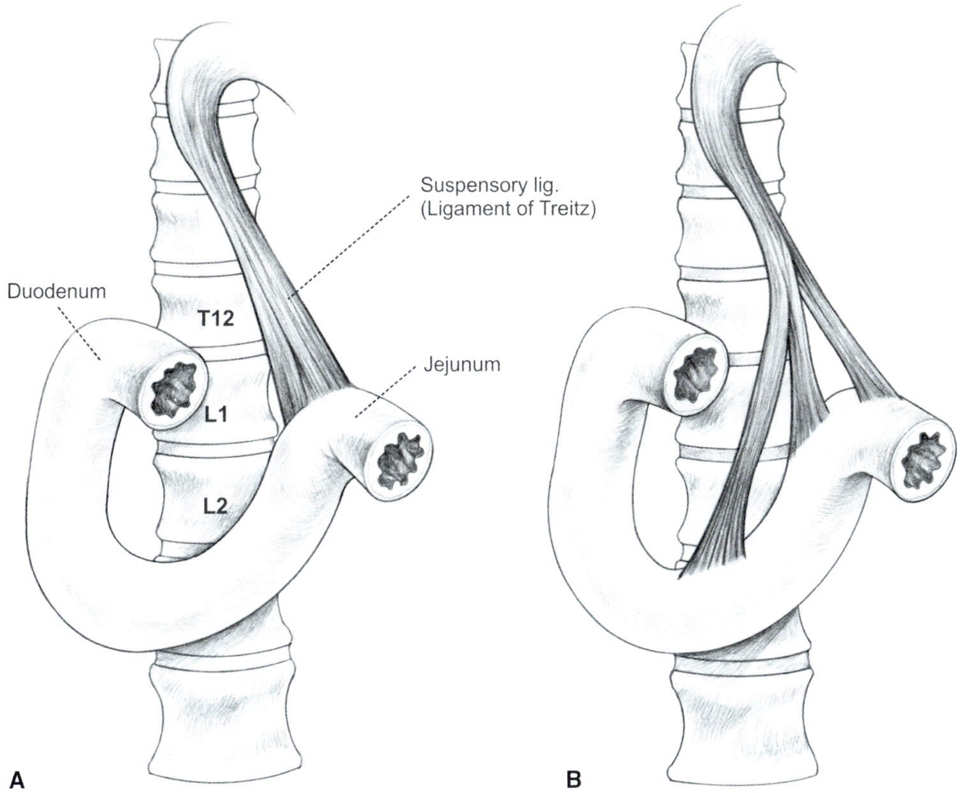

Figure 6.36 **A, B:** Location and patterns of ligament of Treitz.

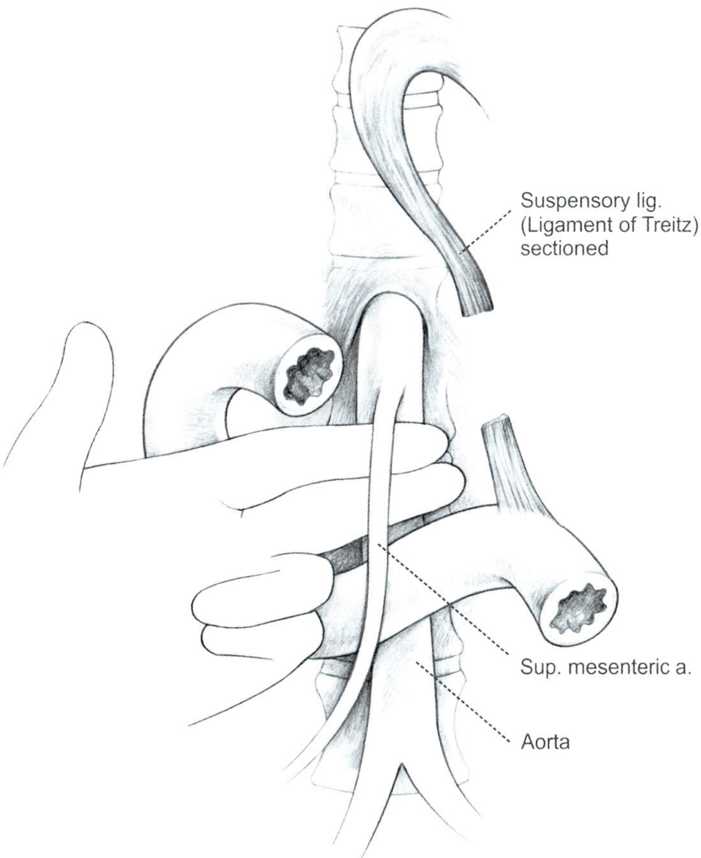

Figure 6.37 Division of ligament of Treitz and repositioning of duodenum for SMA syndrome.

CLINICAL HIGHLIGHTS

- The ligament of Treitz originates from the right crus of the diaphragm.
- The SMA syndrome is compression of the third portion of the duodenum by the SMA.
- The most common clinical presentation of the SMA syndrome is in thin teenage girls where the fat pad is diminished or absent.
- Conservative management of SMA syndrome is the first line of therapy. Diagnosis is aided by lateral aortography. Normally, the angle formed by the SMA and aorta is 25°-60°. When the SMA syndrome is present, the angle between the SMA and aorta is reduced to 6°-15° (Fig. 5.22).
- Doudenal segments, major structures, and pathology (Fig. 6.21):
  - First portion: duodenal bulb, peptic ulcer
  - Second portion: ampulla of Vater, diverticulum
  - Third portion: aortoduodenal fistula s/p AAA graft
  - Fourth portion: ligament of Treitz

GASTROTOMY AND GASTROSTOMIES

INCISION

Full access to the abdomen will facilitate this procedure (Fig. 5.1). Reflect the abdominal wall inferiorly and the anterior chest wall superiorly. Identify the stomach.

OPERATIVE PROCEDURE

Gastrotomies should be performed over the area of suspected pathology. Compromising the lumen by closure of a gastrotomy is unlikely. When performing a gastrotomy, consider underlying pathology and blood supply for healing. Suggested stomach incisions, which may be useful depending on the clinical situation, are demonstrated (Fig. 6.38).

Gastrostomies may be temporary or permanent. The most common temporary gastrostomy performed is a Stamm (Fig. 6.39A–C). An open approach or laparoscopic approach uses a Foley catheter, or a designed gastrostomy tube with a balloon on the internal portion of the tube, or a button device. The gastrostomy

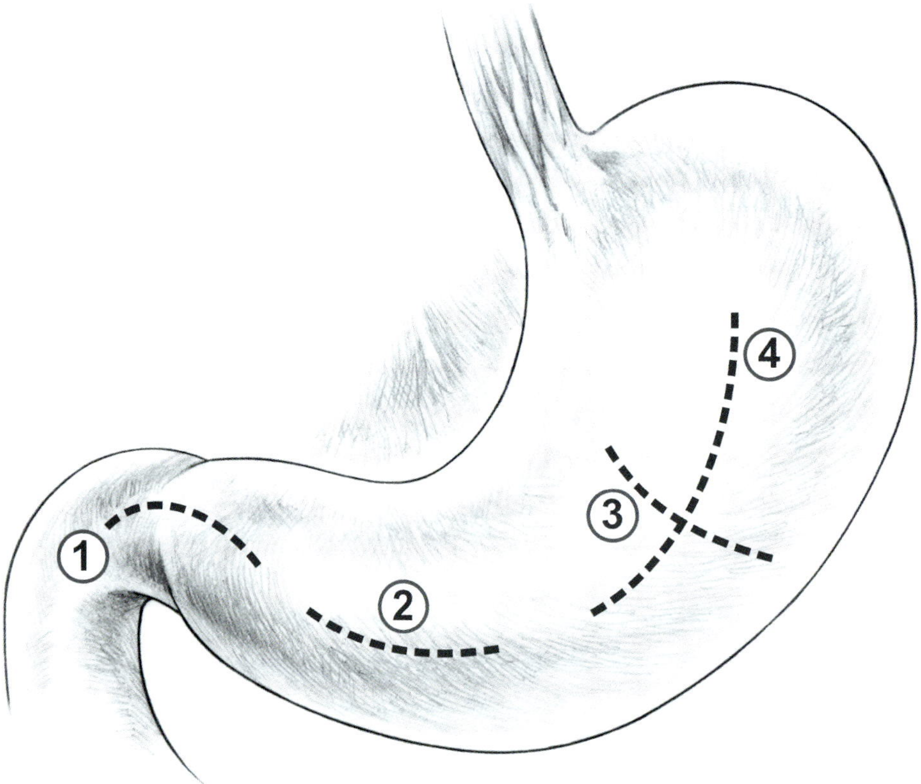

Figure 6.38 Gastrotomy incisions.

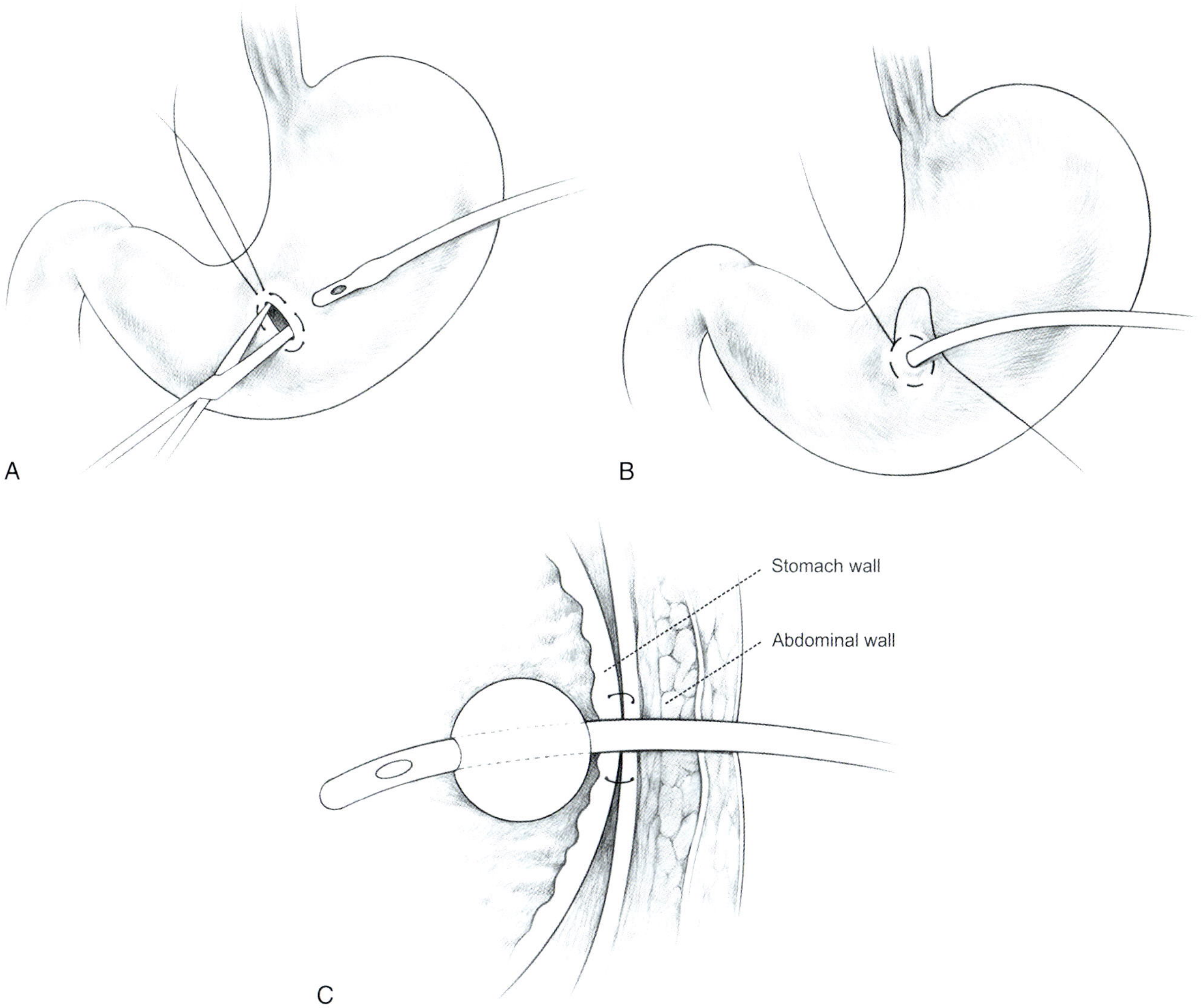

Figure 6.39 **A:** Incision and first purse-string suture for Stamm gastrostomy. **B:** Foley catheter placement for Stamm gastrostomy. **C:** Cross-section of completed Stamm gastrostomy.

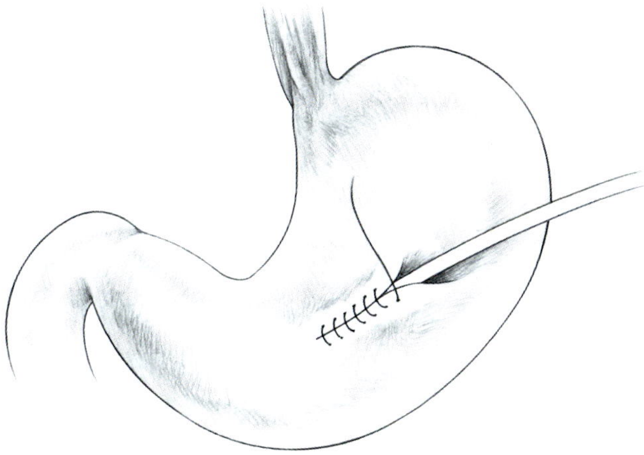

Figure 6.40 Tunnel construction for Witzel gastrostomy.

tube is brought straight through the abdominal wall and through the gastric wall with purse-string sutures placed around the catheter on the gastric wall. Inner and outer concentric rings of purse-string suture are placed. The purse-string suture may be used to tack the anterior gastric wall to the posterior surface of the abdominal wall (Fig. 6.39C). Three other sutures are used to complete the tacking of the gastric wall to the abdominal wall.

Starting with the Stamm method and then using the anterior gastric wall to form a tunnel for the gastrostomy tube creates a Witzel gastrostomy (Fig. 6.40). A running 3-0 Vicryl suture is an appropriate suture to use for the Witzel tunnel. Two layers of anterior gastric wall are noted in classic descriptions of the Witzel procedure.

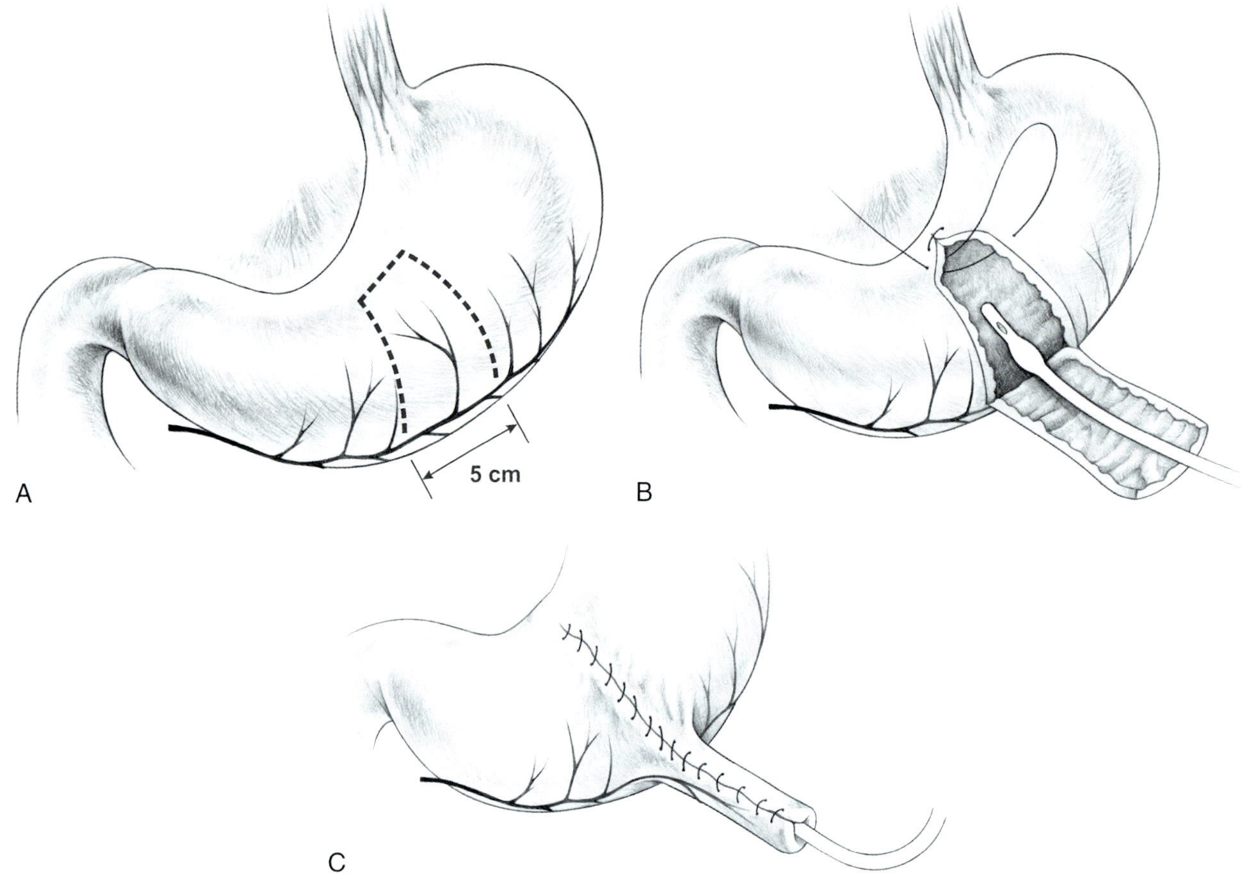

Figure 6.41 **A:** Dimensions for Janeway gastrostomy. **B:** Initial suture placement for Janeway gastrostomy. **C:** Completed Janeway gastrostomy.

One type of permanent gastrostomy is known as a Janeway gastrostomy. Fashion a flap of stomach based on the blood supply from the right gastroepiploic at the greater curve as shown in the figure. Notice the base of the gastrostomy flap is wider than the distal end (Fig. 6.41A–C). Closure of the stomach and the gastric tube may be performed in two layers of running suture over a catheter to serve as a stent. To mobilize the tube fully, divide the gastrocolic omentum interior to the gastroepiploic vessels. The gastric tube is then finished and brought through the abdominal wall.

CLINICAL HIGHLIGHTS

- For gastrostomies used for esophageal obstruction due to malignancy, consider using the Seldinger technique with dilators and a gastrostomy tube placed through a peel-away sheath to avoid drawing the tube through the tumor.
- When nutritional access is needed for obstructing esophageal tumors, consider laparoscopic placement of a jejunostomy tube. The stomach will not be altered so that it may be used in reconstruction after esophagectomy.
- For most other gastrostomies, the transoral technique is commonly used for Stamm-type percutaneous endoscopic gastrostomy (PEG) gastric access.
- Witzel technique for gastrostomies can also be used for placement of jejunostomies. The Witzel technique is particularly useful for needle jejunostomy.
- Permanent gastrostomies described above are particularly useful when permanent gastric access is sought.

ESOPHAGUS (TRANSHIATAL ESOPHAGECTOMY)

INCISION

Access to the chest and abdomen will facilitate this procedure (Figs. 3.2 and 5.1).

OPERATIVE PROCEDURE

The initial steps in performing a transhiatal esophagectomy have been completed in other sections of this chapter. Review Section 3 for mobilization of the stomach and preservation of the right gastroepiploic artery. The Kocher maneuver (Figs. 6.19 and 6.20) and a pyloromyotomy or pyloroplasty, which are necessary in a transhiatal esophagectomy, have been performed previously in Section 5. The esophageal hiatus and vagus nerves have been identified and mobilized in Sections 1 and 2. Take down the Nissen fundoplication previously performed in Section 2. You may find some of your patients have had a previous fundoplication and will need the fundoplication taken down prior to esophagectomy if it does not compromise tumor margins. An upper abdominal lymph node dissection is performed taking the nodal-bearing tissue along the proximal splenic artery, celiac artery, and common hepatic artery (Fig. 6.5). The left gastric artery is ligated close to its origin on the celiac axis (Fig. 7.30).

Mobilization of the esophagus now proceeds in a blunt fashion into the posterior mediastinum. To gain access to the posterior mediastinum, enlarge the esophageal hiatus transversely (Fig. 3.1).

The cervical phase of the procedure is performed through a sternocleidomastoid utility incision beginning at the base of the neck in the suprasternal notch. Prior neck dissection in previous chapters has exposed the trachea and esophagus at the base of the neck. Mobilize the esophagus circumferentially and control it with a Penrose drain (Fig. 6.42). Mobilize as much cervical esophagus as possible using blunt finger dissection inferiorly into the superior mediastinum (Fig. 6.42).

Release tension on the cervical esophagus and insert your hand through the diaphragmatic hiatus and continue dissection as shown (Fig. 6.42). Proximal thoracic esophageal mobilization continues bluntly with traction on the stomach and finger mobilization as shown. The blunt dissection is done by tactile sensation and tearing of the bands of tissue investing the thoracic esophagus. The anterior and posterior planes generally dissect easily in the loose areolar tissue in these planes. The lateral planes can be more difficult due to small communicating vessels (Fig. 10.2) and other structures that are close, including the azygous vein, the trachea, left mainstem bronchus, and the heart (Fig. 6.43). Be very careful when dissecting the esophagus free of the trachea and mainstem bronchi in the superior mediastinum. Remember that the posterior wall of the trachea and left mainstem bronchus is simply a thin mucosal layer. Tearing the posterior tracheal wall will profoundly complicate the rest of the esophagectomy likely necessitating a thoracotomy for repair. A gentle but firm upward pressure on the esophagus distracting it posteriorly toward the vertebral bodies

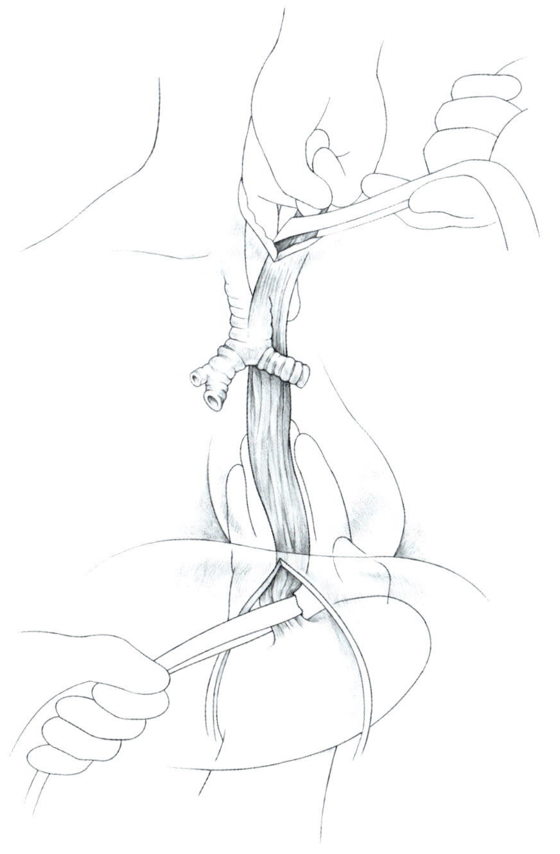

Figure 6.42 Hand placement for first stage of esophageal mobilization for transhiatal esophagectomy.

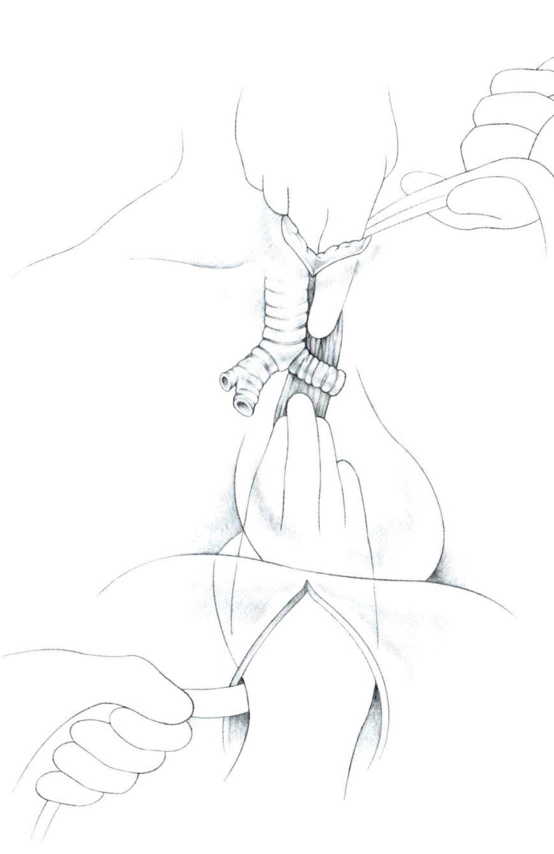

Figure 6.43 Hand placement for second stage of transhiatal esophagectomy.

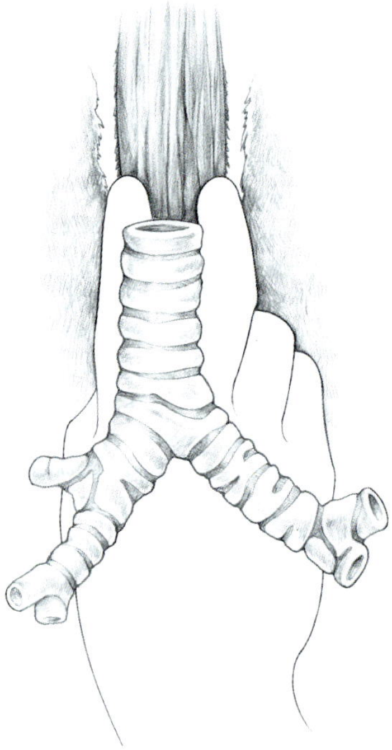

Figure 6.44 Close-up of hand placement during separation of esophagus and trachea for transhiatal esophagectomy.

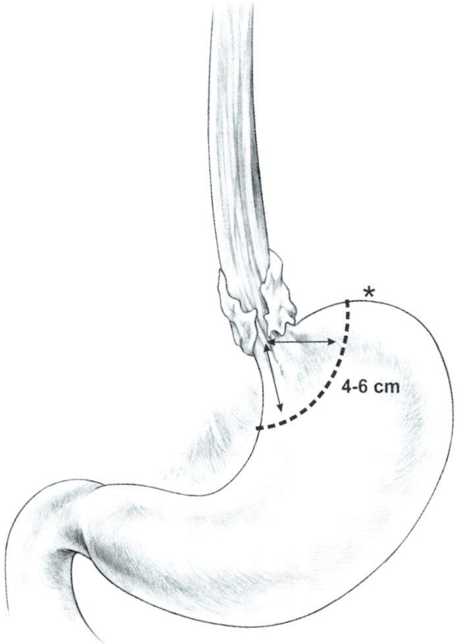

Figure 6.45 Transection line of stomach for tumor and *asterisk* at apex, which will be used for esophagogastrostomy.

while continuing downward traction on the body of the stomach will facilitate the blunt dissection (Fig. 6.44).

At some point, the cervical and abdominal fingers will begin to meet within the thorax (Fig. 6.43). At this point, the stomach can be transected with a stapler taking the cardia and a portion of the lesser curvature. Note that the farthest reach of the stomach as a replacement conduit will be at the apex of the fundus and not at the gastroesophageal junction (Fig. 6.45). The cervical esophagus is then transected in the lower neck at the level of the suprasternal notch to allow sufficient cervical esophagus below the larynx for anastomosis. After transection of the cervical esophagus, the proximal (thoracic) end of the esophagogastric specimen is tied shut with an umbilical tape and then extracted from cranial to caudal through the mediastinum keeping the cervical end of the umbilical tape in the neck. The umbilical tape can be cut on the abdominal side when the specimen is removed and then sutured to the gastric conduit to assist in drawing it superiorly through the posterior mediastinum and guide it to the cervical position for reconstruction. The cadaver may be too stiff to use the posterior mediastinum. If so, lay the stomach in the middle mediastinum.

A gastroesophageal anastomosis can now be performed. At the superior end of the gastric conduit, a gastrotomy is performed. The gastric conduit in the neck and the transected end of the cervical esophagus are then placed in opposition and the anastomosis is performed. It is likely that you will not have access to a good stapler in the lab, so a hand-sewn anastomosis is described and illustrated. A posterior layer of interrupted sutures is placed with the knots tied inside of the anastomosis (Fig. 6.46A). In the operating room at this point, a 46- or 48-French dilator is placed down the pharynx into the gastric conduit under direct vision, but not in the anatomy lab. The anterior anastomosis is completed (Fig. 6.46B). As in all gastrointestinal anastomoses, take care to have mucosa to mucosa apposition.

CLINICAL HIGHLIGHTS

- When performed for malignancy, the patient should be staged preoperatively, both before and after neoadjuvant chemotherapy and radiation. This is generally done with EUS (endoscopic ultrasound) and a CT scan or PET scan pretreatment and a repeat PET scan posttreatment to rule out distant metastases and ensure resectability of the esophagus. If there is any question about the use of the stomach for a replacement conduit, the colon must be prepared as an alternative conduit.

- Currently, neoadjuvant chemotherapy and radiation are the standard of care for all esophageal adenocarcinomas with stage 2 or greater lesions. Esophagectomy is usually performed 4-8 weeks after neoadjuvant therapy is completed.

- Many esophageal centers treat squamous cell cancers of the esophagus with definitive chemotherapy and radiation, withholding esophagectomy for recurrent or residual disease.

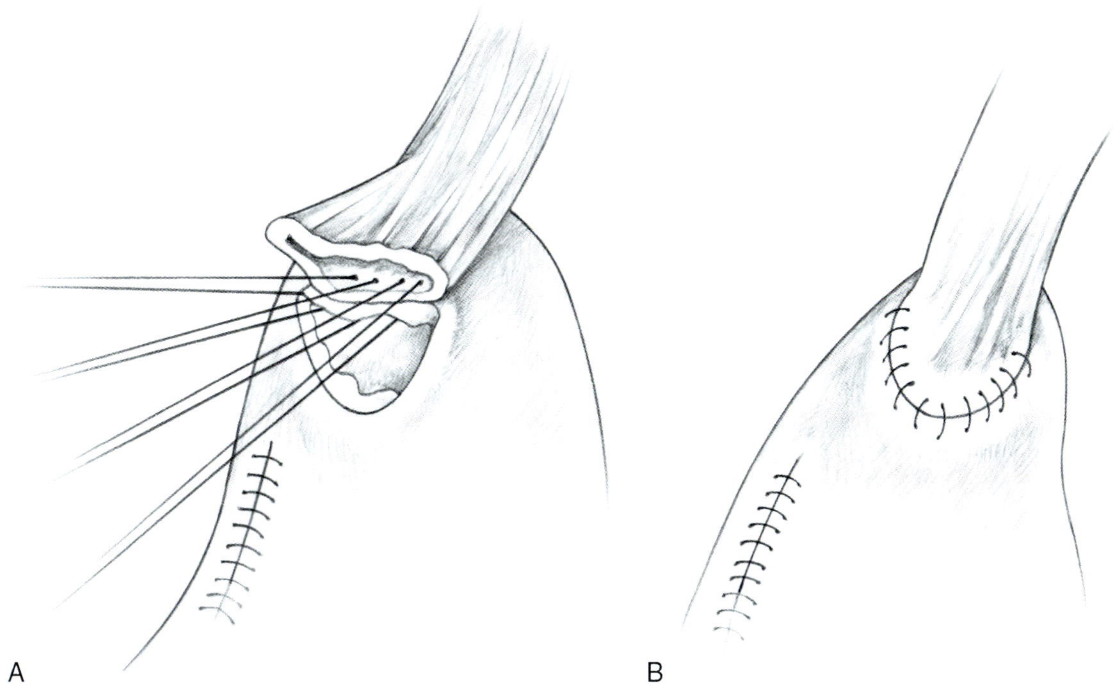

Figure 6.46 **A:** Back wall of esophagogastrostomy completed. **B:** Completion of cervical esophagogastrostomy.

- Blood loss from the blunt mediastinal dissection is usually moderate (300-600 cc) but generally does not require transfusion.

- When performing the blunt mediastinal dissection, it is important to continuously monitor the blood pressure via A-line, as it is common for the presence of the surgeon's hand in the mediastinum to cause hypotension by vena caval and right atrial compression causing decreased venous return to the heart. When this occurs, the surgeon should quickly remove his/her hand from the mediastinum and allow the heart to fill, which will return the blood pressure to normal.

- Preservation and maintenance of an intact right gastroepiploic artery is of critical importance in a transhiatal esophagectomy as it is the primary remaining blood supply to the gastric conduit after mobilization. If the right gastroepiploic artery has been taken in a previous surgery or is injured during gastric mobilization, then the stomach cannot be used as a conduit and an alternative conduit (colon interposition) must be used.

- Leaks or necrosis most commonly occur at superior aspect of the gastric conduit (the fundus of the stomach) where the anastomosis is performed. It is the farthest away from the primary remaining blood supply to the stomach, the right gastroepiploic artery. After gastric mobilization, the vascular supply to this area of the stomach is reliant on the submucosal plexus, which is adequate in most patients but can be compromised in smokers.

- In the OR, the gastroesophageal anastomosis in the neck is stapled with a roticulating 30-mm or 45-mm Endo GIA stapler.

- If a posterior tracheal or mainstem bronchial tear does occur, have anesthesia deflate the balloon, advance it into one of the mainstem bronchi, and reinflate the balloon. The esophagectomy and GI continuity should be completed prior to repair of the tracheal tear.

- A tracheal tear may be exposed and repaired through either an upper sternal extension of the cervical incision or right thoracotomy.

- Consider some type of feeding tube to be used if one has not been previously inserted in before neoadjuvant therapy. In the perioperative setting, the author prefers a nasojejunal feeding tube. Placement is performed in the OR after resection and can be confirmed by direct palpation and visualization.

- Significant unilateral or bilateral reactive pleural effusions are common after transhiatal esophagectomy. The author prefers to insert prophylactic small bore (20- to 24-F) chest tubes bilaterally to prevent subsequent respiratory compromise due to pleural effusions.

- Advantages of a transhiatal approach rather than a transthoracic approach to esophagectomy include avoidance of thoracotomy, cervical anastomosis with easier management of anastomotic problems, decreased pulmonary problems, and shorter recovery. Long-term outcomes have not been found to be significantly different.

GASTRIC RESECTION (ANTRECTOMY, HEMI-, SUBTOTAL, COMPLETE)

INCISION

Full access to the abdomen will facilitate this procedure (Fig. 5.1). Reflect the abdominal wall inferiorly and the anterior chest wall superiorly. Identify the stomach and return it to the abdomen if used for Section 8, transhiatal esophagectomy. Identify the right gastroepiploic artery on the greater curvature of the stomach.

OPERATIVE PROCEDURE

Arbitrary landmarks for gastric resection are as follows: antrectomy, 30% distal gastric resection to include the lesser curvature of the stomach; hemigastrectomy, 50% gastrectomy halfway along greater curvature of the stomach to the right G-E junction to include the lesser curvature of the stomach; subtotal gastrectomy, 75% gastrectomy that is resection of entire lesser curvature of the stomach so that short gastric vessels supply cardia of the stomach (Fig. 6.47). Use a Payr clamp or other long clamp to perform the gastric resection if available; otherwise, simply incise the stomach per the desired resection line.

The following reconstruction procedures may be performed. The author's preference is an interrupted outer layer of 2-0 or 3-0 silk or Vicryl suture and a continuous inner layer of 3-0 chromic catgut or Vicryl using a running locked posterior wall and Connell stitch for the anterior wall of the mucosa.

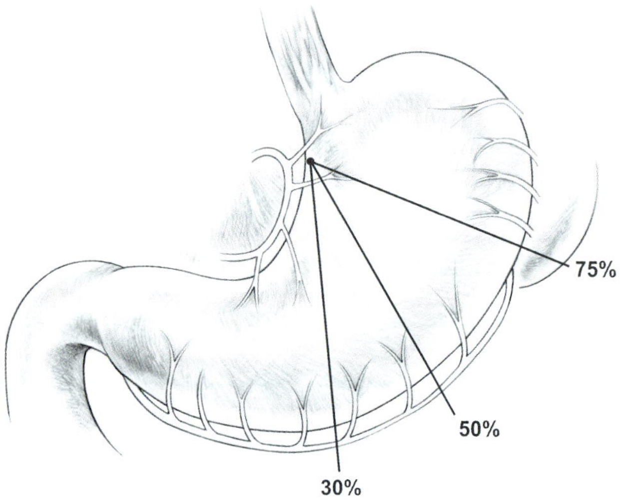

Figure 6.47 Lines of resection for antrectomy, hemigastrectomy, and subtotal gastrectomy; all intersect high on the lesser curve of the stomach.

Polya drainage of the stomach simply uses the entire open end after distal gastric resection to drain into the small bowel (Fig. 6.48A, B).

Hofmeister repair is closure of the open end from the lesser curve toward the greater curve to create a 4- to 5-cm stoma that will drain into the small intestine (Fig. 6.49).

B I: Billroth I drainage of the stomach is a direct reanastomosis of the stomach to the duodenum (Fig. 6.50).

B II: Billroth II drainage of the stomach is an anastomosis of the stomach to the proximal jejunum in either an antecolic (Fig. 6.49) or retrocolic fashion (Fig. 6.51). Be sure to repair the mesenteric defect of the retrocolic repair from the inferior side of the mesentery.

CLINICAL HIGHLIGHTS

- The stomach has a rich blood supply from several sources (right and left gastric arteries, right and left gastroepiploic arteries, short gastric arteries), which allow various types of resections and sacrificing of several of these blood supplies with low risk of causing ischemia in most patients.
- Pay special attention to the blood supply to the "coffin corner" of a Hofmeister closure of the stomach (Fig. 6.52). Postoperative leaks can occur at this site due to necrosis from inadequate blood supply.
- In peptic ulcer disease, inclusion of the lesser curve in the antrectomy specimen is recommended by some authors because of the possible presence of gastrin-producing cells in this area. Also, blood supply to the lesser curve is more tenuous than other areas of the stomach. The lesser curve above the incisura is a common location for gastric ulcer and is eliminated by resection of this area.
- When done for malignancy, antrectomy may be performed from any point on the lesser curvature (typically the incisura), depending on the tumor location. Care is taken to preserve the left gastric artery, which is one of the two remaining major sources of blood supply to the proximal stomach.
- Remember that for a Hofmeister gastric closure, it is necessary to properly size a Billroth I drainage anastomosis (Fig. 6.50).
- Proximal gastric resection may be necessary for proximal gastric lesions. Direct extension of tumor into the spleen or distal pancreas may determine resection of these organs.

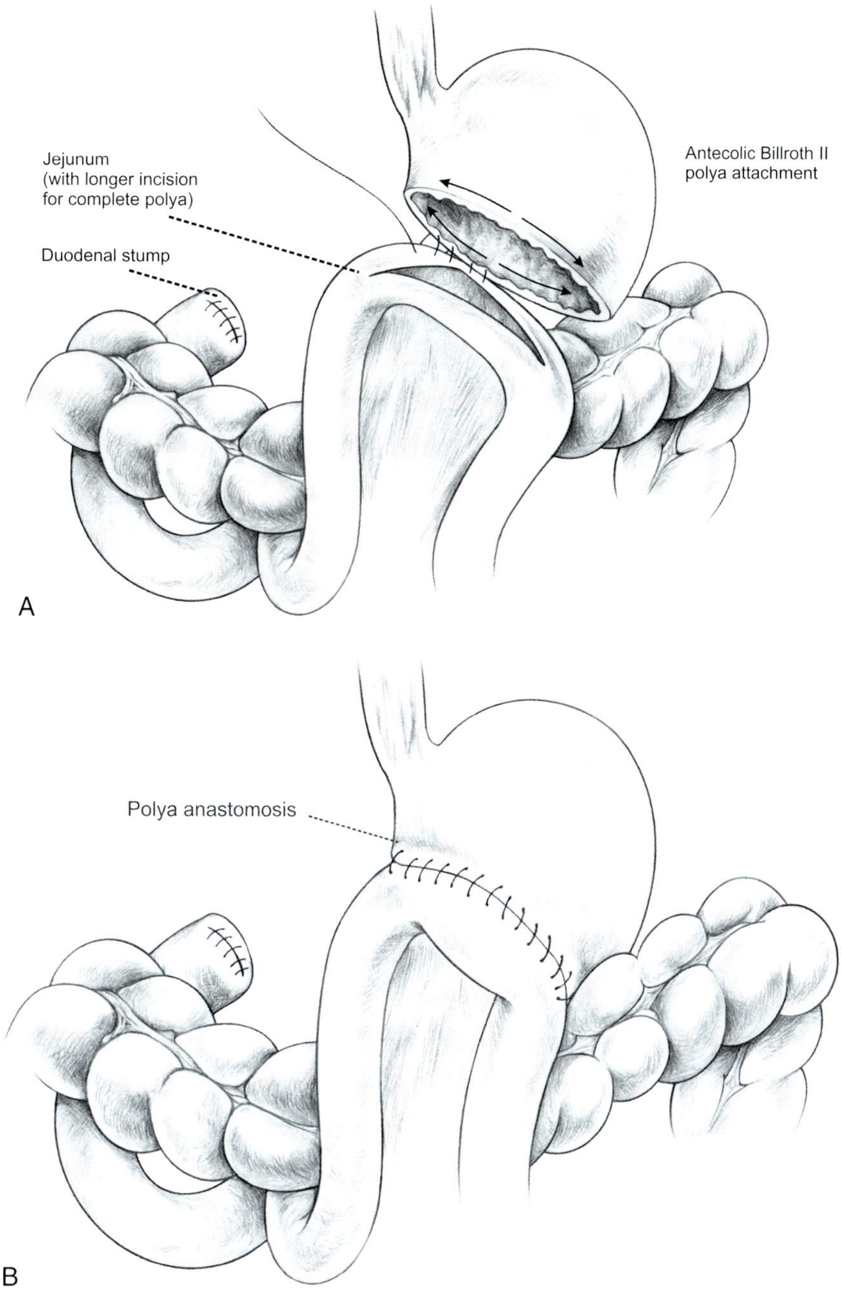

Figure 6.48 **A:** Antecolic Billroth II with back wall construction of Polya anastomosis. **B:** Completed Polya anastomosis.

- Some tumors (GIST, leiomyoma) are amenable to local, nonanatomic gastric resection with narrow 1-cm margins. The intact stomach has several major feeding vessels making it rare to cause ischemia in a local resection. These resections can be done in various locations of the stomach with direct excision and primary hand-sewn repair or a gastrointestinal stapling device. Care is taken to avoid narrowing of the gastric lumen <1.5 cm, which causes obstructive symptoms.

- Complete (total) gastrectomy is required for some malignancies and involves complete gastric resection with esophagojejunal reconstruction.

- Most surgical oncologists perform a D2 lymph node dissection with gastric resection for malignancy, which improves staging accuracy but has not been conclusively proven to improve survival.

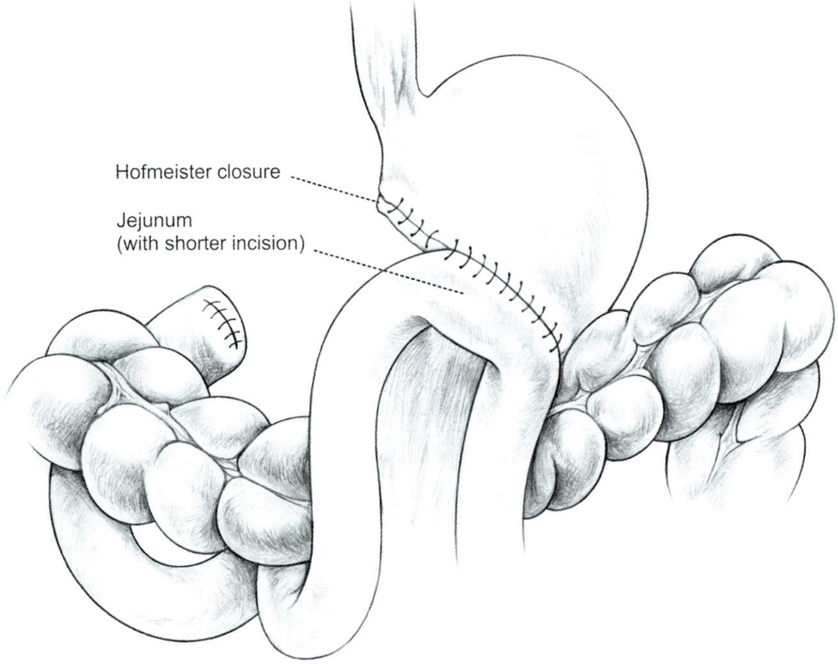

Figure 6.49 Antecolic Billroth II with completed anastomosis using Hofmeister closure of stomach.

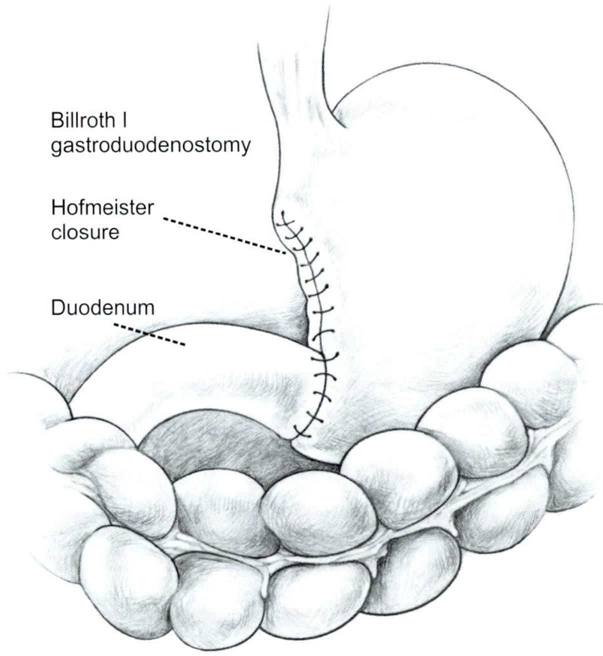

Figure 6.50 Billroth I with Hofmeister anastomosis.

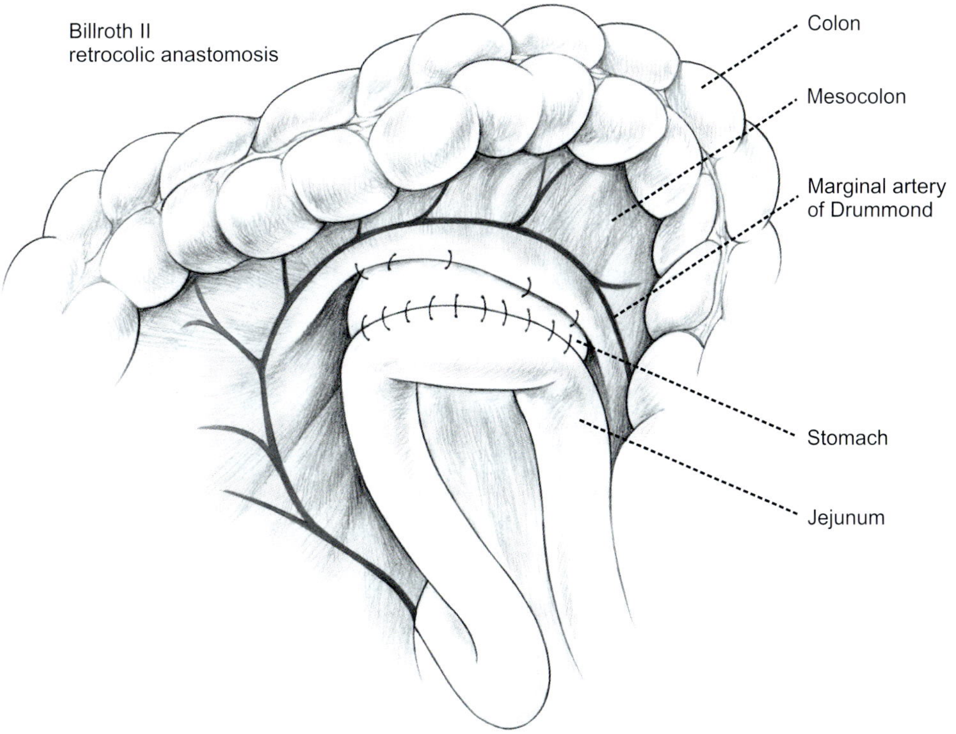

Figure 6.51 Billroth II retrocolic anastomosis through the space of Riolan and closure of mesenteric defect.

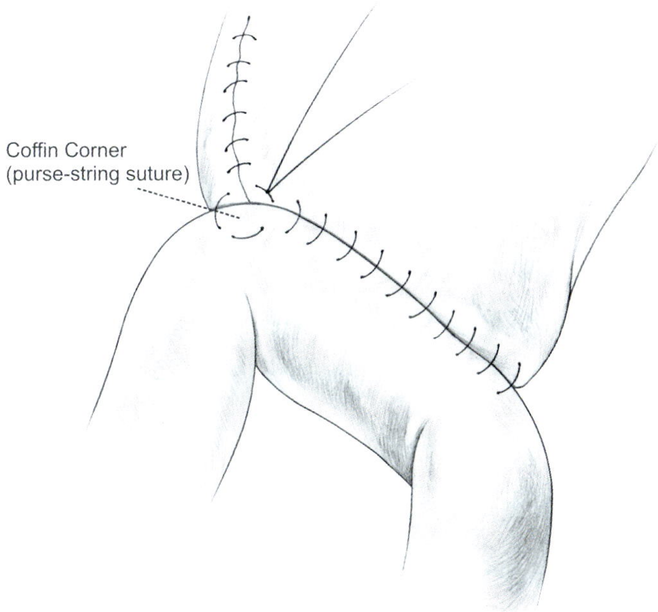

Figure 6.52 Anchor suture for the "coffin corner" to avoid anastomotic leak.

7

Liver, Biliary System, Pancreas, and Spleen

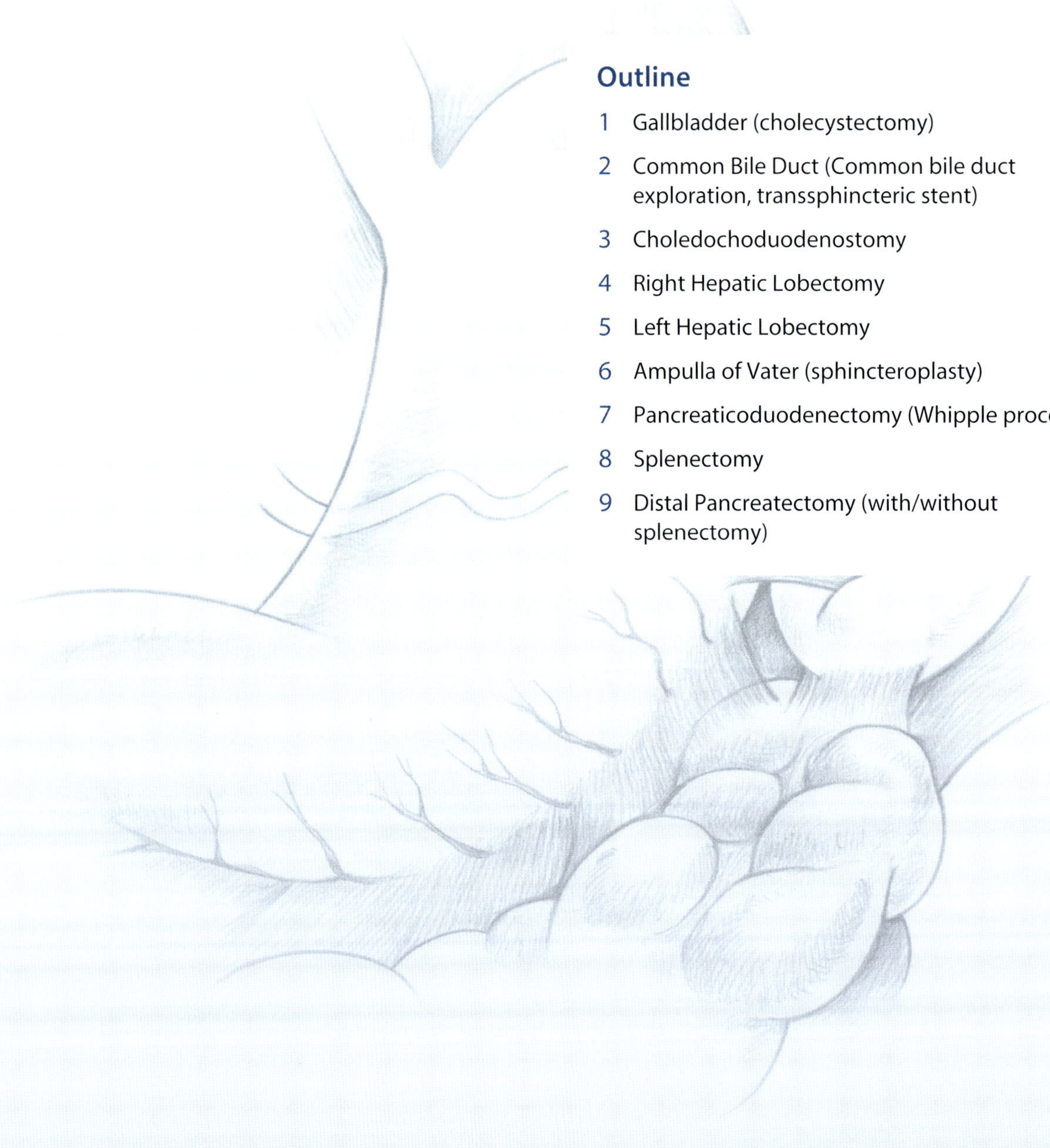

Outline

1 Gallbladder (cholecystectomy)
2 Common Bile Duct (Common bile duct exploration, transsphincteric stent)
3 Choledochoduodenostomy
4 Right Hepatic Lobectomy
5 Left Hepatic Lobectomy
6 Ampulla of Vater (sphincteroplasty)
7 Pancreaticoduodenectomy (Whipple procedure)
8 Splenectomy
9 Distal Pancreatectomy (with/without splenectomy)

GALLBLADDER (CHOLECYSTECTOMY)

INCISION

Open the abdominal wall and chest plate as previously done (Figs. 3.2 and 5.1). An assistant may retract the liver superiorly to expose the gallbladder and porta hepatis.

OPERATIVE PROCEDURE

Begin dissection of the gallbladder by retraction of the infundibulum to the right lateral abdomen and the fundus toward the right shoulder. This is the "critical view" (Fig. 7.1A). Retraction will put the tension on the cystic duct and cystic artery making identification much simpler and less fraught with complications. The area to be dissected is the hepatocystic triangle and triangle of Calot (Fig. 7.1B). Dissect the bile-stained tissues (postmortem bile staining is common) from the cystic duct and artery. Within the hepatocystic triangle, also identify the lateral edge of the common bile duct, the right hepatic artery, and right hepatic duct. After positive identification of the cystic duct and artery, divide these structures and dissect the gallbladder from its hepatic bed.

CLINICAL HIGHLIGHTS

- Some authors advocate routine cholangiography particularly early in a surgeon's training and experience to avoid bile duct injury. This has been shown to be unnecessary with increased experience.
- Transection of the cystic duct should be adjacent to the base of the gallbladder.
- If inflamed, verify the cystic duct by visualizing its direct entry into the base of the gallbladder. Note the widening of the cystic duct as it transforms into the gallbladder.
- Excessive traction may induce common bile duct (CBD) injury by tenting up the CBD leading to incorrect clip placement (Fig. 7.2).
- Wide variation in origin of the cystic artery may lead to pulsatile bleeding during dissection. Some examples can be seen in Figure 7.3 demonstrated by circles.
- Also, note a posterior branch of the cystic artery on the posterior surface of the gallbladder may be present causing bleeding if unrecognized (Fig. 7.1A).

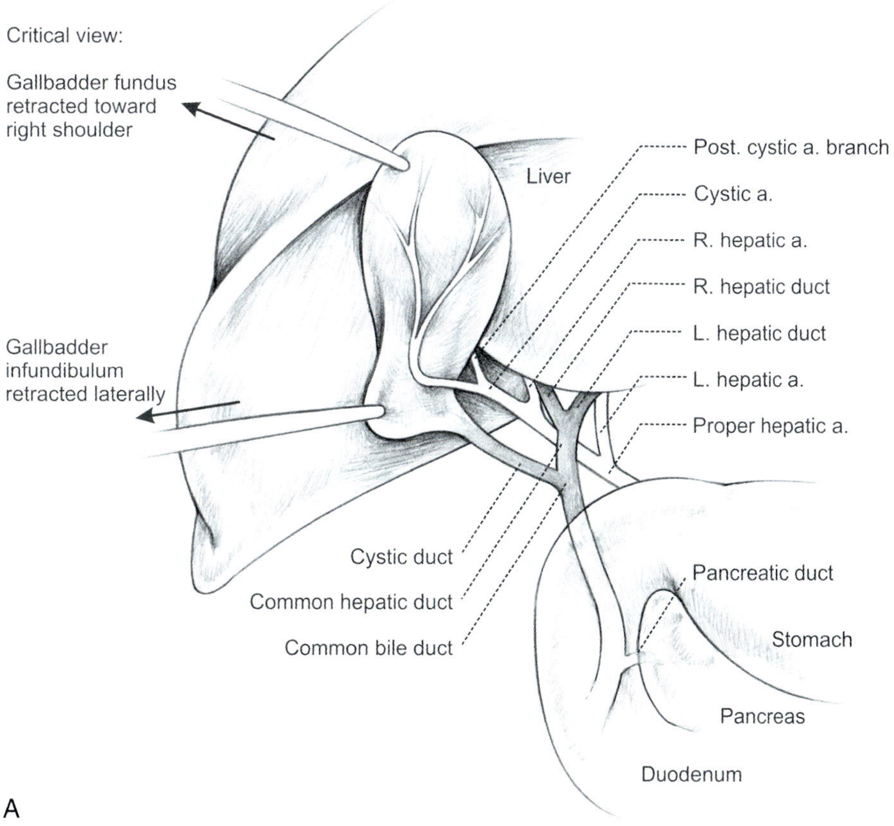

Figure 7.1 A: Begin cholecystectomy with "critical view."

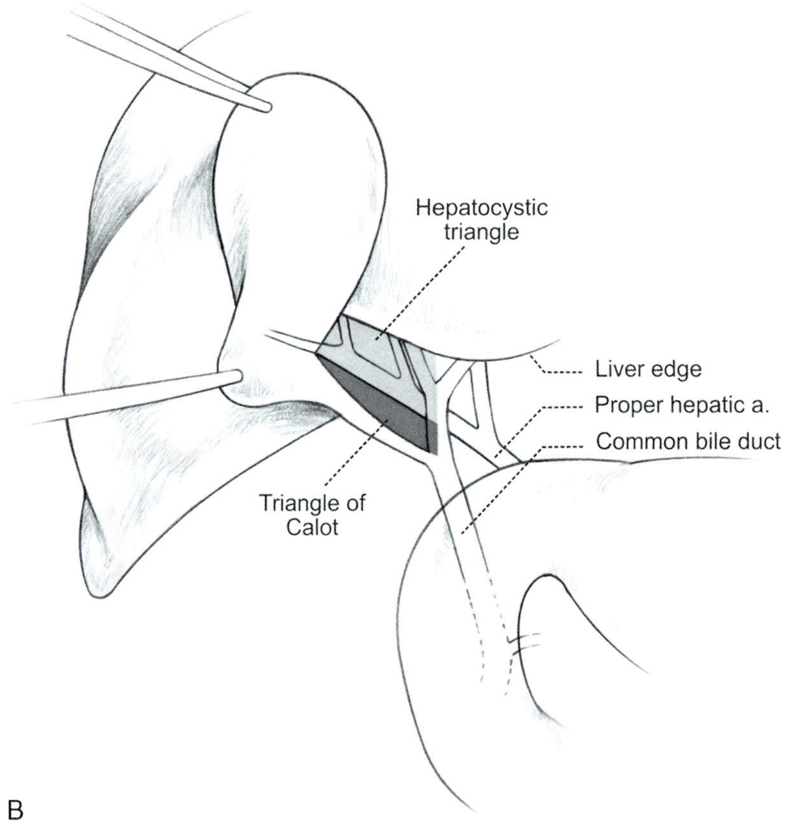

Figure 7.1 (Continued) **B:** Hepatocystic triangle and triangle of Calot.

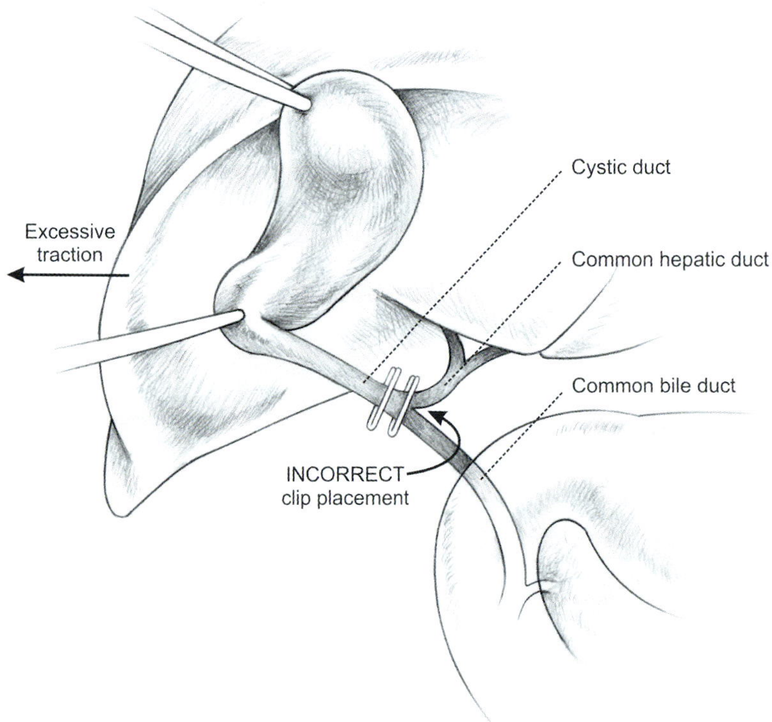

Figure 7.2 Cystic duct traction causing common bile duct injury.

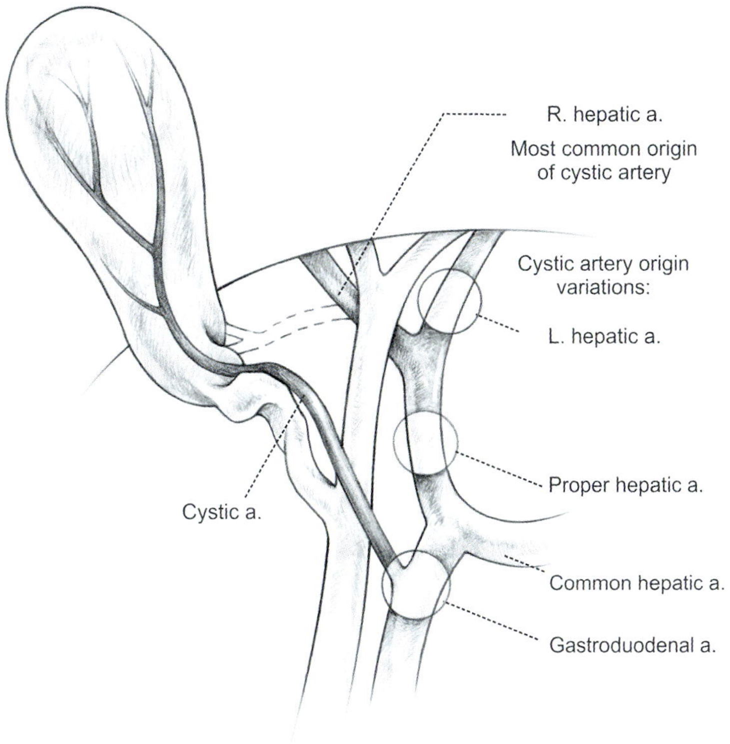

Figure 7.3 Cystic artery variations.

- Wide variation in cystic duct anatomy may occur and may be a source of inadvertent bile duct injury. Parallel duct (Fig. 7.4A), retrohepatic duct (Fig. 7.4B), and short cystic duct (Fig. 7.4C) are some of the most dangerous variations, which may predispose to bile duct injury.
- Duct of Luschka
  - Small biliary ducts may drain directly from the liver bed into the gallbladder and cause post-cholecystectomy biloma. They are called ducts of Luschka (Fig. 7.5).
  - Unlike normal biliary radicals, ducts of Luschka radicals do not have accompanying arterioles and venules.

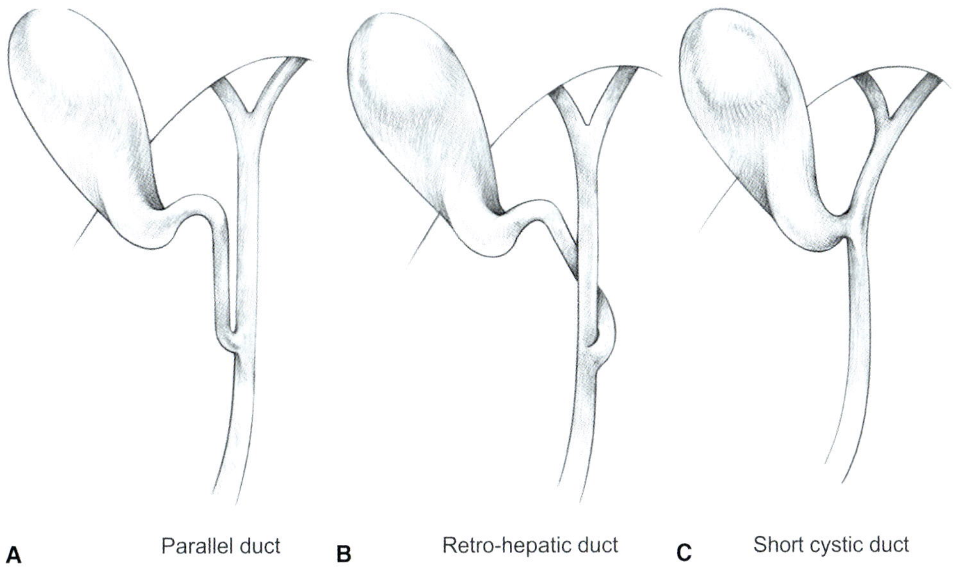

Figure 7.4 **A–C:** Cystic duct variations.

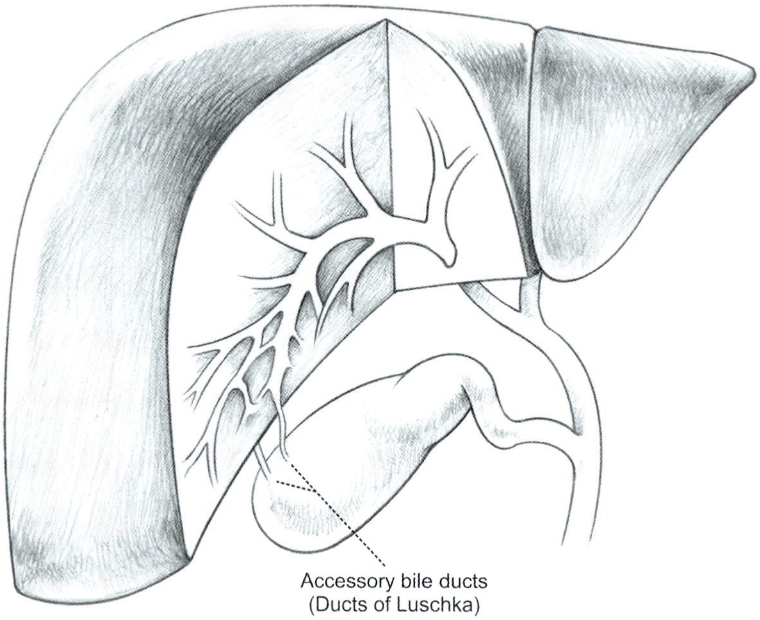

Figure 7.5 Duct of Luschka, common.

- Duct of Luschka size is 1-2 mm. Previous literature stated an incidence of 1 in 500 patients. A modern review estimated incidence to be 1 in 25. That means you will encounter this anatomical variant in your practice.
- Uncommon locations of the duct of Luschka can be from the right hepatic duct or directly from the CBD (Fig. 7.6).

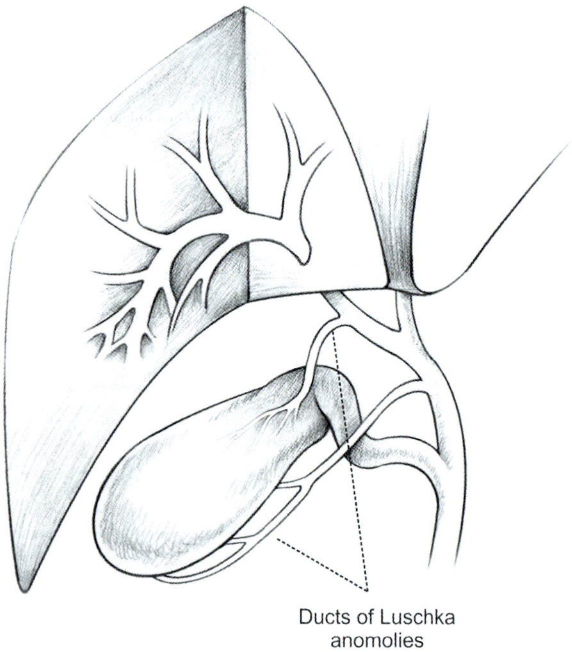

Figure 7.6 Duct of Luschka, uncommon.

COMMON BILE DUCT (COMMON BILE DUCT EXPLORATION, TRANSSPHINCTERIC STENT)

INCISION

Open the abdominal wall and chest plate as for cholecystectomy. Similarly, an assistant may retract the liver superiorly to expose the porta hepatis. Remember that the blood supply to the CBD is segmental and enters from the medial and lateral surfaces of the duct (Fig. 7.7A). The choledochotomy to be performed is demonstrated by a dotted line in the closeup illustration (Fig. 7.7B).

OPERATIVE PROCEDURE

Explore the surface of the CBD and expose the duct. Be sure to have hepatic and duodenal retraction to make the CBD taut. At a point 1-2 cm distal to the cystic duct, place vertical stay sutures in the CBD (Fig. 7.8A). Use an 11 blade and incise the CBD (Fig. 7.8B). Use the Pott scissors or Metzenbaum scissors to enlarge the choledochotomy to ~1.5 cm. Remove stones out of the CBD using compression of the distal CBD and milking stones superiorly to the choledochotomy (Fig. 7.8C). Irrigation and flushing of the CBD, milking of stones retrograde, biliary Fogarty catheter use, choledochoscopy with wire basket retrieval, and Randall stone forceps are all techniques that may be used for CBD stone retrieval. Practice each of these techniques.

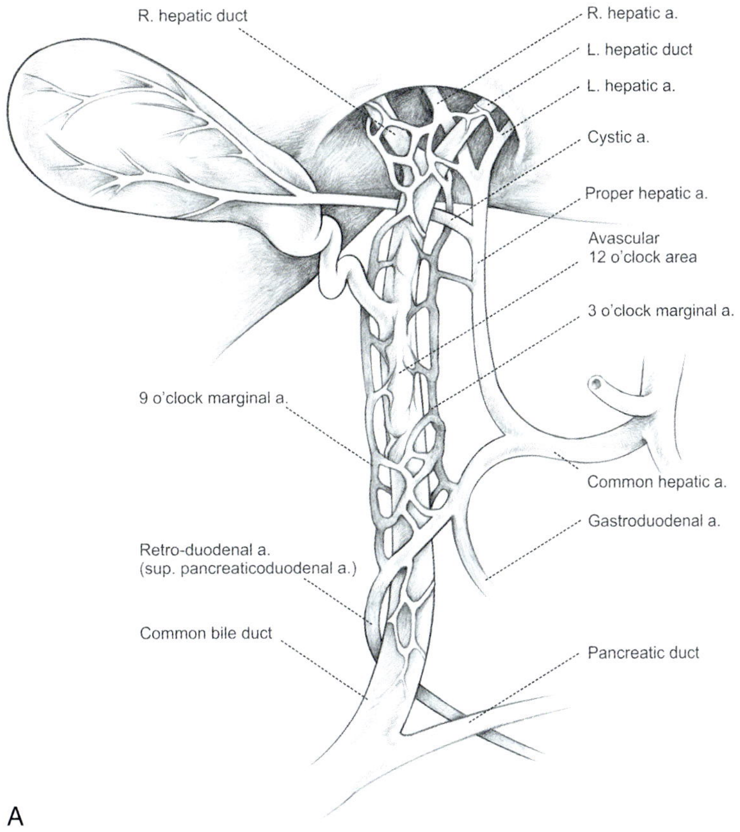

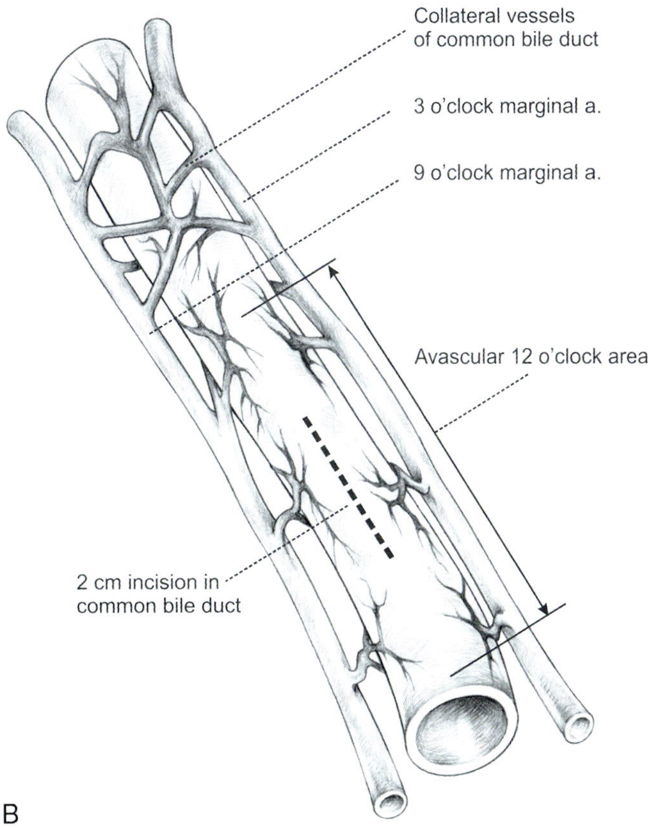

Figure 7.7 **A:** Blood supply to the common bile duct. **B:** Close-up of the common bile duct and location of choledochotomy.

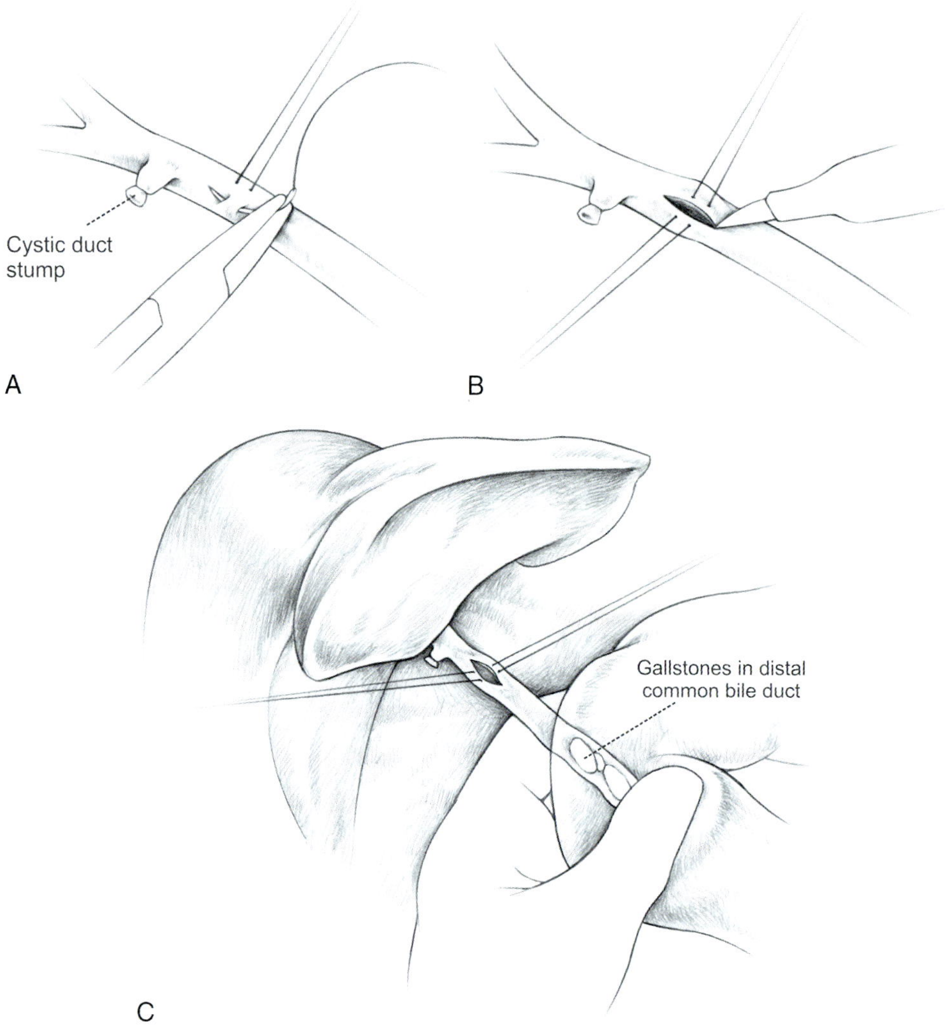

Figure 7.8 **A:** Stay sutures in the common bile duct. **B:** Choledochotomy aligned to avoid blood supply. **C:** Squeezing distal common bile duct stones up to choledochotomy.

CBD explorations typically require T-tube placement. Fashion your T tube as shown (Fig. 7.9A). Place one limb of the T tube into the CBD and then the other limb as shown (Fig. 7.9B).

CLINICAL HIGHLIGHTS

- Blood supply of the CBD is segmental from the following:
    - Proper hepatic artery
    - Gastroduodenal artery
    - Right hepatic artery
    - Left hepatic artery
    - Retroduodenal artery
- The collateral circulation is located on the lateral side of the CBD and requires longitudinal incisions of the duct (Fig. 7.7B).
- Do not skeletonize the CBD. This will lead to ischemia and necrosis of the duct.
- During common bile duct exploration (CBDE), 50% of distal common bile stones will be retrieved using simple saline irrigation through the choledochotomy.
- When using a biliary Fogarty catheter, remember to use air in the balloon. Air is compressible and will not damage the CBD mucosa.
- The biliary Fogarty is also useful for identification of the ampulla of Vater. The catheter balloon may be palpated through the duodenum to confirm patency of the CBD. Randall stone forceps and Bakes dilators have limited use compared to the other techniques mentioned.
- Flexible choledochoscopy with stone basket forceps is a highly effective technique whether used laparoscopically or open.
    - Technique: Pass the scope up into both hepatic ducts and down through the ampulla of Vater to visualize duodenal mucosa.

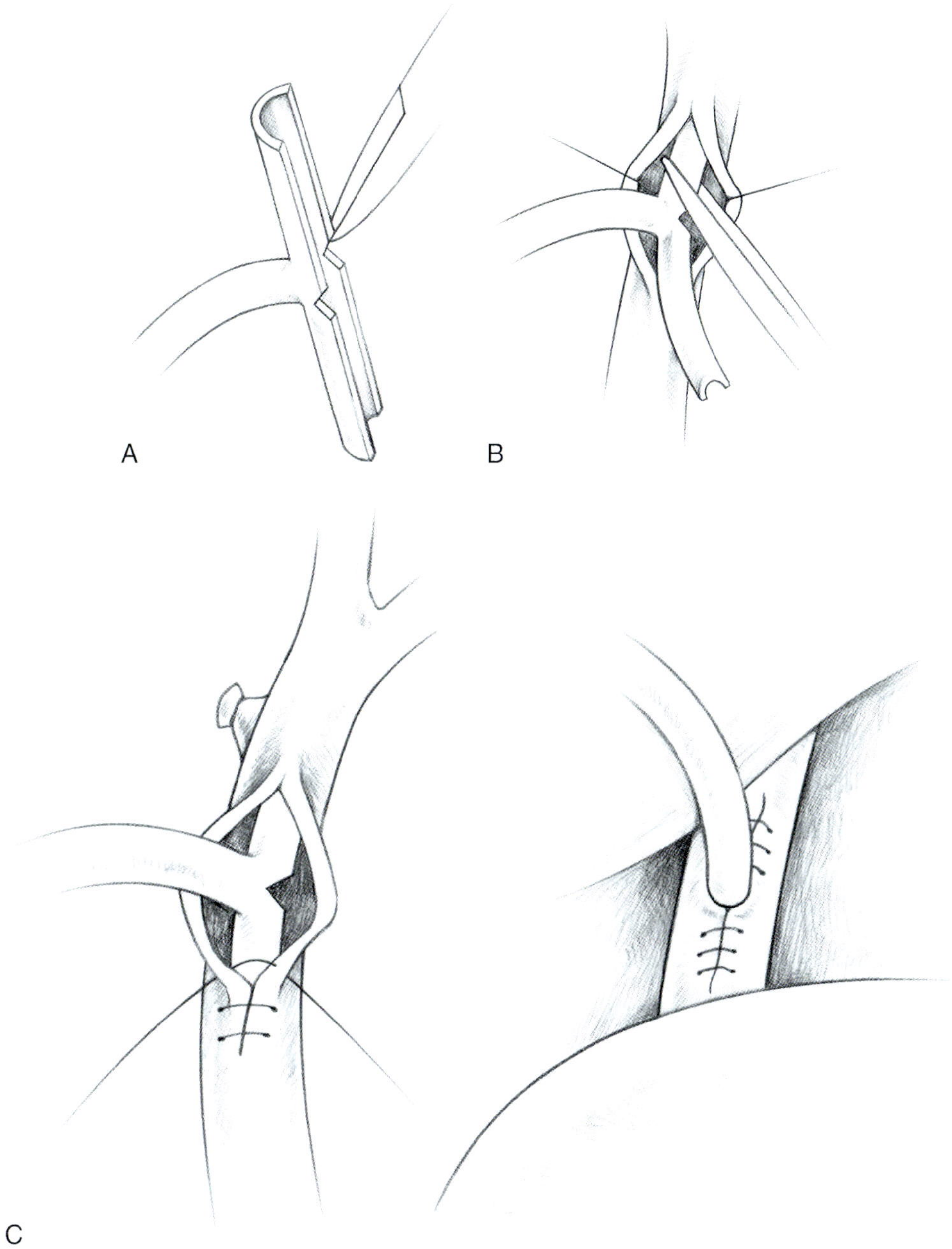

Figure 7.9 **A–C:** Shaping and placement of common bile duct T tube.

- Flexible choledochoscopy is recommended for assure clearance of all stones in the hepatic and CBDs in the OR. However, in the lab due to tissue character, it may be difficult.

- After CBD exploration, placement of a T tube with interrupted monofilament absorbable suture is used to repair the choledochotomy (Fig. 7.9C).

- Completion cholangiography through the T tube may be performed with a C-arm in the OR to ensure that retained stones are not present. Be sure to look superiorly toward and into the liver.

- Practice placement of a transsphincteric stent (Fanelli stent) through the CBD into the duodenum. In the OR, this can be done open or laparoscopically via the cystic duct or the CBD.

CHOLEDOCHODUODENOSTOMY

INCISION

If a T tube is in place, remove the interrupted sutures and the T tube.

OPERATIVE PROCEDURE

The surgeon will use the previously created choledochotomy (Fig. 7.10A). The vertical choledochotomy should be ~2 cm in length. The previously kocherized duodenum is then elevated to this distal choledochotomy. A longitudinal incision in the duodenum for ~2 cm is then performed (Fig. 7.10A). Interrupted monofilament 4-0 or 5-0 suture is used for the choledochoduodenostomy. Place corner sutures first and tag them without tying (Fig. 7.10B). Finish the back wall and tie the sutures (Fig. 7.10C). Complete the choledochoduodenostomy along the front wall (Fig. 7.10D). While doing this choledochoduodenostomy, observe that the opposing incisions will hold each other open in a diamond-shaped or oval pattern (Fig. 7.10E). This will prevent future stricture formation. A single layer of interrupted suture will maximize the size of the anastomosis.

CLINICAL HIGHLIGHTS

- This choledochoduodenostomy may be used for distal CBD obstructions.
- Mucosa-to-mucosa apposition is essential for long-term stricture avoidance.
- Distal CBD obstructions may occur from tumor, benign stricture, impacted stone with scar formation, or pancreatic disease. Primary therapy should, of course, be directed at the underlying disease. However, choledochoduodenostomy is a very effective drainage method when used appropriately.
- For open removal of primary CBD stones, choledochoduodenostomy can be used to ensure complete drainage and prevent recurrent CBD stones.

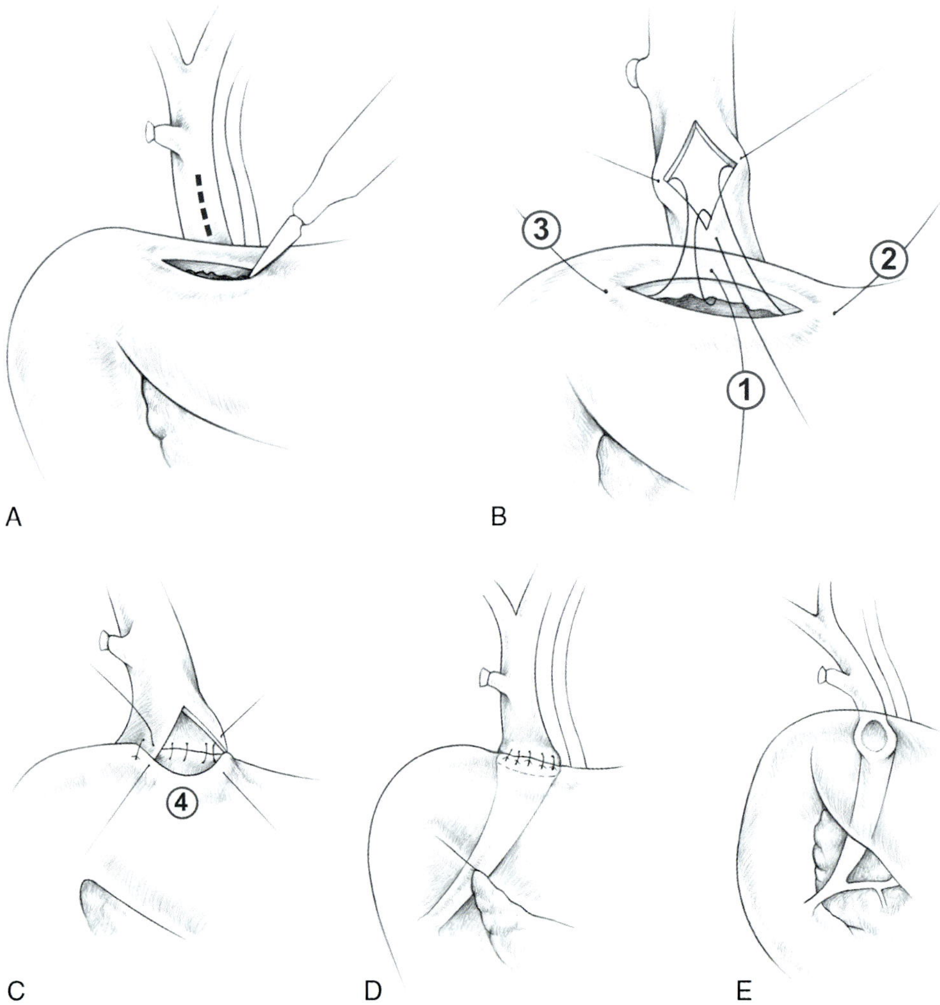

Figure 7.10 **A:** Choledochoduodenostomy—placement of incisions in duodenum and common bile duct. **B:** Choledochoduodenostomy—placement of back wall sutures. **C:** Choledochoduodenostomy—placement of front wall sutures. **D:** Choledochoduodenostomy—anastomosis completed. **E:** Choledochoduodenostomy—ghosted anastomosis demonstrating bile pathway.

- Identification of the CBD in the above disease processes can sometimes be challenging. Remember that a fine gauge (25-gauge or 22-gauge) needle or butterfly catheter may be effectively employed to discover location of the CBD.

RIGHT HEPATIC LOBECTOMY

OPERATIVE PROCEDURE

Begin by conceptualizing the liver segments (Fig. 7.11). Right hepatic lobectomy will remove segments 5-8. Keep in mind the internal hepatic anatomy of the hepatic artery, hepatic duct, portal vein, and hepatic veins (Fig. 7.12). The right triangular and coronary ligaments (Fig. 7.13A) are sharply divided. This dissection is carried down to the inferior vena cava (Fig. 7.13B). The right hepatic vein is then carefully identified and vessel looped (Fig. 7.14). The hepatic veins are very short and the surgeon may need to dissect into liver parenchyma to clearly delineate the hepatic vein. There are from 6 to 20 lower hepatic veins between the right and caudate lobes of the liver and the vena cava (Fig. 7.14). The posterior plane between the liver and vena cava should be carefully dissected to identify and control these veins with either ligation or hemoclips. In the anatomy lab, divide the lower hepatic veins sharply.

Incise the peritoneum overlying the porta hepatis. Remember the hepatic artery is typically on the medial side of the porta, the CBD on the lateral side, and the portal vein deep to these structures. Identify the right and left hepatic arteries. Transect the right hepatic artery (Fig. 7.15). Next, identify the common hepatic duct as well as the right and left hepatic ducts. Follow these ducts proximally up to the liver surface for positive identification. Transect the right hepatic duct (Fig. 7.16). Deep to the bile duct, the large portal vein will be identified. Again, identify the right and left branches of the portal vein. Transect the right portal vein branch (Fig. 7.17). It is very important to realize that early tributaries in the proximal right and left branches of the portal vein may be present. Injury to these small branches may result in a rather brisk bleeding in the OR. To avoid injury to any of the vascular tubular structures discussed, be sure to use good technique by spreading along the wall of the vessel. Learning how to safely dissect along the wall of a structure to define its anatomy is an important technical skill that will save the surgeon a great deal of trouble in the operating room.

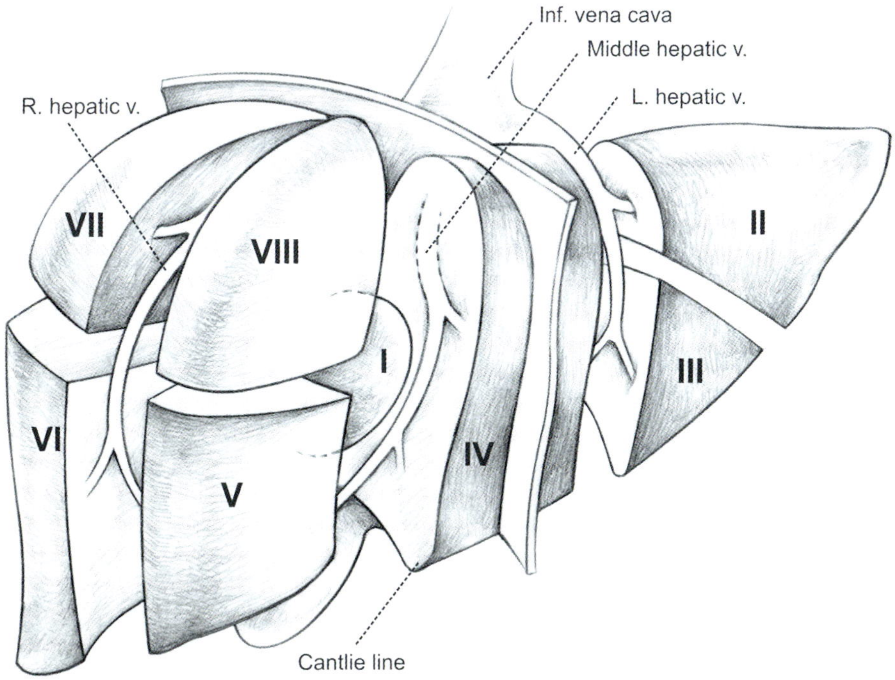

Figure 7.11 Liver segments and the Cantlie line.

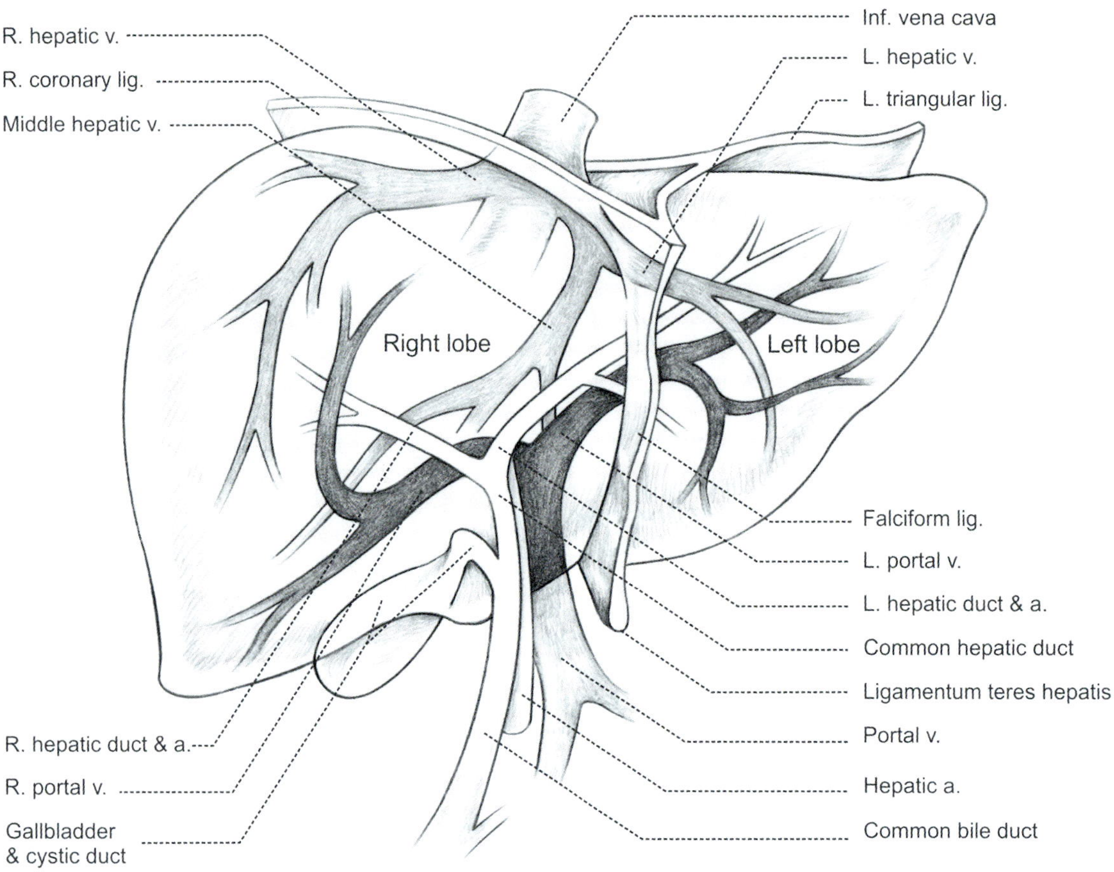

Figure 7.12 Liver segment distribution of hepatic artery, hepatic vein, portal vein, and bile duct. Portal triad.

Once the right portal vein branch is identified, place straight clamps upon it and use a scalpel blade be to divide the vein. In the operating room, use horizontal mattress suture followed by removal of the clamp. Then, use the same suture to oversew in a running fashion back to the origin of the suture. This is an effective and secure technique to control these structures. A 5-0 or 6-0 Prolene is an appropriate suture to consider for this purpose (Fig. 7.17). Divide the right hepatic vein (Fig. 7.14). Stiffness of cadaveric tissues may necessitate delay of right hepatic vein division until the liver parenchyma is divided.

In the OR, demarcation of the liver will have occurred after occlusion of the right hepatic artery due to ischemia of the right lobe. Lobar resection is then carried out according to the needs of that particular patient (Fig. 7.18). Note that a classic right hepatic lobectomy includes resection in a line from the gallbladder bed to the inferior vena cava, the Cantlie line. For purposes of cadaveric dissection, use a number 10 scalpel blade and sharply divide the liver parenchyma staying just to the right lateral side of the Cantlie line previously described.

Identify the middle hepatic vein. It may be found in the right hepatic lobe anteriorly draining segment 5 (Fig. 7.18; ghosted in the figure). In the OR, this middle hepatic vein is the surgeon's guide to the division of the right and left anatomic lobes of the liver. The right hepatic lobe is now nearly free of its attachments. Elevate the right hepatic lobe and identify the inferior vena cava. The right hepatic lobe may now be removed.

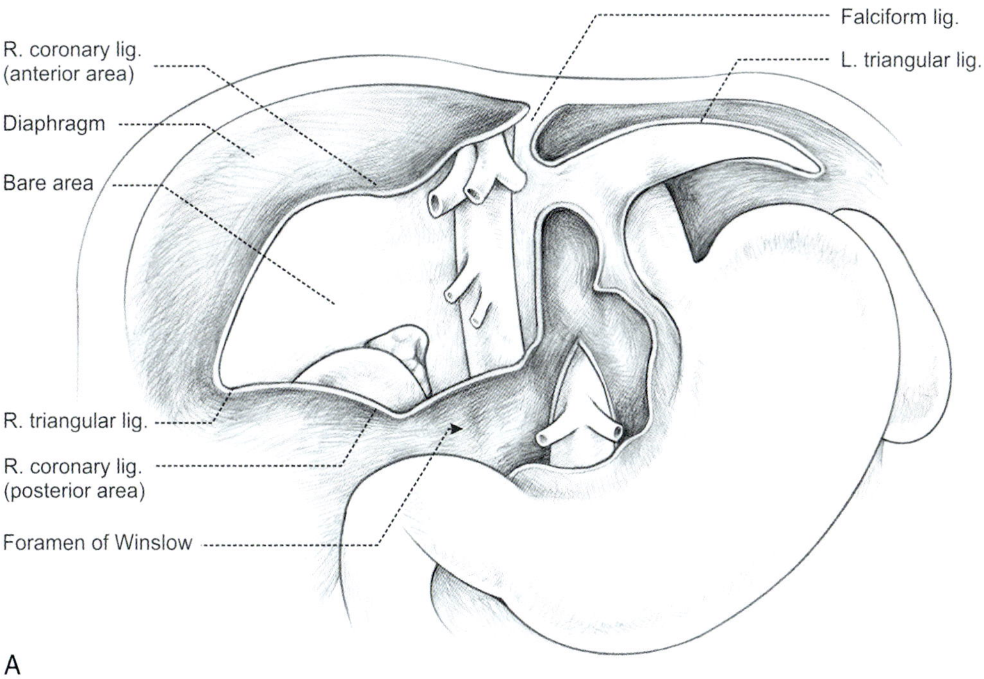

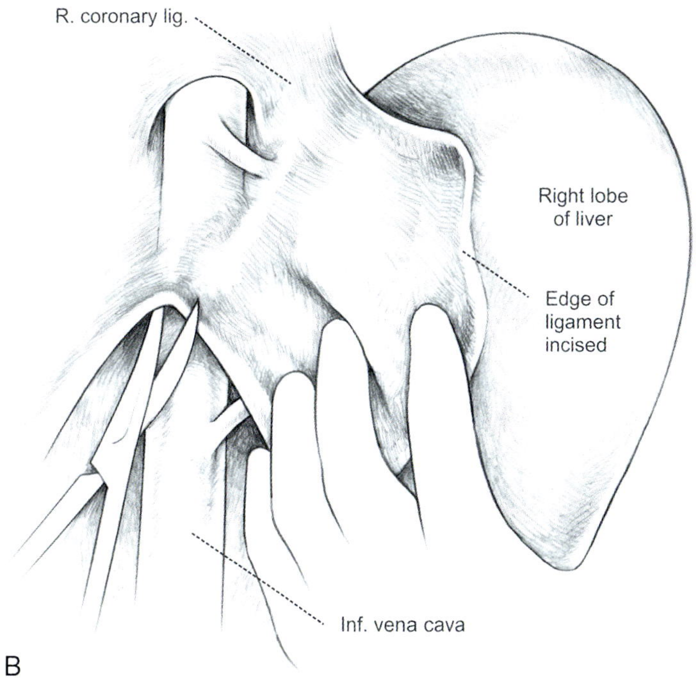

Figure 7.13 **A:** Peritoneal attachments of liver (liver removed) and posterior position of IVC behind liver. **B:** Peritoneal mobilization of right lobe of the liver.

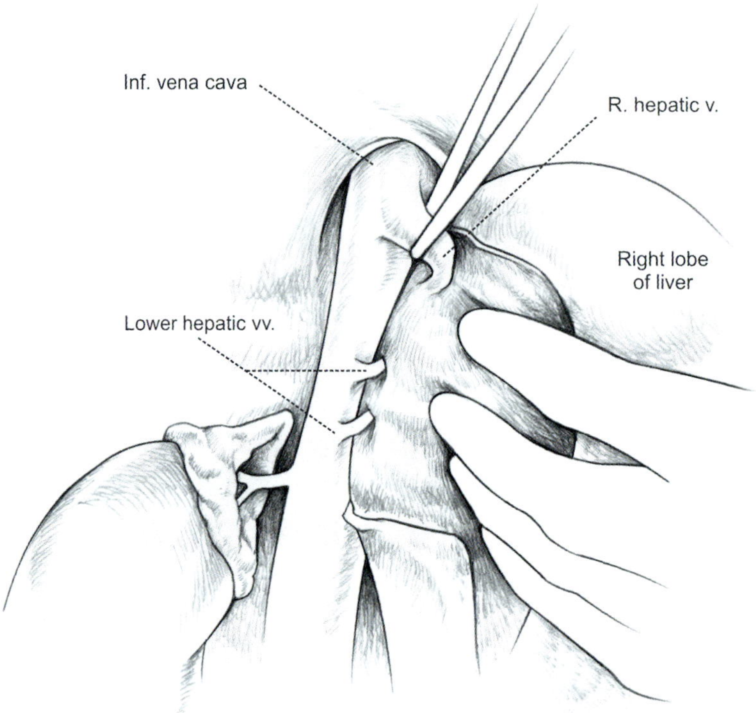

Figure 7.14 Vessel loop on right hepatic vein and demonstration of direct venous connections between liver and IVC.

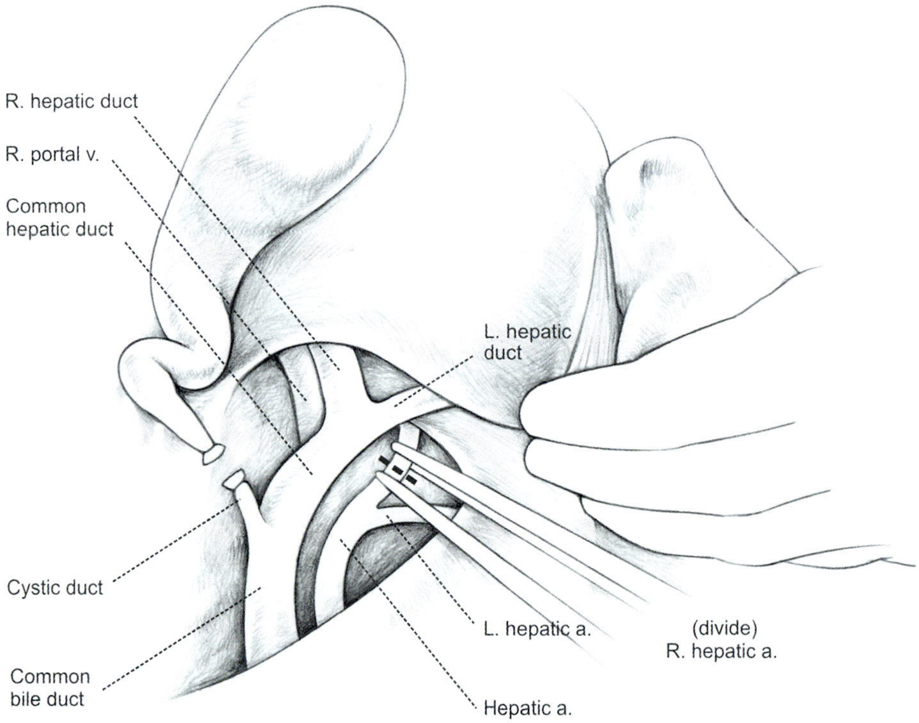

Figure 7.15 Division of right hepatic artery.

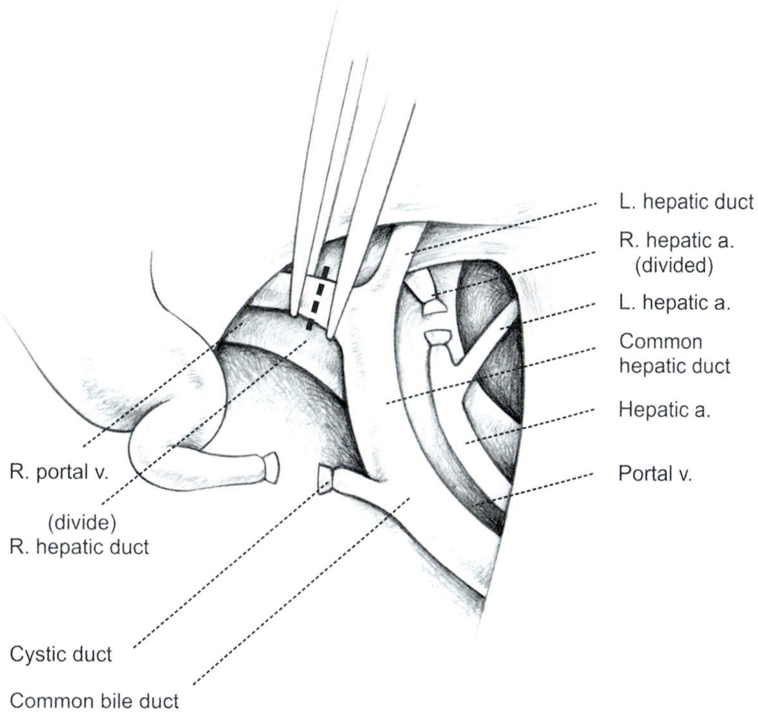

Figure 7.16 Division of right hepatic duct.

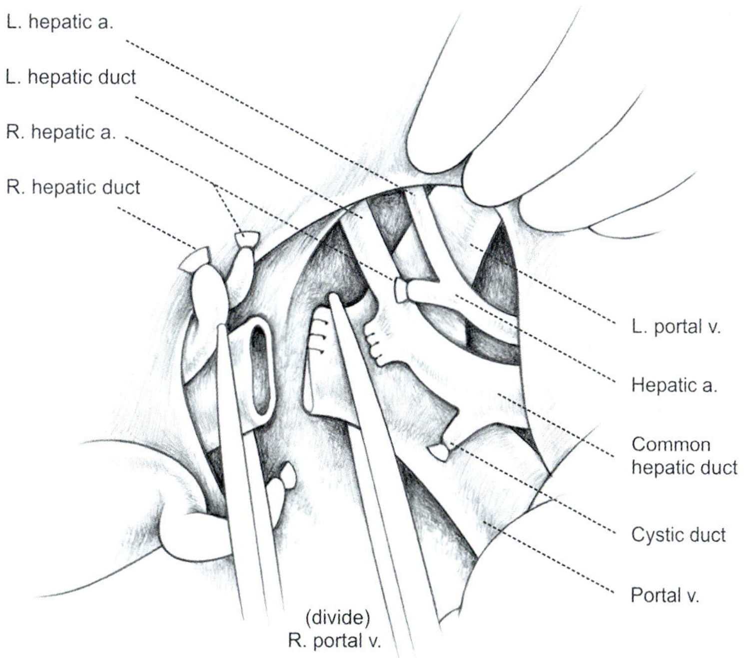

Figure 7.17 Division of right portal vein.

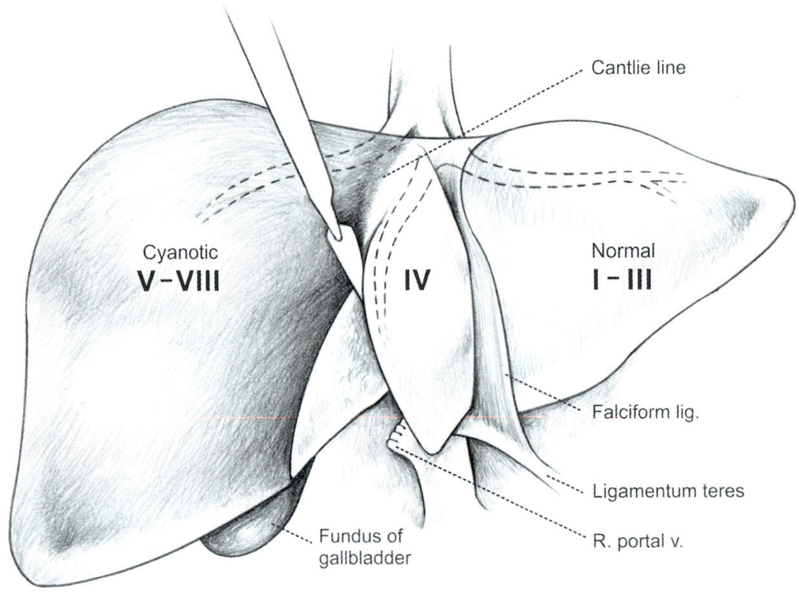

Figure 7.18 Parenchymal division along the Cantlie line and orientation of hepatic veins.

LEFT HEPATIC LOBECTOMY

OPERATIVE PROCEDURE

Identify the left hepatic artery. Trace the left hepatic artery back to the common hepatic artery originating from the celiac axis. Identify the takeoff of the gastroduodenal artery, right gastric artery, and left gastric artery (Fig. 5.2). Note the close proximity of the celiac axis to the superior mesenteric artery origin off the aorta (Fig. 6.21).

Identify the origin of the left hepatic artery from the hepatic artery and transect the left hepatic artery (Fig. 7.19). Identify the common hepatic duct and left hepatic duct. Transect the left hepatic duct (Fig. 7.20). Finally, identify the portal vein and the left portal vein branch. Remember to use good technique dissecting down the walls of the portal vein branch and transect the left portal vein (Fig. 7.21).

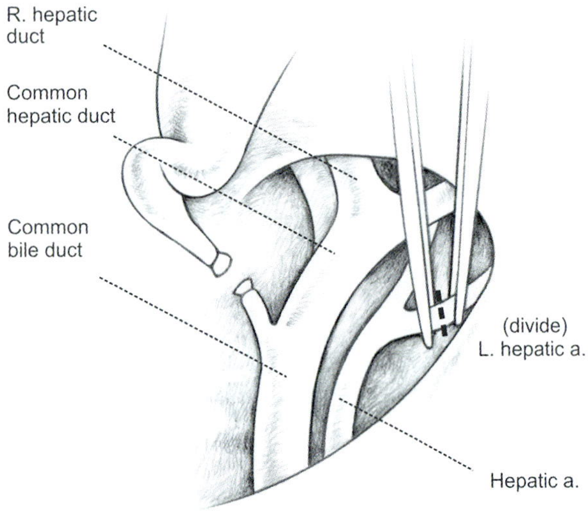

Figure 7.19 Division of left hepatic artery.

At this point, the falciform ligament and the left coronary ligament should be sharply divided and dissected to the inferior vena cava. Carefully isolate the left hepatic vein and middle hepatic vein using good technique. These are large veins and need to be carefully handled to avoid bleeding. They may arise from the IVC from a common trunk or individually. Transect the middle hepatic vein and left hepatic vein once completely identified and dissected (Fig. 7.22). Remove the left hepatic lobe from the inferior vena cava noting the several small branches between the liver and IVC. Note that there are direct branches from the caudate lobe (segment 1) to the portal vein and biliary duct system. These need to be carefully identified and transected to avoid damage.

CLINICAL HIGHLIGHTS

- Segmental resections of the liver may be accomplished if compromised hepatic function is suspected (see Fig. 7.11).
- In the setting of normal liver function, at least 20% of liver parenchyma is necessary for survival post resection.
- 75% of hepatic blood supply is via the portal system. 25% of blood supply is via the hepatic arterial system. Oxygen supply to the liver is 50% portal and 50% hepatic arterial.

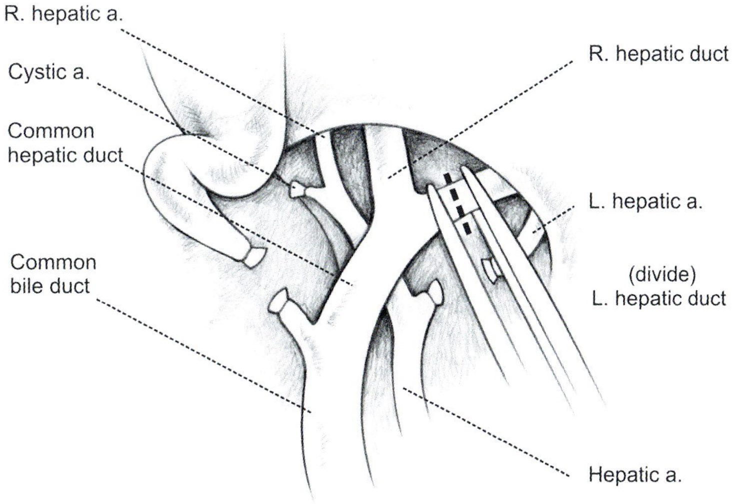

Figure 7.20 Division of left hepatic duct.

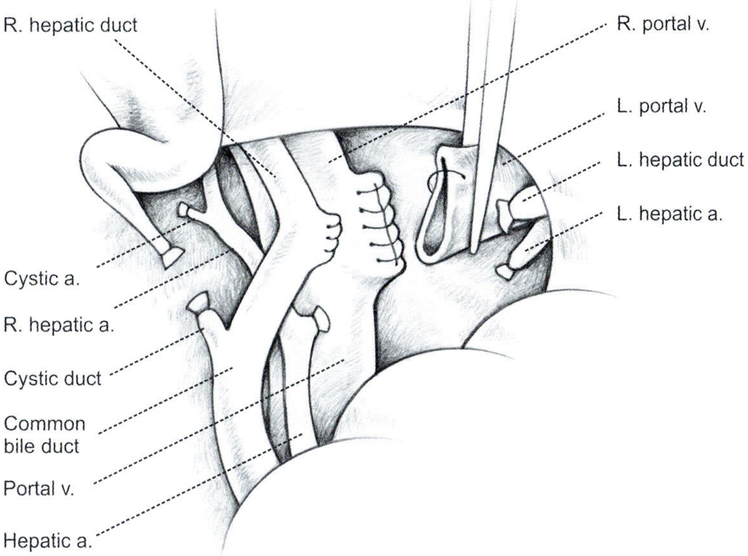

Figure 7.21 Division of left portal vein.

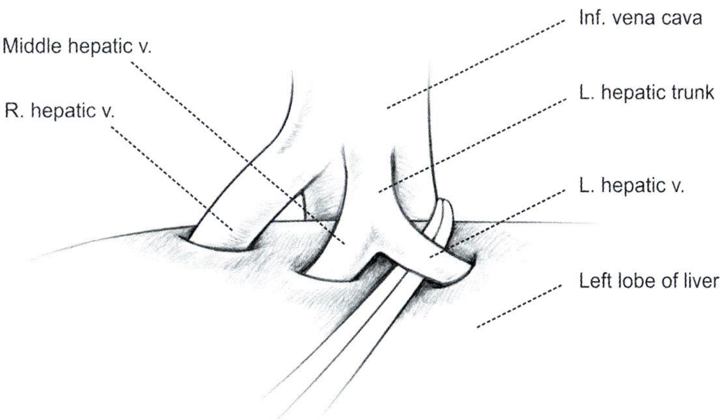

Figure 7.22 Orientation of hepatic veins and IVC.

- Once the hepatic artery is ligated to the appropriate segment, dark discoloration of the affected segment will occur within the next few minutes. Remember that portal venous supply is deoxygenated, dark blood.

- A wide variation in hepatic duct anatomy and hepatic artery anatomy is common.

- "Replaced" aberrant hepatic arteries are end arteries. "Accessory" hepatic arteries are additional hepatic arterial blood supply (see Fig. 6.5).

- Anatomic variations, though multiple, can be identified by careful dissection and identification of the structures into the appropriate hepatic segment.

- Pretreatment of the liver with a Pringle maneuver for 5 minutes followed by 10 minutes of blood flow back to the liver may improve postoperative hepatic function. Preconditioning may allow the surgeon to occlude all vascular supply during the parenchymal dissection for hepatic lobectomy.

- During parenchymal dissection, the middle hepatic vein is the guide to a classic hepatic lobectomy (see Fig. 7.18).

- Failure to ligate all lower hepatic veins can result in significant hemorrhage, which can be difficult to control both on the hepatic and caval surfaces (see Fig. 7.14). Suture ligation is the most secure and is preferred by the authors.

- In the OR, there are several transection techniques for the hepatic parenchyma including radiofrequency ablation, bipolar cauterization, ultrasonic dissection, and finger fracture techniques. Current technology for hemostasis allows transection of the hepatic parenchyma to minimize blood loss.

- A 1 cm liver margin is desired. Histologically negative margins are required for satisfactory tumor resection.

AMPULLA OF VATER (SPHINCTEROPLASTY)

INCISION

If a T tube was previously placed in the CBD, remove the interrupted sutures and the T tube at this time. Open the lateral wall of the duodenum in a longitudinal fashion and identify the ampulla of Vater containing the sphincter of Oddi (Fig. 7.23).

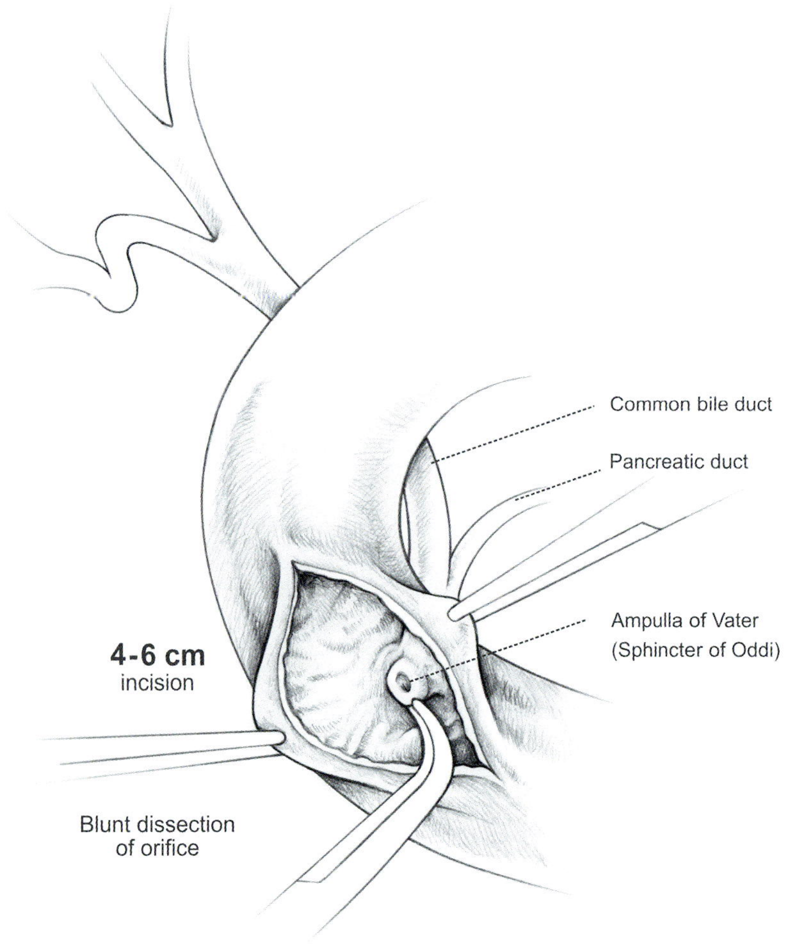

Figure 7.23 Longitudinal duodenal incision and exposure of ampulla of Vater.

OPERATIVE PROCEDURE

Attempt to insert a soft catheter through the choledochotomy, down the CBD, and through the ampulla of Vater. If the soft rubber catheter will not work, use a metal dissection probe. In the cadaveric specimen, it may be necessary to cannulate the ampulla of Vater retrograde beginning from the duodenal side. There may be an anatomic anomaly. Be aware of these three ampullary configurations (Fig. 7.24B).

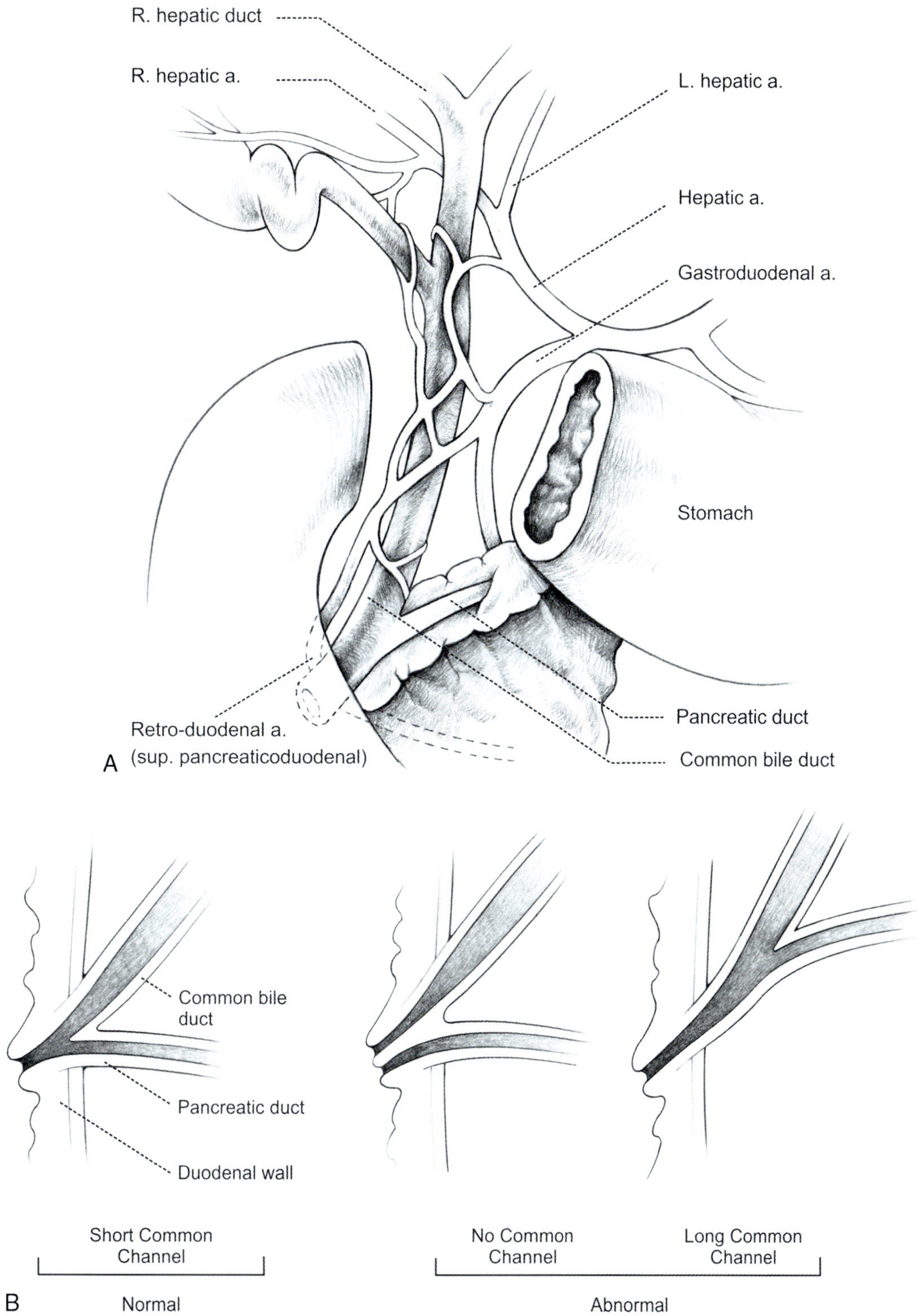

Figure 7.24 **A:** Location of retroduodenal artery. **B:** Ampulla of Vater configurations.

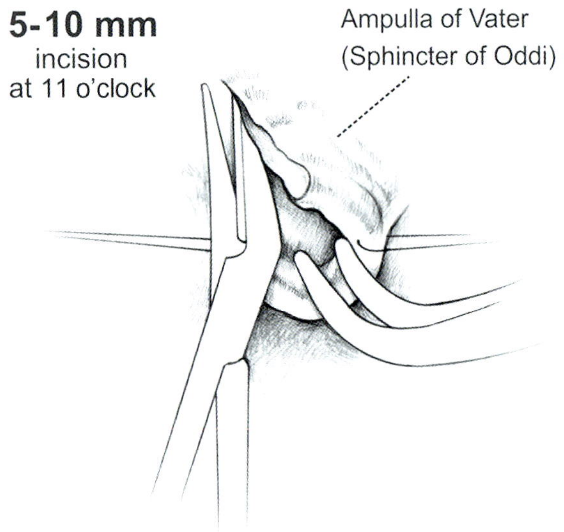

Figure 7.25 11 o'clock incision for sphincterotomy of sphincter of Oddi.

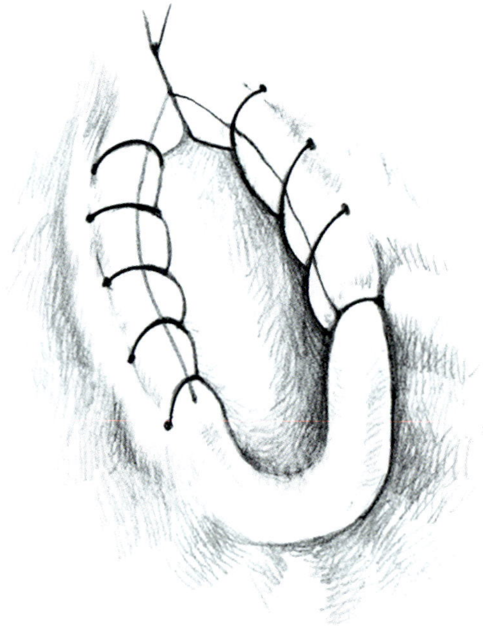

Figure 7.27 Completed sphincteroplasty.

Once a catheter, probe, or right angle clamp has been placed through the sphincter of Oddi, use a Pott scissor or scalpel to create a 15-mm sphincterotomy at the 11 o'clock position (Fig. 7.25). Examine the sphincterotomy and note that the ampulla of Vater is now very patulous and will allow free biliary flow. Start with the apical suture (#1) and continue with interrupted sutures (#2) (Fig. 7.26). Completion of the sphincteroplasty is accomplished using 4-0 or 5-0 interrupted absorbable suture (Figs. 7.26 and 7.27). Also note that the posterior superior pancreaticoduodenal artery (also known as the retroduodenal artery) is anatomically close (Fig. 7.24). If that artery is damaged by sphincteroplasty or ERCP, suture closure of the sphincterotomy will control the bleeding. Transverse repair of the duodenotomy using an inner layer of running absorbable suture and an outer layer of interrupted Lembert suture or single-layer closure will minimize long-term risk of duodenal stricture (Fig. 7.28).

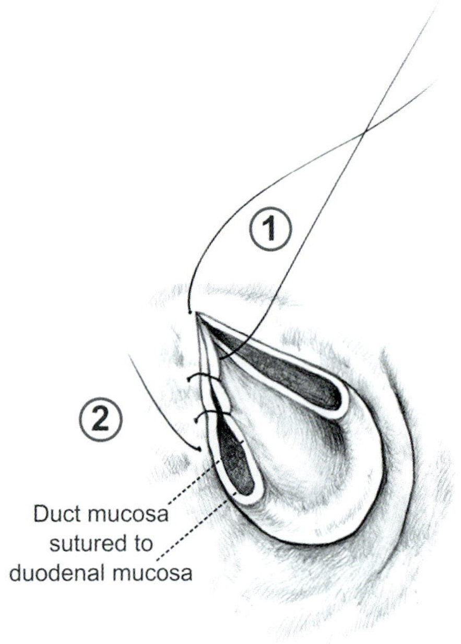

Figure 7.26 Mucosal repair of sphincteroplasty.

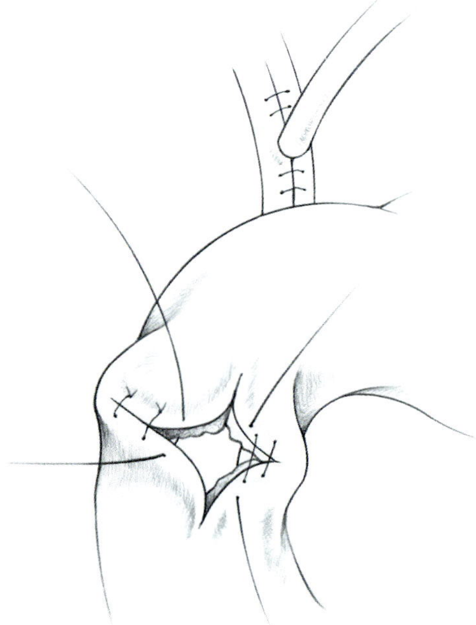

Figure 7.28 Transverse closure of duodenum.

CLINICAL HIGHLIGHTS

- Distal CBD stone or stricture may be observed on cholangiography.
- Endoscopic retrograde cholangiopancreatography (ERCP) is another method of performing sphincteroplasty.
- ERCP and open operative sphincteroplasty may cause rapid hemorrhage if the posterior superior pancreaticoduodenal artery (retroduodenal artery) or its branch is damaged (see Fig. 7.24). If this should occur, remember that completing the sphincteroplasty with suture will control the hemorrhage and insure long-term patency of the sphincteroplasty (see Fig. 7.27).
- Mucosa-to-mucosa apposition will minimize long-term risk of stricture.
- Normal common channel of CBD and pancreatic duct is located in the wall of the duodenum (see Fig. 7.24A).
- Diseases associated with (see Fig. 7.24B):
  - Anomalous pancreaticobiliary ductal configurations—congenital cystic dilatation of CBD
  - Long common channel—carcinoma of the gallbladder
  - No common channel—alcohol-induced chronic pancreatitis

PANCREATICODUODENECTOMY (WHIPPLE PROCEDURE)

INCISION

Remove any previously placed T tubes or sutures in the CBD. Kocherize the duodenum if not previously done (Figs. 6.19 and 6.20).

OPERATIVE PROCEDURE

Many modifications of the Whipple procedure are presented in texts and current literature. A pyloric preserving pancreaticoduodenectomy (PPPD) is described. The classic Whipple procedure includes excision of the following structures:

1. Distal stomach (pyloric preservation may be possible)
2. Duodenum
3. Head, neck, and uncinate process of the pancreas
4. Distal CBD
5. Regional lymph nodes

After kocherization of the duodenum and taking down of the gastrocolic omentum, divide the duodenum beyond the pylorus (Fig. 7.29). Complete dissection of

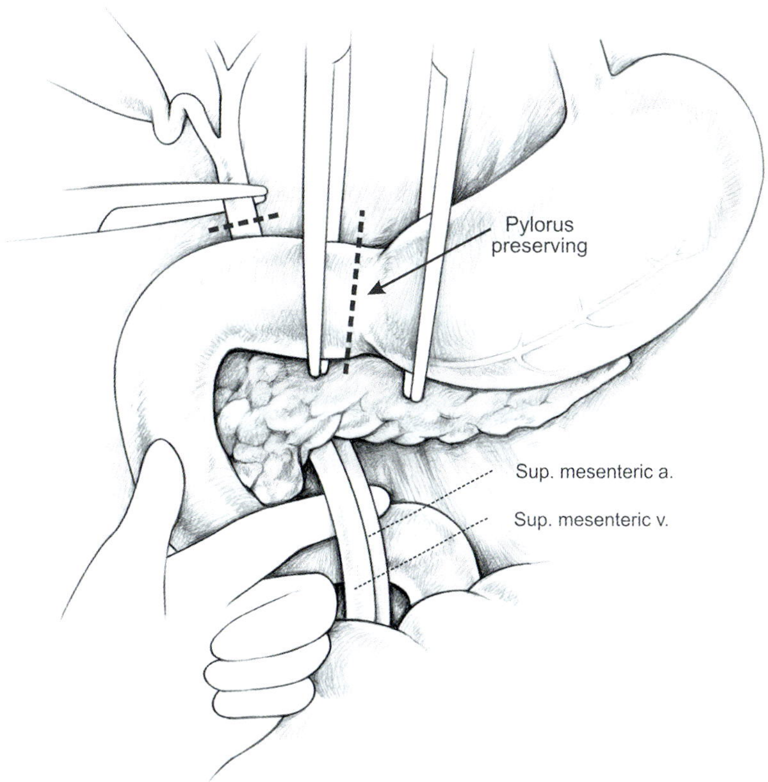

Figure 7.29 Whipple—pylorus-preserving division of duodenum, division of common bile duct, and identification of SMA and SMV.

the porta hepatis is performed by identifying and isolating the three major structures in the porta. On the medial side of the porta, the common hepatic artery, gastroduodenal artery, proper hepatic artery, and right and left hepatic artery branches are identified and dissected free. On the lateral side of the porta, the CBD is identified and dissected free (Fig. 7.24). Between and deep to the hepatic artery and CBD, the portal vein is identified and exposed. The areolar plane anterior to the portal vein is dissected inferiorly under the neck of the pancreas as far inferior as is safely possible (Fig. 7.30).

The inferior border of the pancreas is identified and the SMV is located and dissected free (Fig. 7.29). The anterior areolar plane along the SMV is dissected superiorly under the neck of the pancreas to join the dissection of the portal vein from the porta hepatis (Fig. 7.31A). Divide the gastroduodenal artery (Figs. 7.30 and 5.2). Divide the CBD (Fig. 7.29). Remove the gallbladder if not already removed (Fig. 7.1A, B). Divide the jejunum 15-20 cm beyond the ligament of Treitz. Divide the vascular attachments of the proximal jejunum retrograde until the dissection reaches the SMA (Fig. 7.32). Divide the ligament of Treitz and mobilize the fourth portion of the duodenum. The plane posterior to the third and fourth portions of the duodenum is dissected under the root of the small bowel mesentery. The remnant of proximal jejunum is passed under the mesentery to the right side of the abdomen.

Close the mesenteric/retroperitoneal defect where the ligament of Treitz and the duodenum exited the retroperitoneum. Divide the neck of the pancreas (Fig. 7.31B). Carefully dissect the portal vein and SMV from the head of the pancreas. Lateral venous tributaries between the pancreas and portal vein are divided (Fig. 7.30). Divide the remaining retroperitoneal attachments taking all retroperitoneal tissue as far medially as possible, skeletonizing the SMA. The inferior pancreaticoduodenal vessels (anterior and posterior) branching from the SMA will be in this retroperitoneal tissue (Fig. 7.33). The head of the pancreas forms an offset "C" shape. The superior element of the "C" is anterior to the SMA/SMV. The inferior element

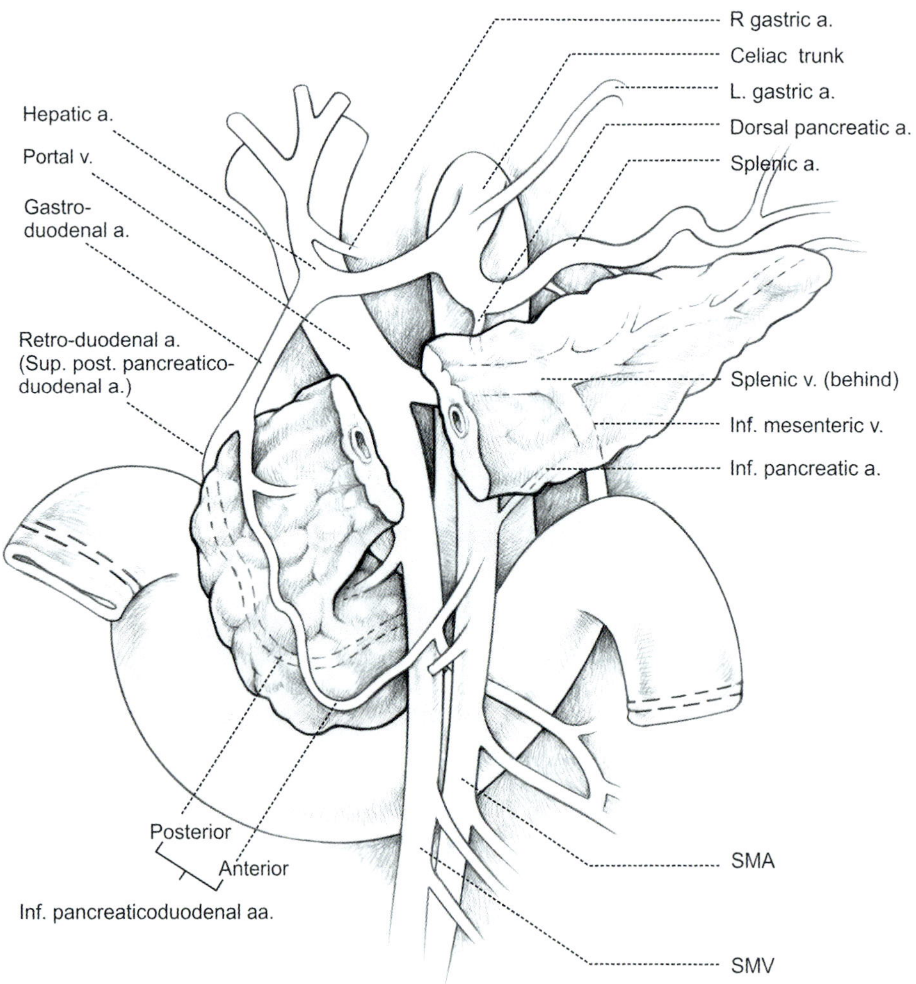

Figure 7.30 Whipple—arterial supply to pancreas, venous drainage to SMV from the pancreas, and orientation of portal venous drainage.

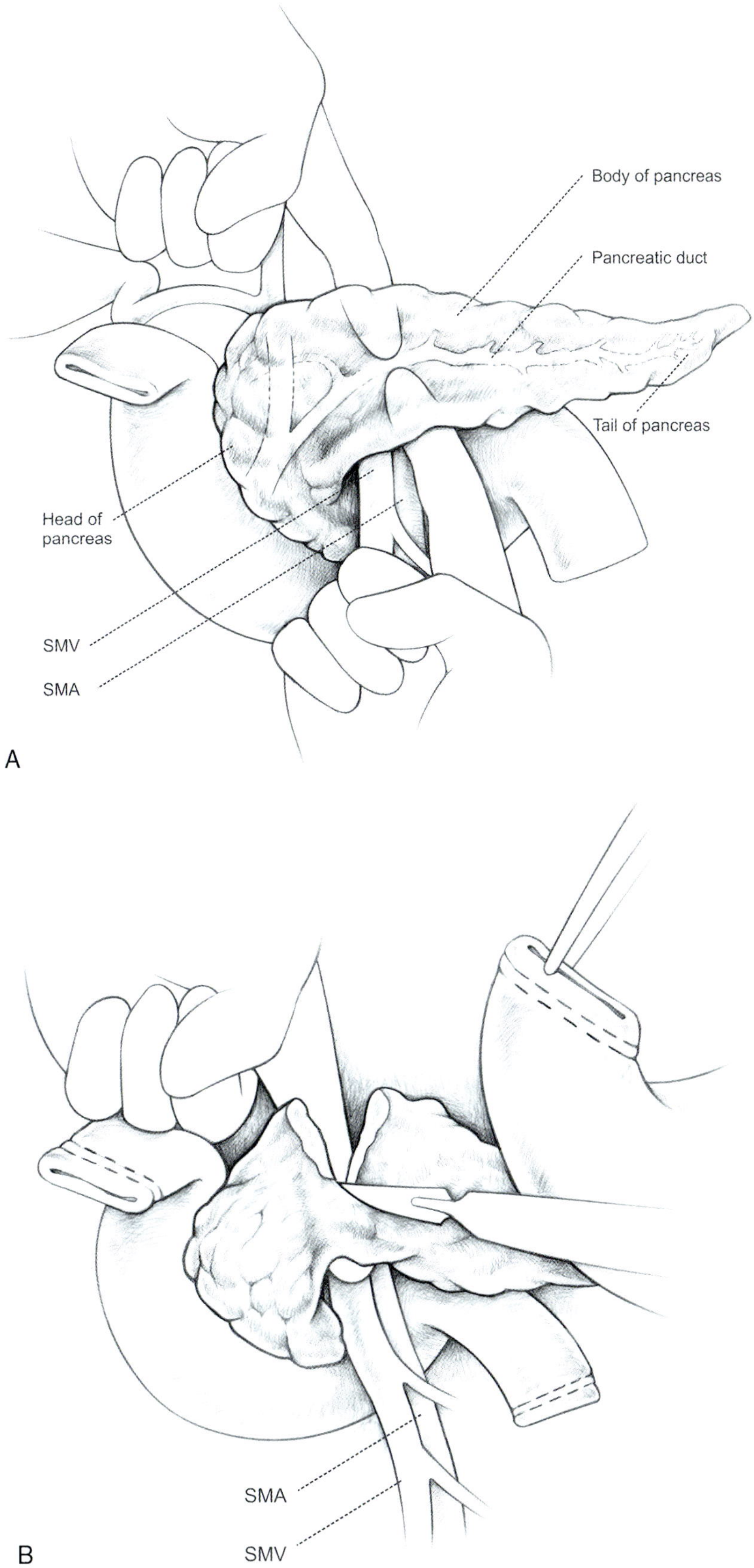

Figure 7.31 **A:** Whipple—releasing body of the pancreas from SMV. **B:** Whipple—division of the body of the pancreas.

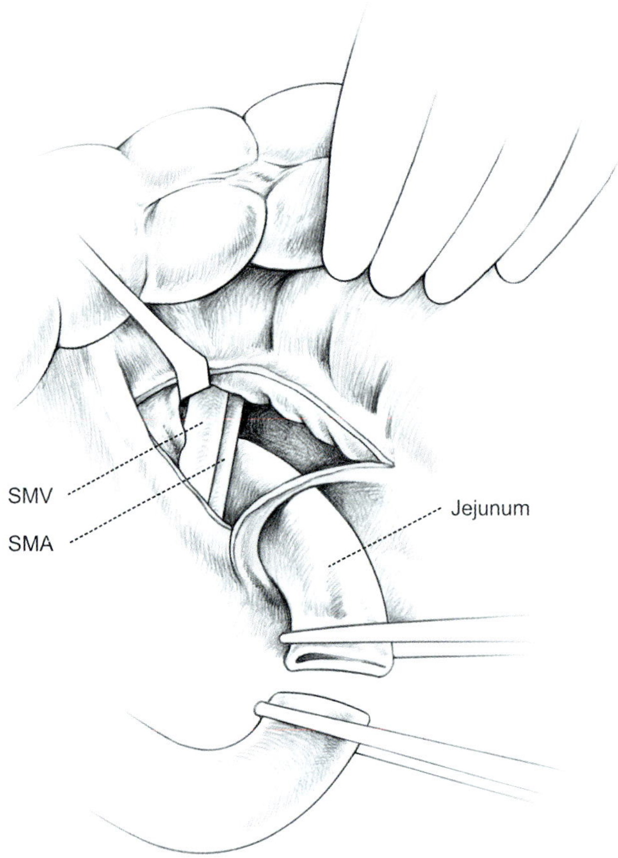

Figure 7.32 Whipple—jejunal division. Colonic mesenteric window only for demonstration of location of SMV and SMA.

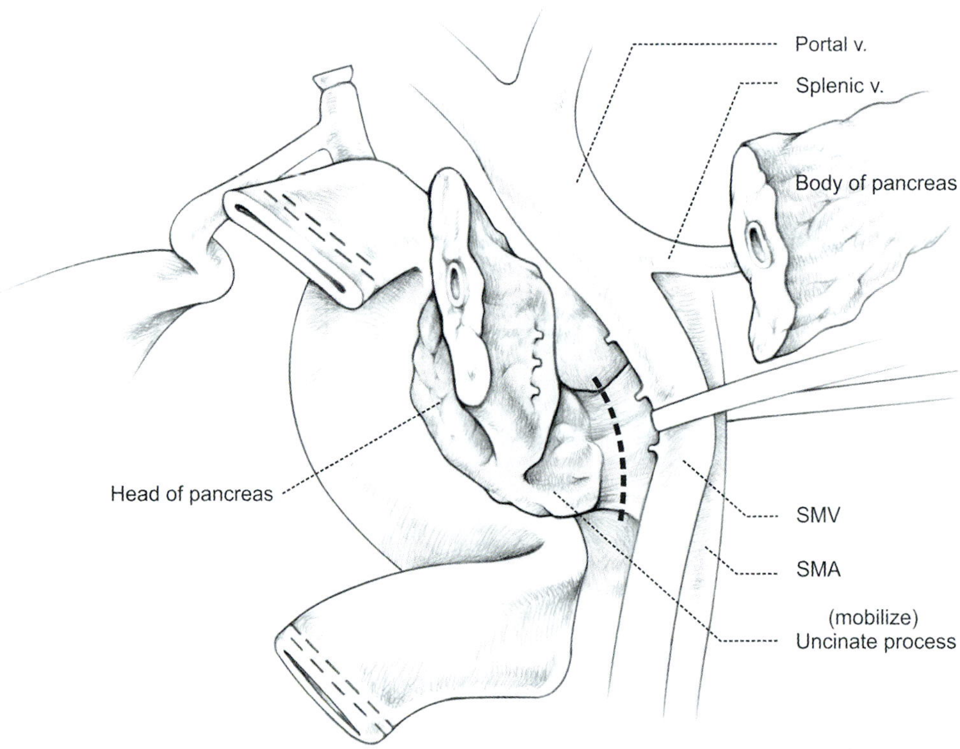

Figure 7.33 Whipple—mobilization of uncinate process of the pancreas.

of the "C" is posterior to the SMA/SMV and is called the uncinate process. Complete the excision by dividing the uncinate process attachments to the medial vascular structures of the SMV and SMA (Fig. 7.33).

There are many variations in the reconstruction of gastrointestinal continuity. All include three anastomoses: pancreaticojejunostomy, choledochojejunostomy, and gastrojejunostomy. The distal transected end of the jejunum is brought up into the RUQ through a defect created in the transverse colon mesentery (Fig. 7.32). To prevent leaks and avoid pancreatic duct stenosis, a short segment of pediatric feeding tube or 7F round Jackson-Pratt drain is placed in the pancreatic duct (Fig. 7.34A). Suture the jejunum with 3-0 suture to the edge of the pancreatic parenchyma. The pancreas can be inverted into the small bowel loop with an additional layer of interrupted sutures between the body of the pancreas and the jejunal serosa to complete the pancreaticojejunostomy.

A single-layer hepaticojejunostomy is performed with interrupted 4-0 suture, ~10 cm distal to the pancreaticojejunostomy on the jejunum. If the CBD is not dilated, a T tube may be placed superiorly in the CBD, with the inferior arm positioned through the anastomosis to stent the choledochojejunostomy. Finally, a gastrojejunostomy is performed in two layers to complete the reconstruction. This is done ~10-15 cm distal to the hepaticojejunostomy. An outer layer of 3-0 Lembert interrupted sutures and an inner layer of running 3-0 suture are placed (Fig. 7.34B). The defect where the jejunum is brought through the transverse colon mesentery is closed with interrupted 3-0 sutures to prevent internal herniation. A drain or drains are placed posterior to the pancreaticojejunostomy and hepaticojejunostomy.

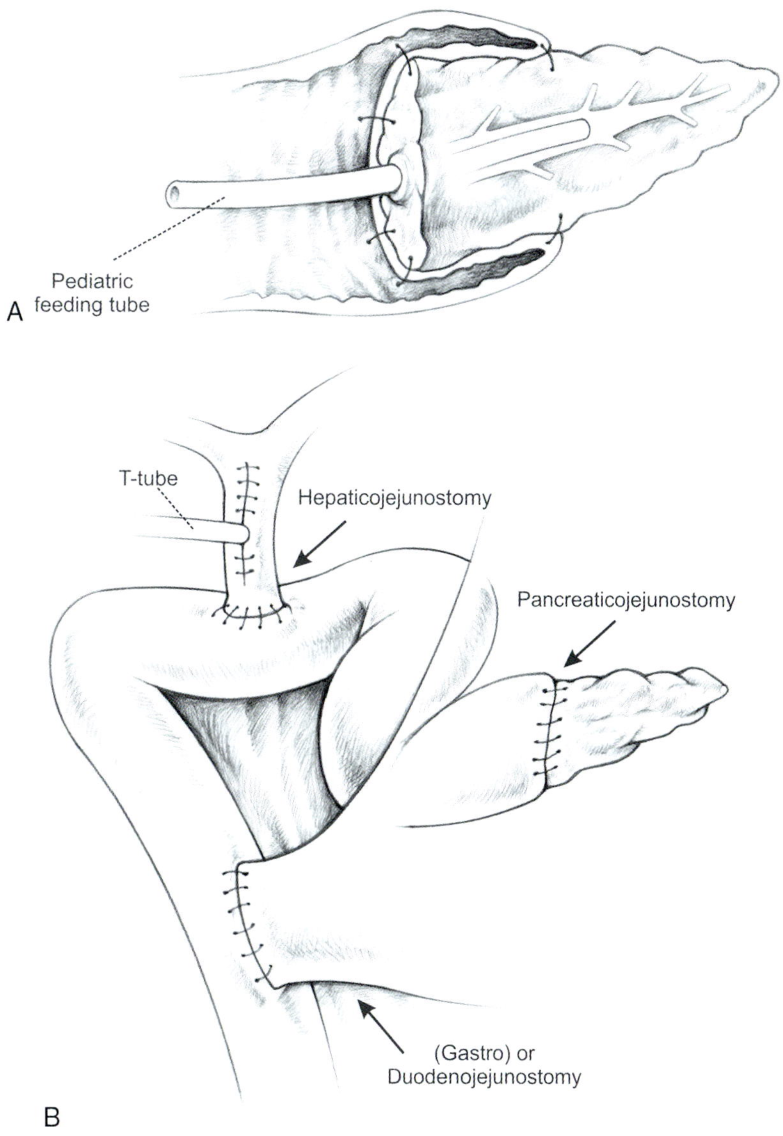

Figure 7.34 **A:** Whipple—pancreaticojejunostomy. **B:** Whipple—completed reconstruction after pancreaticoduodenectomy.

CLINICAL HIGHLIGHTS

- Most Whipple procedures are performed for malignant (ductal adenocarcinoma, neuroendocrine tumors) or premalignant lesions (IPMN, cystic neoplasms) of the pancreas.
- The ability to remove pancreatic malignancies has been greatly improved in the past decade due to the development of effective chemotherapy regimens, which can be used to shrink marginally resectable tumors and convert them to resectable lesions.
- Many acceptable alternative reconstruction methods for reconstruction after pancreaticoduodenectomy may be found in the literature.
- The many steps to pancreaticoduodenectomy are performed in a stepwise fashion.
  - The resection is performed in a clockwise sequence (kocherize duodenum and gastrocolic omentum division, porta hepatis and SMV dissection, cholecystectomy, CBD and gastric division, jejunal transection and mobilization, pancreatic transection, and removal of specimen from portal vein and retroperitoneum).
  - The reconstruction is performed in three counterclockwise steps (pancreaticojejunostomy, hepaticojejunostomy, gastrojejunostomy).
- The key step in pancreaticoduodenectomy is the ability to dissect the neck of the pancreas from the portal vein and SMV. With current diagnostic techniques of endoscopic ultrasound and high-resolution CT scan, invasion can be determined preoperatively and resection of the portal vein or SMV can be planned.
- Most surgeons prefer a pylorus-preserving procedure (PPPD) due to improved GI function postoperatively. Large tumors involving the proximal duodenum or pyloric area require a classic pancreaticoduodenectomy plus antral resection.
- Historically, bleeding has been a problem with the Whipple procedure. Meticulous hemostasis is very important to avoid postoperative hemorrhage. Remember the blood supply as shown (see Fig. 7.30).
- The most common leak is from the anastomosis between the pancreas and the bowel. Careful technical attention must be paid to this anastomosis to insure proper healing.
- Prior to proceeding with the Whipple procedure for tumor in the head of the pancreas, careful abdominal exploration and biopsy of any suspicious nodal tissue, hepatic masses, or any other intraperitoneal abnormality should be done to insure the best possible outcome.
- A great deal of lymphatic tissue will be found in the specimen from a Whipple procedure. During dissection of this tissue, fine vessels may be encountered and must be controlled.
- Watch for an aberrant right hepatic artery (Type III), which arises from the SMA, courses lateral to and supplies the CBD.

SPLENECTOMY

OPERATIVE PROCEDURE

Identify the spleen, gastrosplenic ligament, and splenocolic ligament (Fig. 7.35). Divide the short gastric vessels found in the gastrosplenic ligament and then the splenocolic ligament. Watch for gastroepiploic vessels in the inferior portion of the gastrosplenic ligament and possibly a superior polar splenic vessel (Figs. 7.36 and 5.2). Begin to mobilize the spleen out of the abdomen by taking down the splenophrenic and then the splenorenal ligaments. Rotate the spleen medially and examine the posterior splenic hilum (Fig. 7.37). The splenic artery and vein will be apparent as well as the tail of the pancreas. Divide the splenic artery and then the splenic vein. In the operating room, these will be controlled with suture ligature and stick-tie suture or a vascular load on a linear stapler. Be sure to identify and separate the tail of the pancreas to avoid damage to the pancreas.

CLINICAL HIGHLIGHTS

- During takedown of the short gastric vessels, be aware that one or two peritoneal layers may be encountered (Fig. 6.15).
- Accessory spleens are found in ~20% of cases.
- In 60% of cases of accessory spleens, one accessory spleen is identified.
- In 20% of cases of accessory spleens, two accessory spleens are discovered.
- In 20% of cases of accessory spleens, three or more accessory spleens may be found.
- Three-fourths of accessory spleens are found in the hilum and 20% are embedded in the tail of the pancreas. To remember the location of where accessory spleens may be identified, take your left hand and place the heel of your hand at your left costal margin and spread your fingers. This should remind you that accessory spleens are most commonly found in the hilum (1) but may also be found (in descending frequency) in the tail of the pancreas (2), the omentum (3), splenic artery (4), splenocolic ligament (5), bowel mesentery (6), or even the gonad (7) (Fig. 7.38).

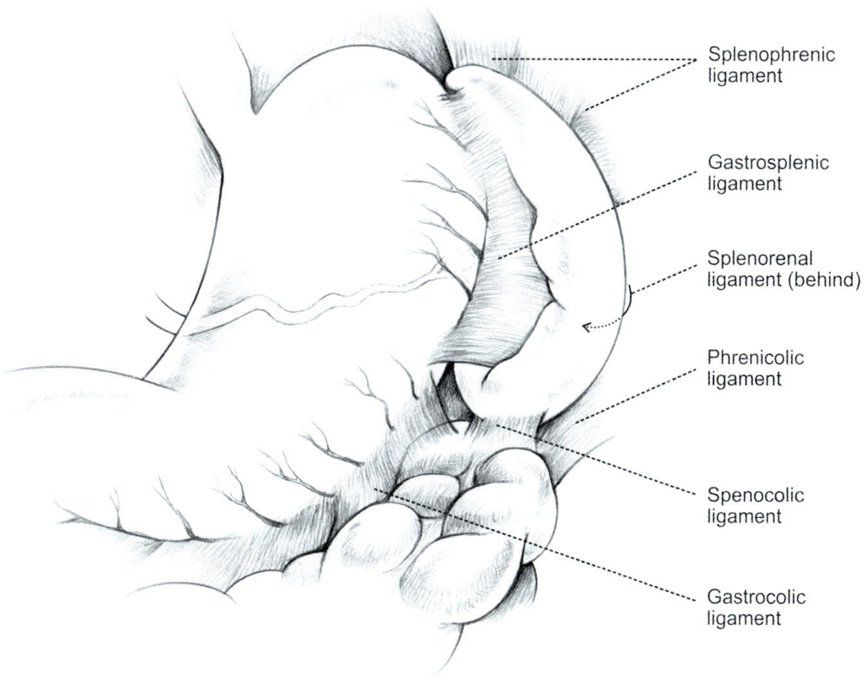

Figure 7.35 Splenic ligaments.

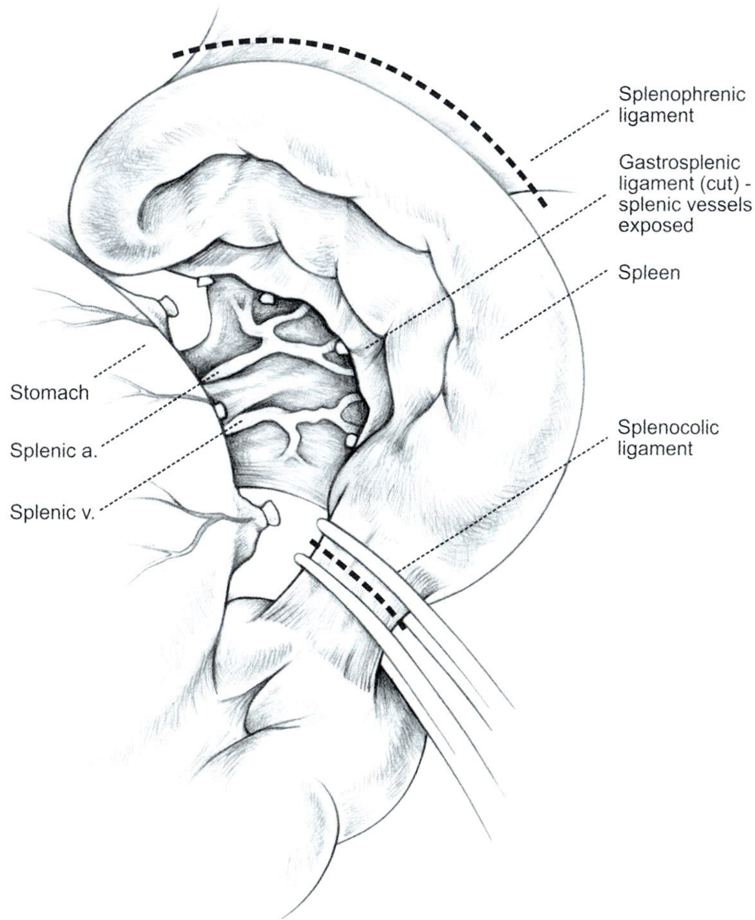

Figure 7.36 Division of splenocolic and splenophrenic ligaments and identification of splenic artery and vein.

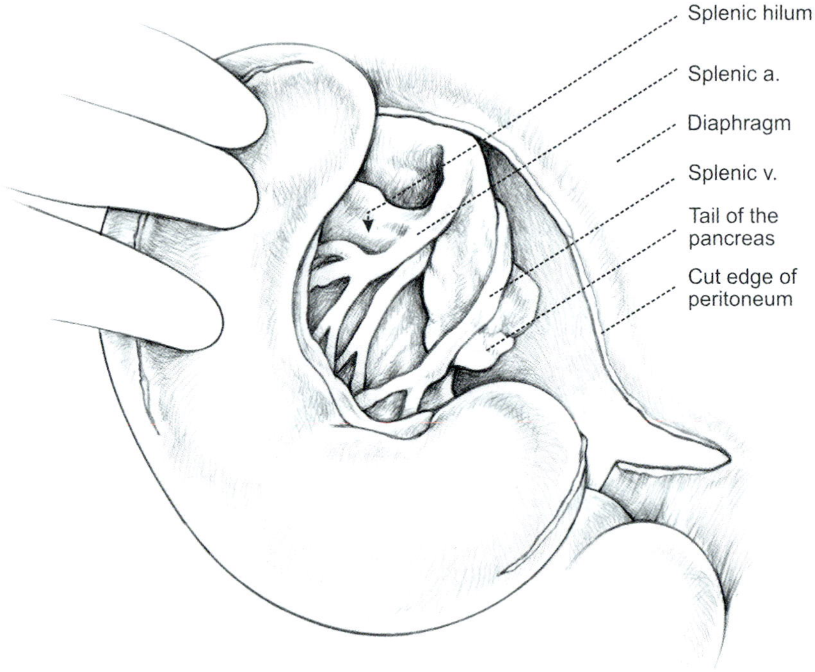

Figure 7.37 Spleen mobilized up and posterior look at splenic vessels.

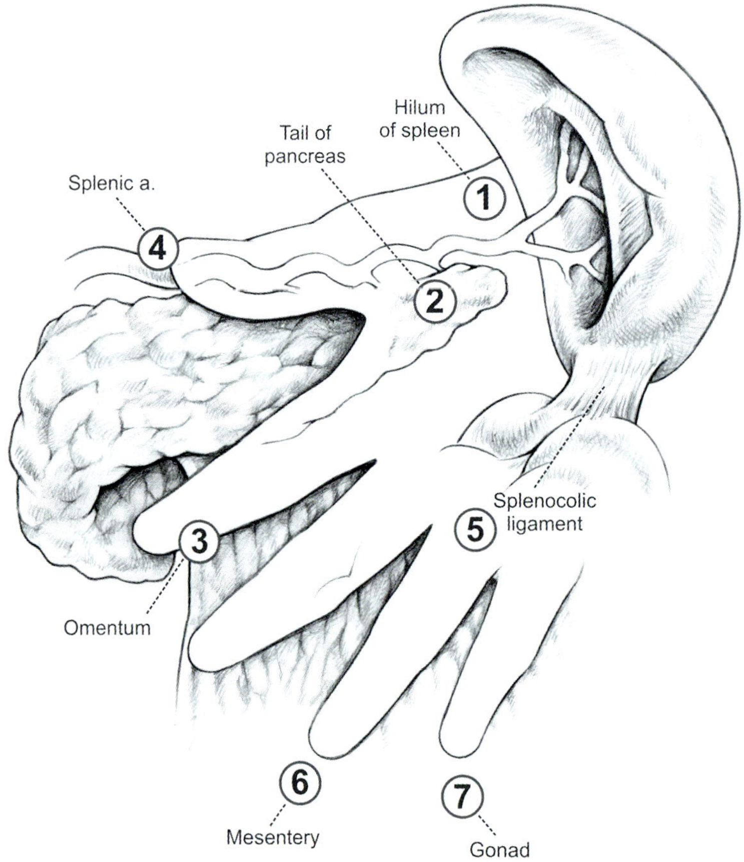

Figure 7.38 "Handy" method of remembering location of accessory spleens.

- Trauma splenectomy—blunt or sharp division of ligaments by palpation without direct visualization allows rapid medial rotation of the spleen. Place a curved clamp across entire splenic hilum.

- Elective splenectomy for splenomegaly—divide the presplenic fold and the gastrosplenic ligament with short gastric vessels (Fig. 6.15). Divide the splenic artery and vein (Fig. 7.36) prior to mobilizing the splenic ligaments. The splenic ligaments may need to be divided by palpation only.

DISTAL PANCREATECTOMY (WITH/WITHOUT SPLENECTOMY)

OPERATIVE PROCEDURE

Enter the lesser sac by dividing the gastrocolic ligament if this has not been previously performed (Fig. 7.35). Tumor location is evaluated and the appropriate approach decided upon.

To remove the body or tail, identify the portal vein in the porta hepatis (Fig. 7.12) and the SMV in the retrogastric retroperitoneum (Fig. 7.29). Finger dissect the areolar plane behind the neck of the pancreas (Fig. 7.31A). Identify the celiac axis and splenic artery (Figs. 7.30 and 5.2). Divide the splenic artery and divide the neck of the pancreas anterior to the portal vein/SMV (Fig. 7.31B). Elevate the body of the pancreas to the left (Fig. 7.39). Identify and mobilize the splenic vein and divide it (Fig. 7.39). Divide the splenophrenic ligaments and mobilize the tail, the pancreas, and the spleen from the retroperitoneum medially to remove (Fig. 7.37).

To remove the pancreatic body with splenic preservation, divide the retroperitoneal attachments and the small branches from the splenic artery and vein (Fig. 7.39). Remove the body and tail of the pancreas from the retroperitoneum, preserving the spleen and its primary blood supply.

To remove the tail of the pancreas with splenic preservation, the site of transection is chosen 1-2 cm proximal to the lesion. Identify and elevate the superior and inferior borders of the pancreas. Dissect the body of the pancreas from its posterior attachments and the splenic vein (Fig. 7.31A). Transect the pancreas and ligate the small branches from the splenic artery and vein (Figs. 7.31A and 7.30). Transect the remaining retroperitoneal attachments removing the tail of the pancreas, preserving the spleen and its primary blood supply.

CLINICAL HIGHLIGHTS

- Distal pancreatectomy is an increasingly common procedure due to the increased incidental identification of lesions of the body and tail of the pancreas in diagnostic imaging for other reasons.

- Distal pancreatectomy for invasive ductal adenocarcinoma requires an en bloc resection of the distal pancreas which may include:
  - Wide resection of the adjacent retroperitoneum
  - Spleen and adjacent stomach and left kidney

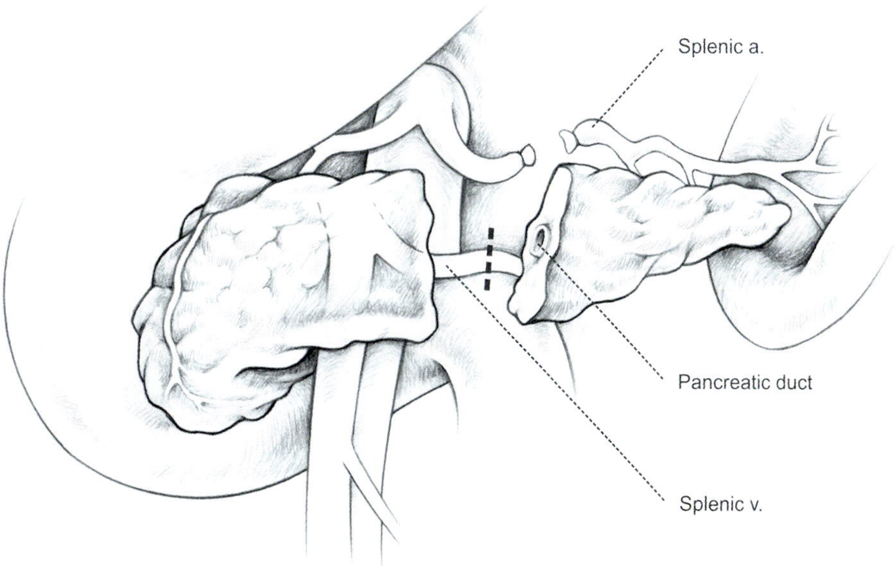

Figure 7.39 Division of splenic artery and vein for distal pancreatectomy and splenectomy.

- Distal lesions that are low grade (ie, neuroendocrine tumors), benign (some cystic neoplasms and islet cell tumors), or premalignant (IPMN) may only require resection of the pancreas 1-2 cm proximal to the lesion.

- For benign disease, splenic preservation should be attempted. Splenic arterial and splenic venous branches are individually divided and ligated when preserving the spleen (Fig. 7.30). Careful dissection and ligation of small branches from the splenic artery and vein are required to avoid significant hemorrhage.

- The pancreatic parenchyma is often transected with a linear stapler with 4.5-mm staples reinforced with figure-of-8 nonabsorbable sutures to ensure pancreatic duct ligation.

- Drainage of the pancreatic bed is recommended after distal pancreatectomy as postoperative fluid collections are common.

- Pancreatic leaks can occur from the transected end of the pancreas, but if properly drained, they often will close spontaneously.

- Reactive left pleural effusions are common after distal pancreatectomy and splenectomy sometimes requiring thoracentesis or chest tube.

- Mobilization of the splenic vein is a useful technique to be remembered when a distal Warren splenorenal shunt is needed for portal decompression in liver cirrhosis.

8

Kidney, Adrenal, Ureters, and Bladder

Outline

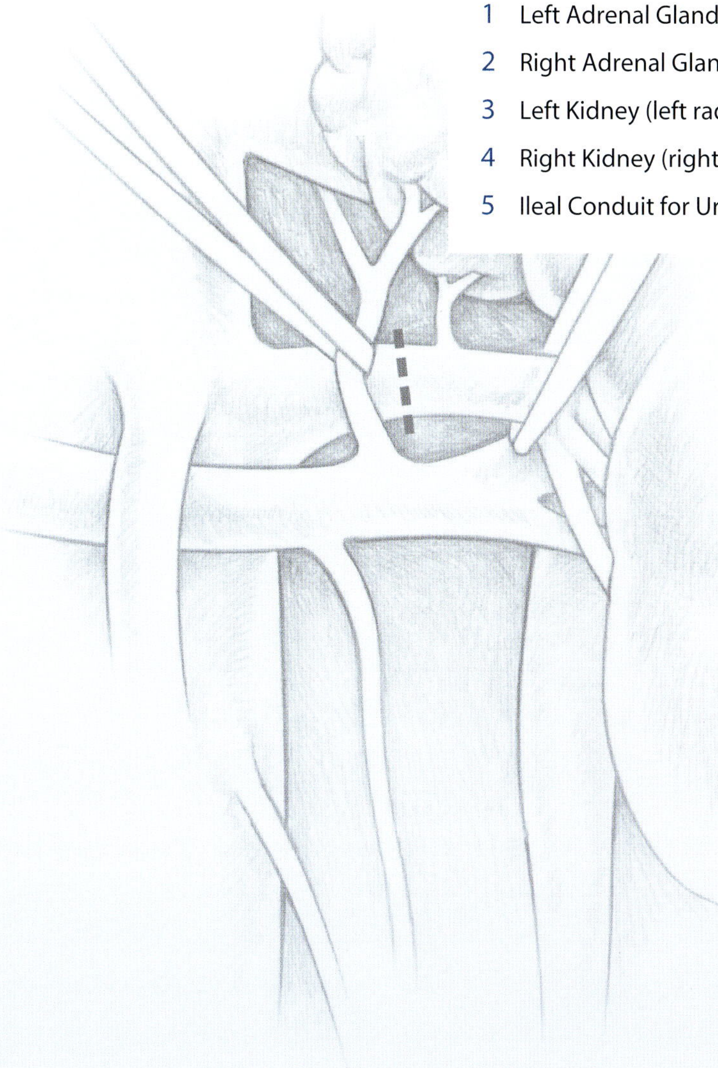

LEFT ADRENAL (LEFT ADRENALECTOMY)

INCISION

Reflect the chest plate superiorly and the abdominal wall inferiorly (see Figs. 3.2 and 5.1), remove the previously dissected organs, and reflect the bowel superiorly and to the right. This will give a good view of the retroperitoneum. Before entering the retroperitoneum, consider adrenal venous connections and lymphatic drainage (Fig. 8.1). Lymphatic drainage is shown by *arrows*. Notice proximity of the pancreas.

OPERATIVE PROCEDURE

Identify the posterior peritoneum as it lays over the adrenal glands and kidneys. Incise the peritoneum and identify the left adrenal gland. In the operating room, the adrenal glands are easily identifiable by a very bright yellow appearance. The surrounding fat is a dull yellow in comparison. In a cadaveric specimen, the adrenal and the surrounding fat do not have the same degree of color contrast.

Identify the adrenal by its anteromedial location to the kidney and firm consistency. Its color is a dull gray and is rather unremarkable in a cadaver. The adrenal glands are a retroperitoneal structure and lie inside a slip of Gerota's fascia separate from the kidney (Fig. 8.2). Dissect the gland off of the kidney. Arterial supply is varied and is very rich. The adrenal glands and the thyroid gland have the greatest blood supply per gram of human tissue. Note that the lymphatic drainage of the left adrenal gland is superiorly to the posterior mediastinal nodes, medially to lateral periaortic nodes, and inferiorly to renal hilar nodes (see Fig. 8.1).

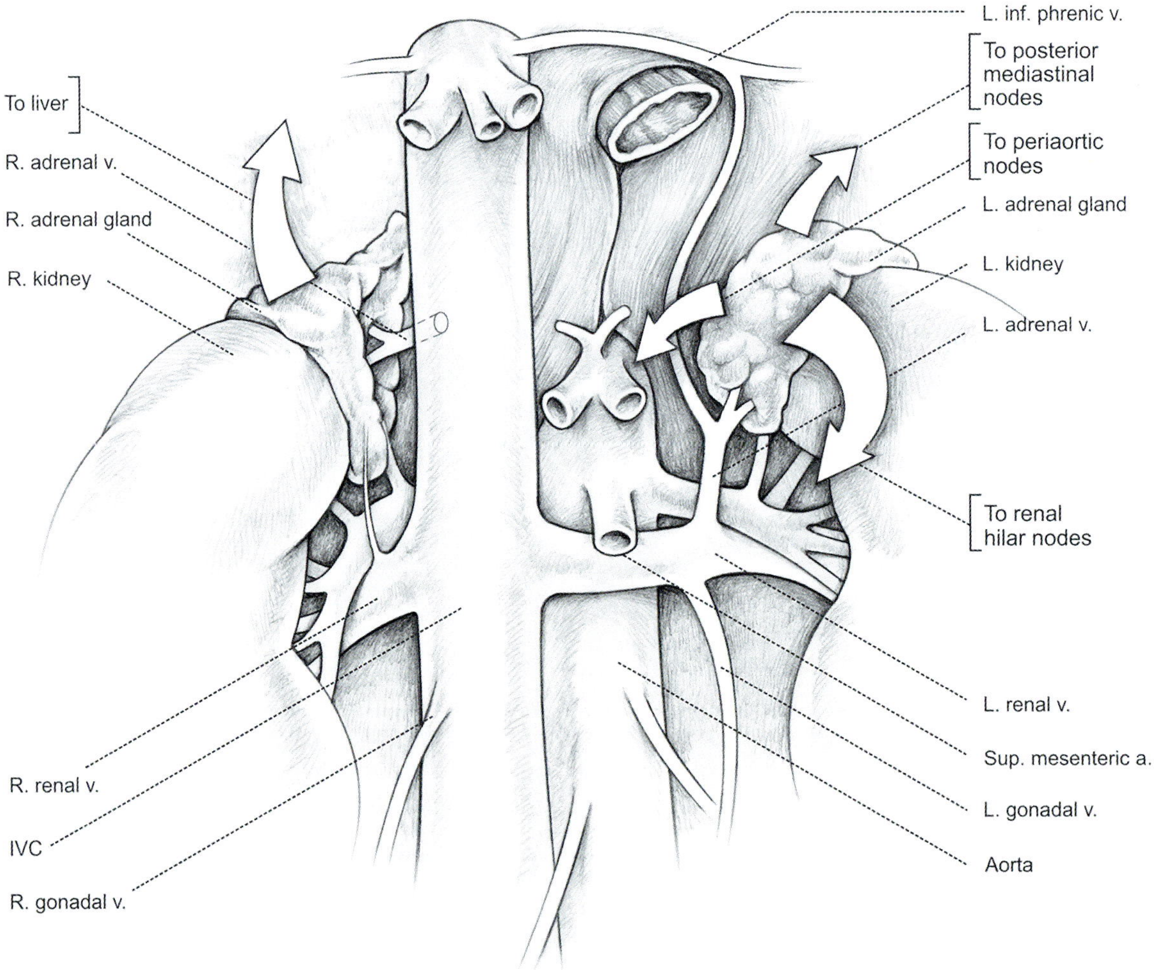

Figure 8.1 Adrenal location, venous and lymphatic drainage. Renal location, arterial supply and venous drainage.

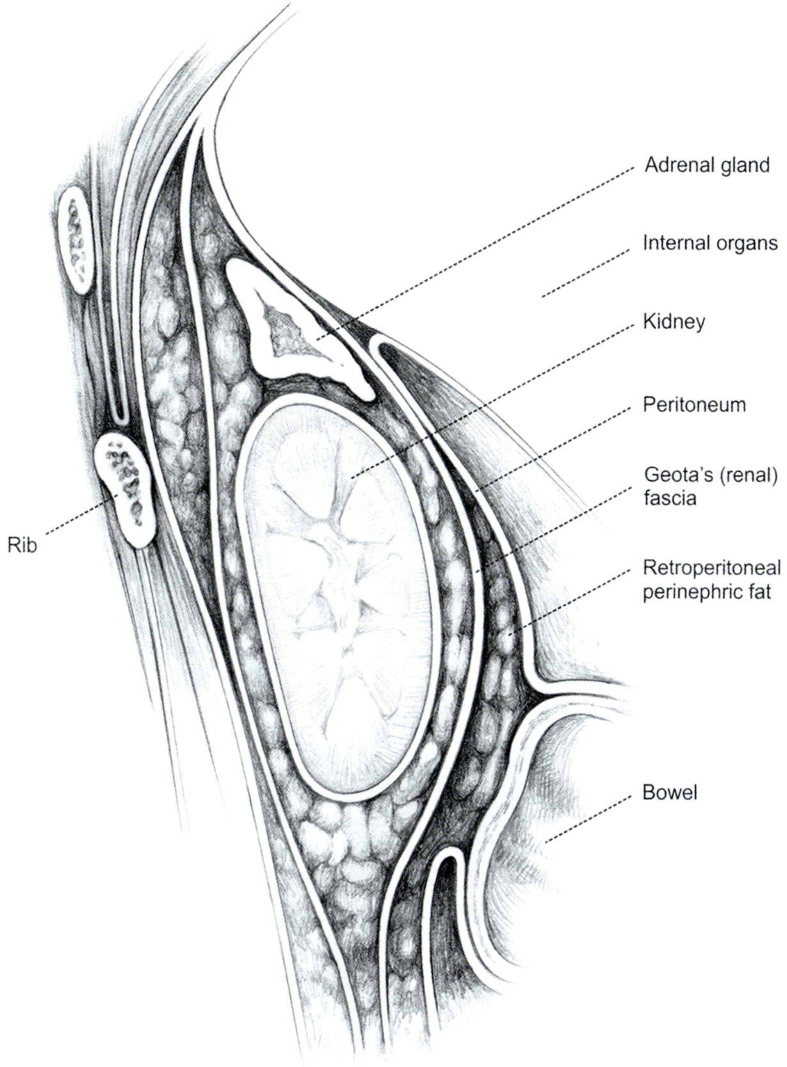

Figure 8.2 Cross-section demonstrating renal and adrenal location within Gerota's fascia.

Note from the diagram (Fig. 8.3) that the arterial supply to the left adrenal comes as follows:

1. Superiorly from inferior phrenic vessels
2. Medially from the aorta
3. Inferiorly from the renal artery

The adrenal arteriolar vessels are drawn large in the diagram to make identification easy; however, in situ, these arterioles are filamentous and may be difficult to identify.

Identify the adrenal vein that drains into the left renal vein (see Fig. 8.1). This is typically a single vein but minor accessory veins to the inferior phrenic vein may occur. After identification of the left adrenal vein, transect the vein as well as its filamentous arterial blood supply and remove the adrenal using blunt and sharp dissection with downward traction on the kidney.

Note the medial location of the left adrenal and that it extends inferiorly into the renal hilum within Gerota's fascia (Fig. 8.2). Exposure is essential to successful adrenal surgery. Control of any bleeding is difficult without stable and excellent exposure with identification of surrounding structures.

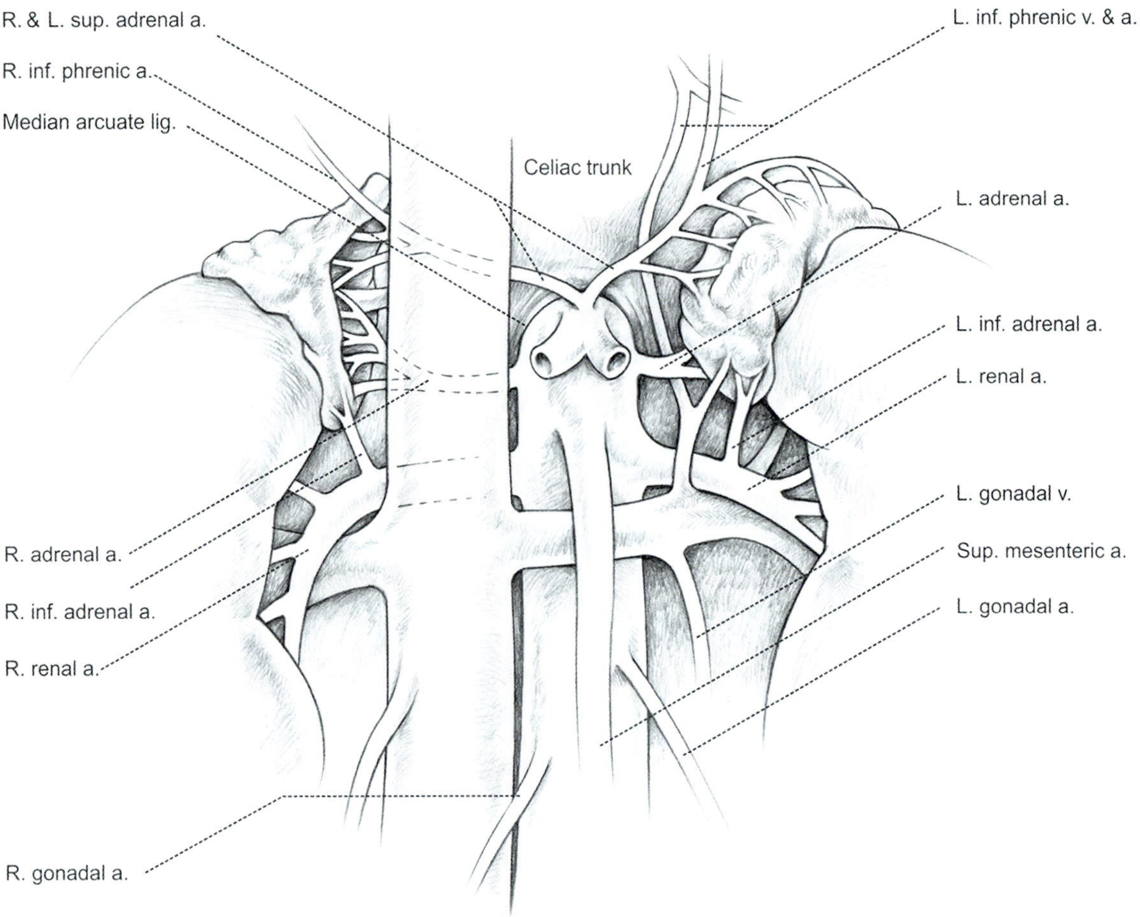

Figure 8.3 Adrenal location and arterial supply.

RIGHT ADRENAL GLAND (RIGHT ADRENALECTOMY)

INCISION

See incision under the section, "Left Adrenal (Left Adrenalectomy)," and prepare the cadaveric specimen similarly.

OPERATIVE PROCEDURE

Identify the peritoneum as it covers the adrenal and kidney on the right side (see Fig. 8.2). Proceed as on the left side with incision of the peritoneum to gain access to the retroperitoneal space. Notice proximity of the duodenum.

Identify the right adrenal gland. As on the left, it will appear a dark gray unremarkable color; however, on palpation, it will have a different consistency than the surrounding retroperitoneal fat. The right adrenal lies in a medial location and is generally hidden behind the inferior vena cava (IVC) (see Fig. 8.3). Retract the IVC to the left gently. Note that the lymphatic drainage of the right adrenal gland travels superiorly toward the liver (see Fig. 8.1).

The arterial supply to the adrenal glands is very rich and on the right comes:

1. From the inferior phrenic artery
2. From the aorta coursing behind the IVC
3. From the renal artery inferiorly (Fig. 8.3)

Retract the right adrenal gland laterally using thumb and forefinger. Gentle medial retraction of the IVC and careful medial dissection of the gland will identify a short, large right adrenal vein draining into the lateral-posterior IVC (see Fig. 8.1). Identification of this vein is essential and often difficult due to its posterior position. Control of the right adrenal vein should be done with care to avoid injury to the IVC. Note that the arterial supply has many branches. Divide the small arterial branches and the large right adrenal vein. The gland may now be removed.

CLINICAL HIGHLIGHTS

- The adrenal glands lie inside a separate sheath of Gerota's fascia in the retroperitoneum. There is a small interspace of connective tissue between the adrenal and corresponding kidney, which can be entered and the organs separated (see Fig. 8.2).
  - The left adrenal is shaped like a watch cap. Imagine it tipped medially on the left kidney.
  - The right adrenal is shaped like a triangular pointed hat. Imagine it tipped medially on the right kidney positioned partially behind the IVC.
  - Dissection of the adrenal arterial supply in vivo generally does not display large arterial feeder vessels unless a substantial adrenal tumor present (see Fig. 8.3).
- The in vivo color of the adrenal glands is a very bright yellow in contrast to the surrounding dull yellow of retroperitoneal fat.
- When looking for the adrenal glands, remember their superior-medial location in relation to the kidney.
- During an adrenalectomy for hormone-secreting tumors (eg, pheochromocytoma and aldosteronoma), ligation of the adrenal venous drainage should be carried out as early as possible during the procedure.
- Pheochromocytomas are the "10% tumor": 10% bilateral, 10% malignant, and 10% extra-adrenal.
- Extra-adrenal tissue may be found in periadrenal tissue, under the capsule of the kidney or liver in a periaortic location, peri-iliac artery location, or near the bladder.
- Adrenalectomy may be accomplished via the open technique, laparoscopic technique, or posterior approach.
- The consistency of the adrenal gland is friable. In the operating room, the gland may crumble with vigorous retraction. This may cause remnants of adrenal tissue to be left behind and may cause an inadequate adrenalectomy.

LEFT KIDNEY (LEFT RADICAL NEPHRECTOMY)

INCISION

Exposure is gained as noted in the section, "Left Adrenal (Left Adrenalectomy)," above.

OPERATIVE PROCEDURE

Remember in the OR that isolation and control of the renal arteries and veins is accomplished early in the procedure. The left renal vein has already been identified in the left adrenal dissection. Now identify the renal arteries and note their position posterior to the left renal vein (see Fig. 8.3). Identify the left gonadal vein as it descends from the left renal vein. Follow this vein in its retroperitoneal path parallel and becoming lateral to the ureter at the pelvic brim. Keep this relationship in mind when dissecting the right or left pelvis for a colorectal procedure. Identify the ureter below the ureteropelvic junction and follow it as it descends inferiorly in the retroperitoneal space (Fig. 8.4). This is a good time to look at the ureter, gonadal artery, and vein. The gonadal vessels originate medial to the ureter and end up *lateral* to the ureter at the pelvic brim.

Once the renal vessels have been identified and controlled, the renal arteries may be transected adjacent to the aorta (see Fig. 8.3). Divide the left renal vein and the left gonadal vein (Fig. 8.5), and then divide the left ureter at the pelvic brim (see Fig. 8.4). A radical nephrectomy done for cancer must be performed excising the retroperitoneal fat and the intact fascia of Gerota. Most often, the corresponding adrenal is removed within Gerota's fascia (see Fig. 8.2). Perform this retroperitoneal dissection bluntly. Note that there is a definite tissue plane, which allows the kidney to be mobilized freely from the posterior body wall (see Fig. 8.2).

Place the specimen on a paper towel. Open Gerota's fascia and identify the kidney within it. Note the branching of the renal vessels in the hilum of the kidney. Bivalve the kidney in a craniocaudal fashion. Note the relationship of the renal cortex, medulla (pyramids), renal papilla, and calices (Fig. 8.6).

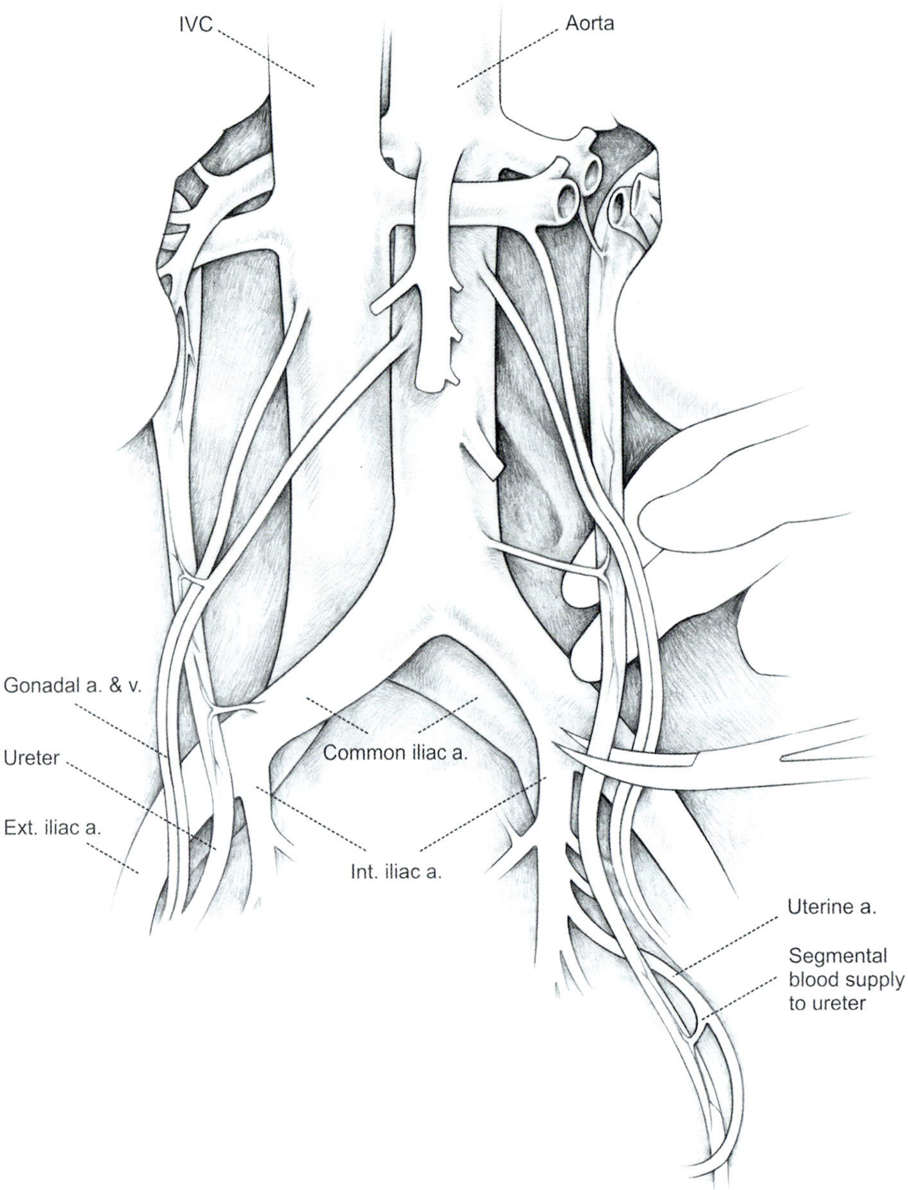

Figure 8.4 Blood supply of ureters. Gonadal vessels origin and course.

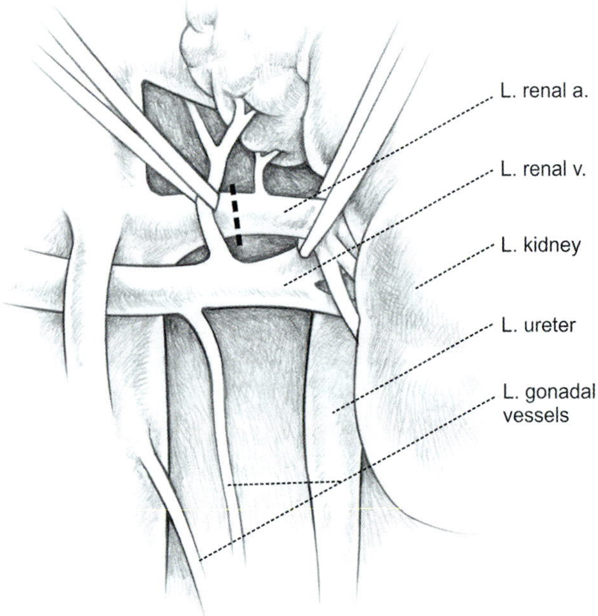

Figure 8.5 Left renal artery and vein.

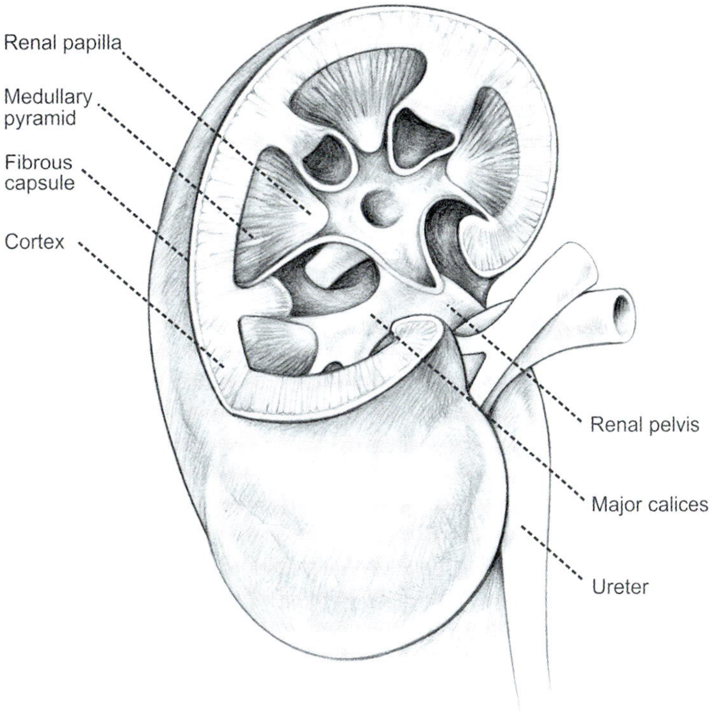

Figure 8.6 Cross-section of the kidney.

RIGHT KIDNEY (RIGHT RADICAL NEPHRECTOMY)

INCISION

Exposure is gained as noted in the section, "Left Adrenal (Left Adrenalectomy)," above.

OPERATIVE PROCEDURE

Note that the right kidney is somewhat inferior in its relationship to the left kidney. For a radical nephrectomy, the renal vessels are dissected and controlled early in the procedure. Dissect the right renal arteries and vein, which may be single or multiple (Fig. 8.7). The renal arteries are posterior to the renal vein and are also posterior to the IVC. Note that the right gonadal vein drains directly into the IVC. Follow the gonadal vessels and ureter to the pelvic brim (see Fig. 8.4). Find the plane outside of Gerota's fascia, which will include the perinephric fat for dissection of the kidney for the radical nephrectomy (see Fig. 8.2). Develop this plane to completely free up the kidney from its retroperitoneal attachments. With the kidney completely freed, place a hand behind the kidney and note the

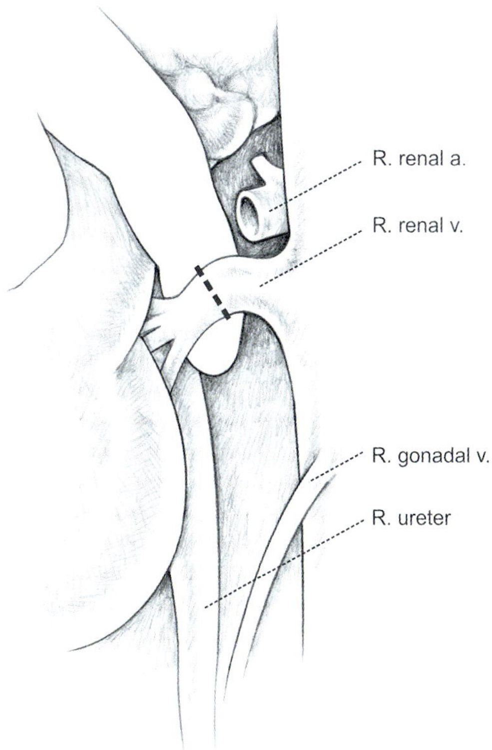

Figure 8.7 Right renal vein after transection of the right renal artery.

relationship of the anterior vein, posterior arteries, and collecting system and ureter posterior to all of the vessels. Divide the renal vessels. Divide the right ureter at the pelvic brim (see Fig. 8.4).

Open Gerota's fascia and identify the branches of the renal vein and renal arteries at the hilum of the kidney. Note the divisions and the presence of polar (superior or inferior polar vessels) arteries if they should be present. Bivalve the kidney in a craniocaudal fashion. Note the relationship of the renal cortex, medulla (pyramids), renal papilla, and calices (Fig. 8.6).

CLINICAL HIGHLIGHTS

- Renal tumors tend to be clinically silent. For this reason, they may be quite large upon detection although modern imaging has resulted in earlier diagnosis of small renal tumors.

- Small tumors in elderly and frail patients may be followed expectantly. With the diagnosis of small tumors occurring more frequently, partial nephrectomy is more common in recent years.

- Be sure a thorough preoperative workup to include estimated glomerular filtration rate, urinary creatinine clearance, and evaluation of contralateral renal anatomy and function. Consultation concerning dialysis may be appropriate if renal function is marginal.

- Remember that the renal artery and vein may be single or multiple on either side (see Figs. 8.1 and 8.3).

- Lymphatic drainage is typically to the periaortic lymph nodes. These nodes should be included with the radical nephrectomy dissection.

- Tumor thrombus may extend into the renal vein and from there into the IVC. Ultrasound of the IVC preoperatively may indicate tumor presence or it may be evident on computed tomography. Complete excision of this tumor extension is essential for an adequate oncologic procedure. Be aware that IVC tumor extension can extend all the way to the right atrium of the heart. In this case, median sternotomy and control of the distal extent of the tumor will be important to avoid tumor embolization.

- Remember that the gonadal vessels are found lateral to the ureter at the pelvic brim.

- Testicular venous drainage: right to IVC and left to left renal vein (see Fig. 8.4).
  - Unilateral varicocele on the right may indicate right-sided retroperitoneal tumor.
  - Left-sided varicoceles are very common and may be associated with infertility or chronic pain. If the left varicocele is new, it may indicate a left renal tumor.

- Blood supply to the ureter is *medial* in the abdomen and *lateral* in the pelvis.

- Practice placing a double J stent in the ureter at the point of division to simulate a ureteral injury. Spatulate the opposing ureteral ends and close with interrupted 5-0 chromic.

ILEAL CONDUIT FOR URINARY DIVERSION

INCISION

Incision and access to the intestinal conduit have been previously obtained.

OPERATIVE PROCEDURE

In the event of a radical cystectomy for bladder cancer, appropriate urinary drainage will need to be constructed. The ureters have been previously identified in the retroperitoneum, followed into the pelvis and transected (see Fig. 8.4).

Identify the terminal ileum and preserve 15-20 cm of the terminal ileum to avoid metabolic complications of malabsorption. A section of bowel 12-18 cm proximal to that is selected and isolated. Continue mobilization into the mesentery preserving blood supply to this segment (Fig. 8.8A). An ileoileostomy is performed usually side to side with sutures or staples taking care to preserve the bowel vascular supply. The anastomosis should be performed to allow the isolated segment of bowel to freely come to the anterior abdominal wall without compromise. Close the mesenteric defect to avoid internal herniation and obstruction (Fig. 8.8B).

The ureters are prepared by spatulating the end of each ureter (Fig. 8.9A). The ureters are joined together with a single layer of running or interrupted suture (Fig. 8.9B). The back wall of the ureters is now approximated to the back wall of the intestine (Fig. 8.9C). Completion of the anastomosis to the bowel is demonstrated (Fig. 8.9D).

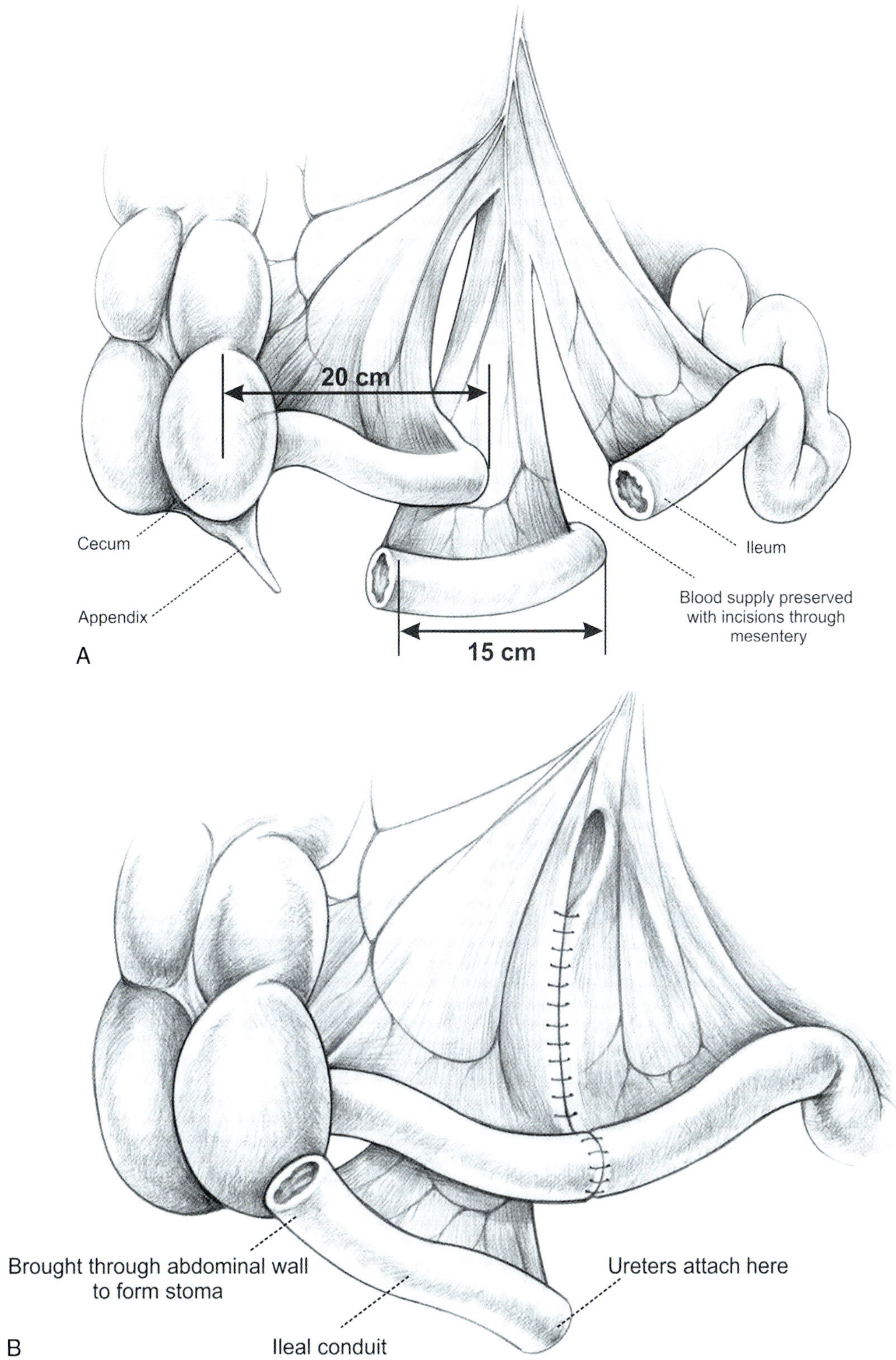

Figure 8.8 **A:** Design of ileal segment for urinary diversion. **B:** Ileal loop ready for ureteral implantation.

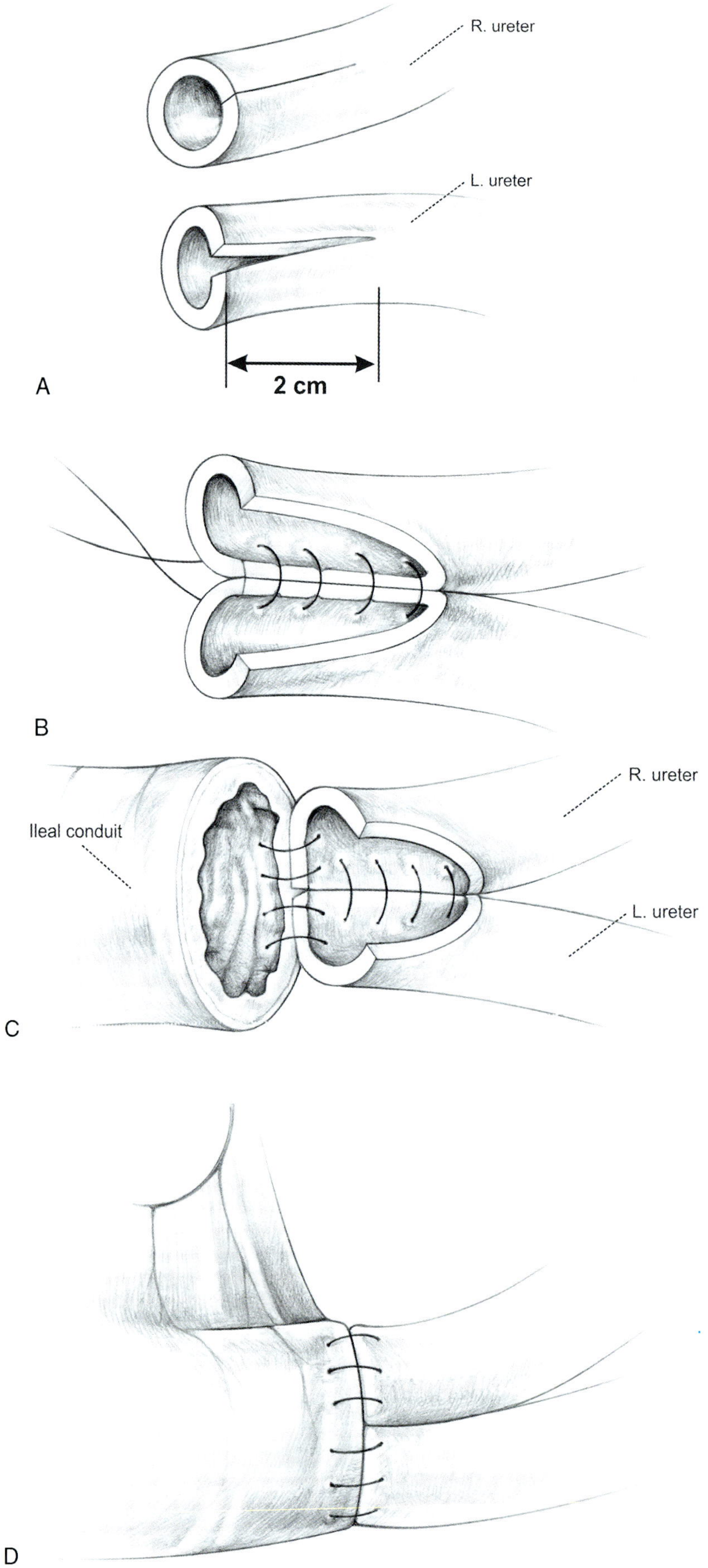

Figure 8.9 **A:** Ureteral spatulation. **B:** Joining ureters. **C:** Back wall of ureteroileal anastomosis. **D:** Completed ureteroileal anastomosis.

Take care to provide a mucosa to mucosa watertight anastomosis reinforced with sutures in the muscular layer. The ileum can be anchored to the posterior abdominal wall to avoid tension on the ureteroileal anastomosis. The ileal conduit is finished by bringing out the bowel through the right lower quadrant and maturing the stoma. As an alternative, each ureteroileal anastomosis may be done separately.

CLINICAL HIGHLIGHTS

- A general surgeon might find himself/herself involved with the urologic service to assist in this procedure. Operative urinary continuity is good for any general surgeon to understand in practice.
- Many urologists will have other methods of handling the ureteral anastomosis to the bowel.
- A drain and temporary ureteral stent may be appropriate to avoid the postoperative development of a urinoma.
- Remember to not disturb the final 15-20 cm of native ileum to avoid problems of malabsorption unique to this section of small intestine.
- Orthotopic urinary diversion utilizing an ileal reservoir and the native urethra is an alternative for continent urinary diversion. This technique provides urinary continence in those patients whose urethral margins are negative by frozen section.

9

Pelvis

Outline

1 Obturator Canal and Hernia
- See Chapter 4 for Obturator Hernia

2 Rectum (proctectomy)
- See Chapter 5, "Intestines," for Total Mesorectal Excision

3 Bladder and Ureters
- See Chapter 8

4 Uterus and Ovaries (hysterectomy and bilateral salpingo-oophorectomy)

5 Urinary Diversion
- See Chapter 8

6 Testicles (orchiectomy)

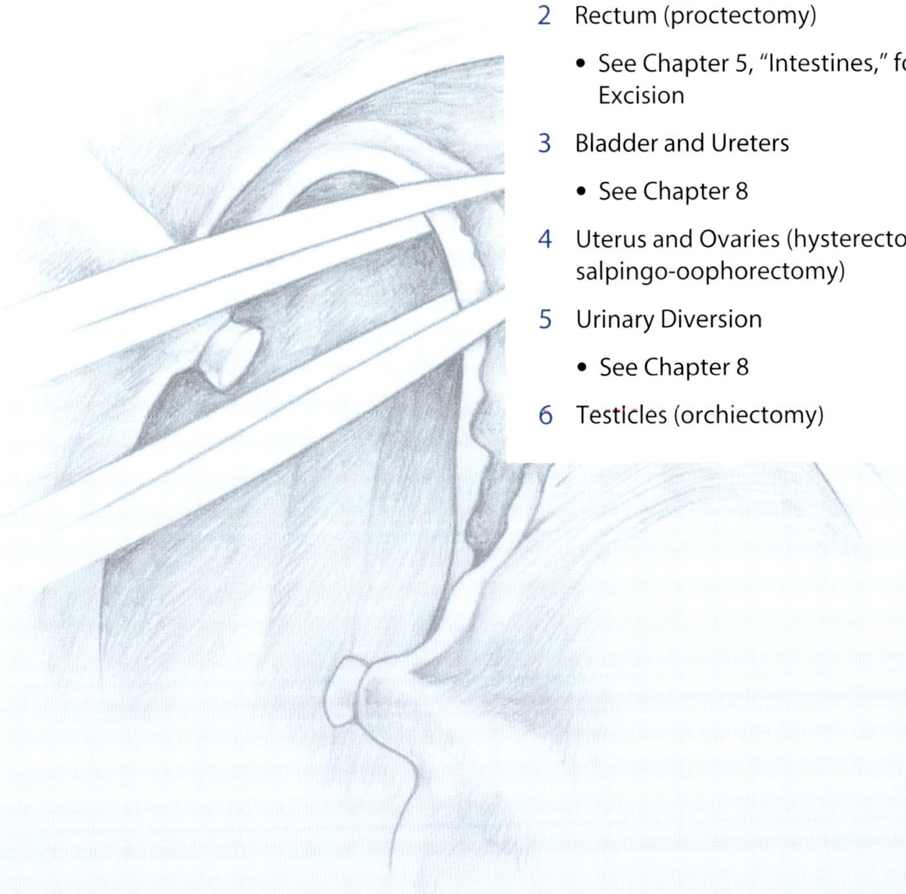

OBTURATOR CANAL AND HERNIA

- (See Chapter 4 for obturator hernia)

RECTUM

- (See Chapter 5 For TME)

BLADDER AND URETERS

- (See Chapter 8)

UTERUS AND OVARIES

- (Hysterectomy and Bilateral salpingo-oophorectomy)

INCISION

Reflect the abdominal wall inferiorly; see Chapter 5, Section 1, "Omentum." Remove the sigmoid colon and rectum if they have been previously dissected.

OPERATIVE PROCEDURE

Identify the round ligaments, broad ligaments, suspensory ligament of each ovary (which contains the ovarian vessels), uterosacral ligaments, and ureters (Fig. 9.1). Place long Pean clamps across each uterine cornu incorporating the round ligaments and fallopian tubes. Utilize these for upward traction on the uterus for improved visualization and dissection throughout the procedure.

Divide the round ligaments bilaterally, which will open the anterior and posterior layers of the broad ligaments (see Fig. 9.1). Dissect the avascular space between these two layers with blunt and sharp technique to expose the uterine vessels and isolate the suspensory ligament of the ovary. Extend this dissection of the anterior broad ligament layer to incise the vesicouterine peritoneum (Fig. 9.2A).

Create a window in the posterior layer of the broad ligament beneath the suspensory ligament/ovarian vessels on either side. Be certain to identify the ureter as it crosses over the pelvic brim and runs beneath the suspensory ligament. Place two clamps across this (now isolated) vascular pedicle and suture ligate both (Fig. 9.2B).

Utilizing upward traction on the uterus, dissect the bladder off the anterior surface of the lower uterine segment, cervix, and upper vagina. Keep the dissection

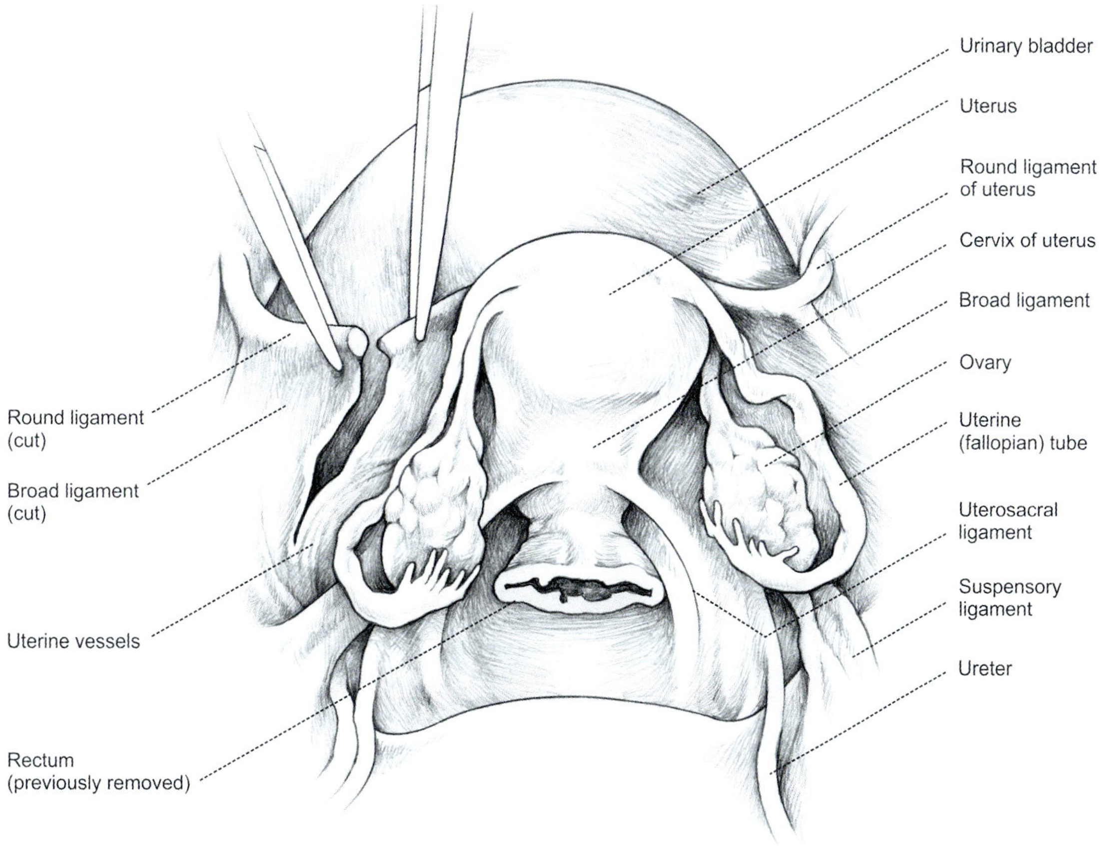

Figure 9.1 Identify pelvic organs: round, broad, suspensory, and uterosacral ligaments; ovaries; ureters.

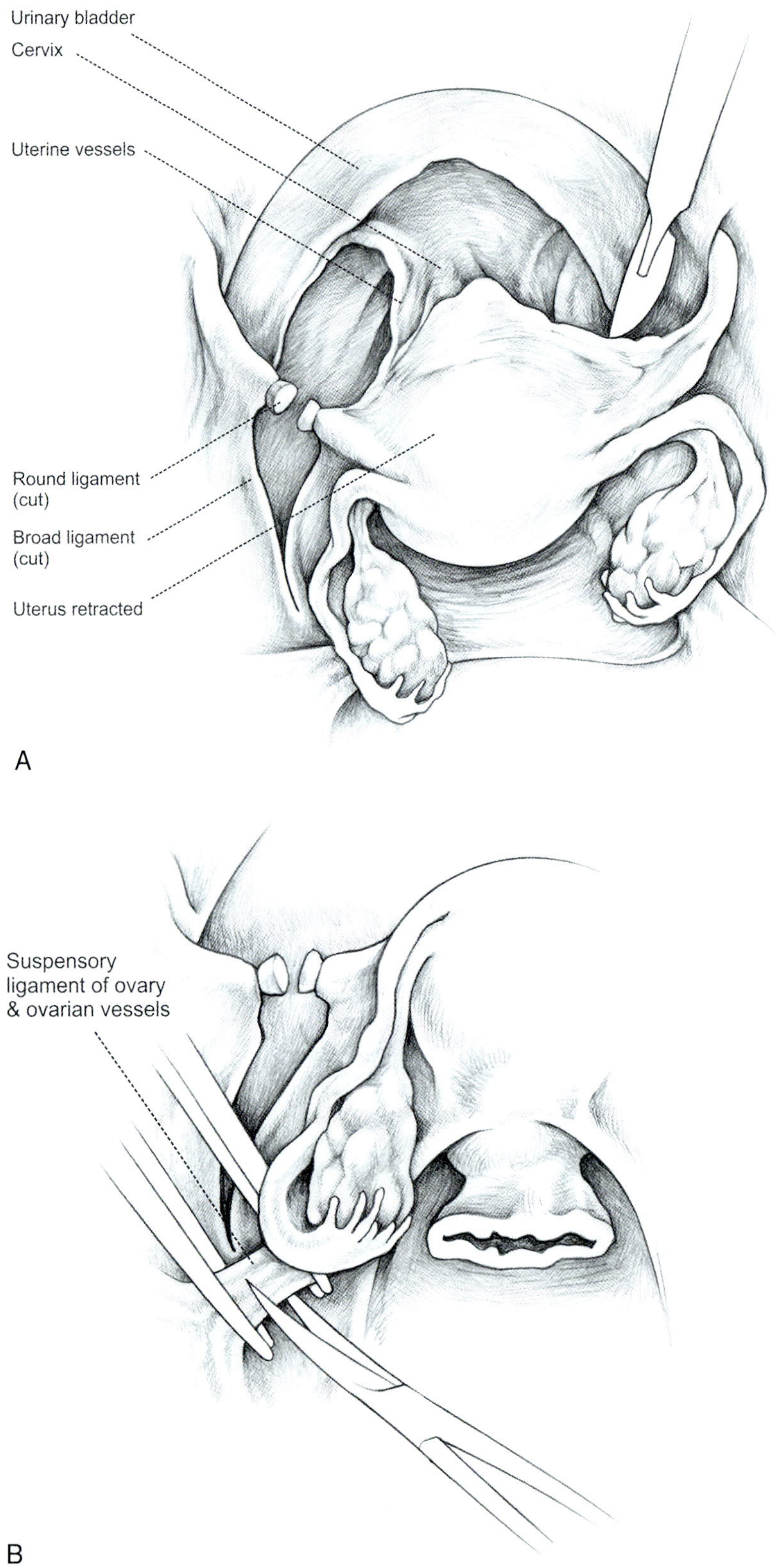

Figure 9.2 **A:** Incise the broad ligament. **B:** Suspensory ligament of the ovary contains the ovarian vessels.

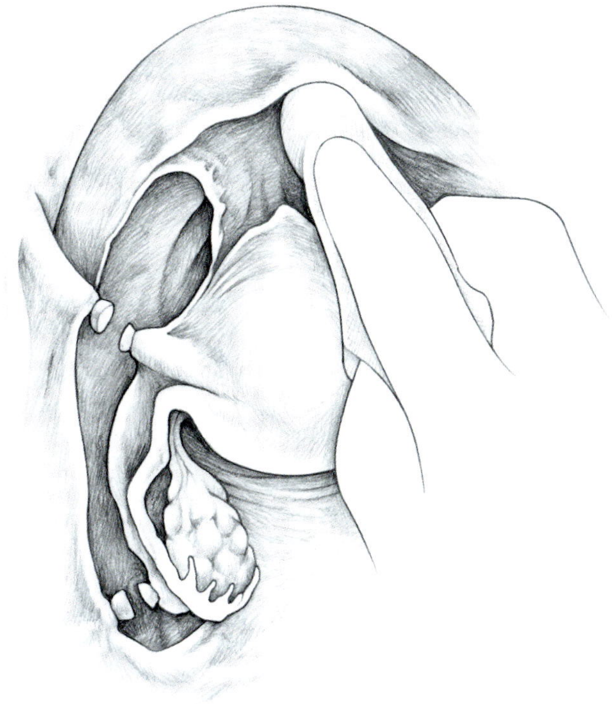

Figure 9.3 Dissection of the bladder off lower uterine segment, cervix, and upper vagina.

directed toward the uterus to avoid bladder injury. This can be accomplished sharply with Metzenbaum scissors or bluntly with a sponge-covered finger (Fig. 9.3).

With continued upward traction on the uterus, deviate to one side to expose the uterine vessels. Place two Heaney clamps at right angles to the vessels with tips tightly against the uterus. Note the location of the ureter below this pedicle (Fig. 9.4). Care should be taken to leave 5-7 mm between the clamps to allow transection (Fig. 9.5). Place a Heaney stitch around the uterine artery pedicle, tightening the suture as the clamp is slowly released (Fig. 9.6). Repeat on the contralateral side.

Palpate the distal end of the cervix by placing a hand around the lower uterine segment while maintaining upward traction utilizing the pean clamps across the cornua. This will identify the top of the vaginal cuff when the specimen is removed. Any additional dissection of the bladder from the cervix and upper vagina should be completed. Identify, clamp, and transect the cardinal ligaments bilaterally. Identify the uterosacral ligaments posterolaterally (Fig. 9.7). Place curved Heaney clamps across the vaginal cuff angles bilaterally taking care to

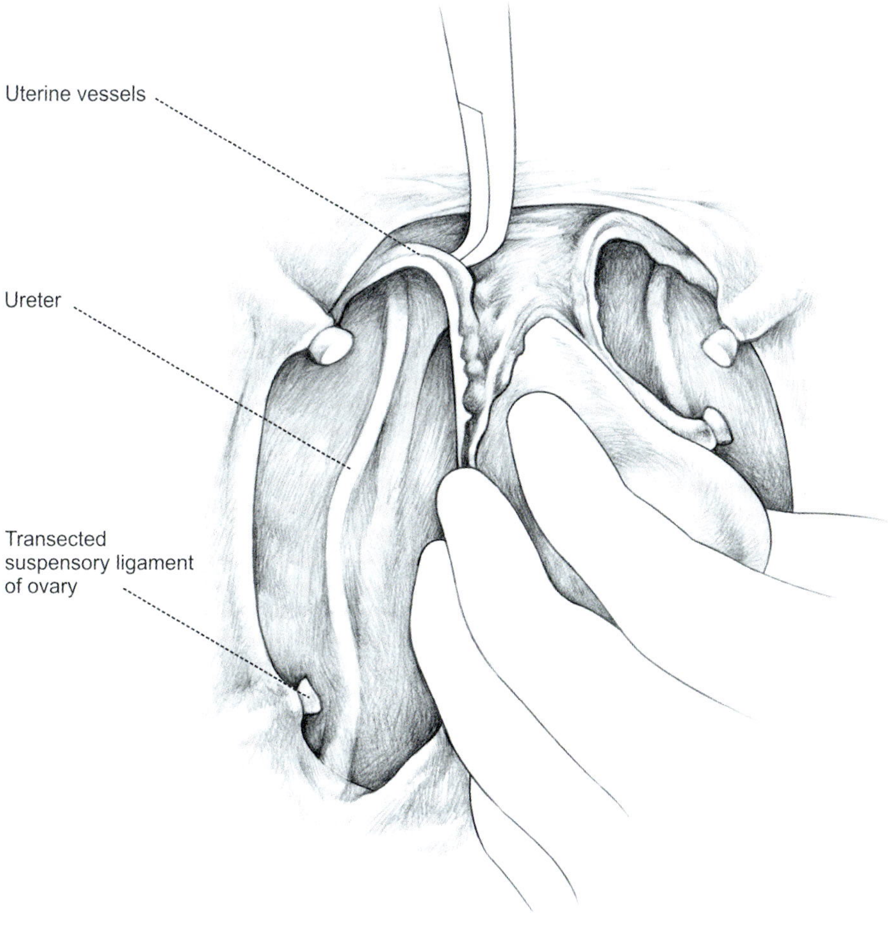

Figure 9.4 Identify the ureter, separate, and divide the uterine vessels.

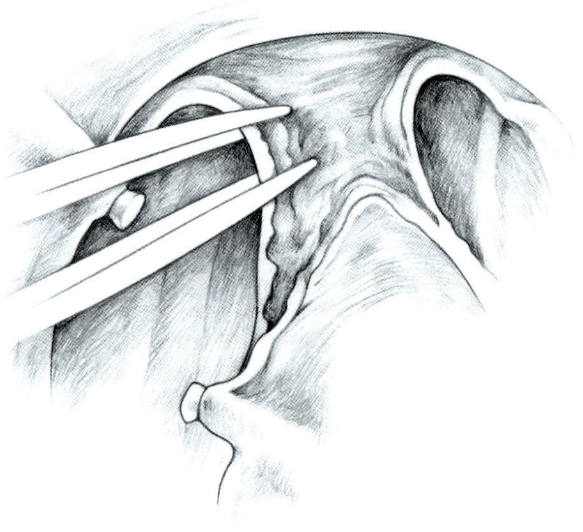

Figure 9.5 Detail of uterine vessel control prior to division.

Figure 9.6 Detail of Heaney stitch to secure uterine vessels.

incorporate the uterosacral ligaments in these pedicles. This is important for cuff suspension and prevention of future vaginal prolapse. The Heaney clamps should be placed snuggly beneath the cervix so as not to foreshorten the vagina (see Fig. 9.7). Divide the uterosacral ligaments (Fig. 9.8A).

Remove the uterus, cervix and attached adnexal structures with a heavy, curved scissors using the cuff angle clamps as a guide (Fig. 9.8B). Once the specimen is removed, grasp the middle of the cuff (between the Heaney clamps) with a long Kocher clamp ensuring both the anterior and posterior vaginal mucosal layers are included. The vaginal cuff is then closed and suspended by placing a Heaney stitch around each angle clamp incorporating the uterosacral ligament (see Fig. 9.8A). The middle of the cuff can then be closed with figure of eights sutures.

CLINICAL HIGHLIGHTS

- Clinically, location and type of incision are based on (a) operative procedures planned concurrent with hysterectomy, (b) size of the uterus to be removed, (c) cosmesis, and (d) strength of healed incision and subsequent risk of hernia formation.

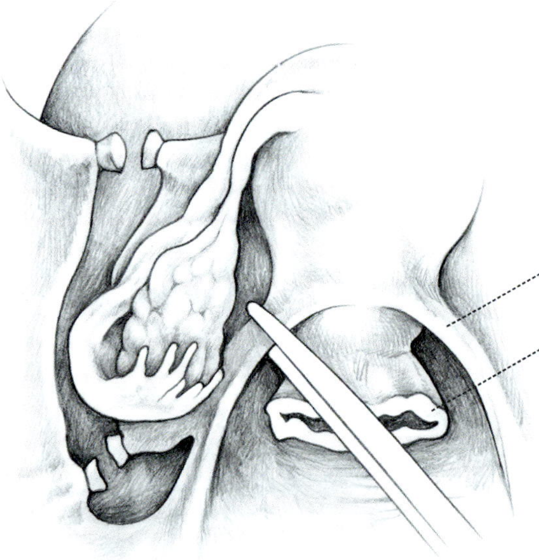

Figure 9.7 Identify uterosacral ligaments.

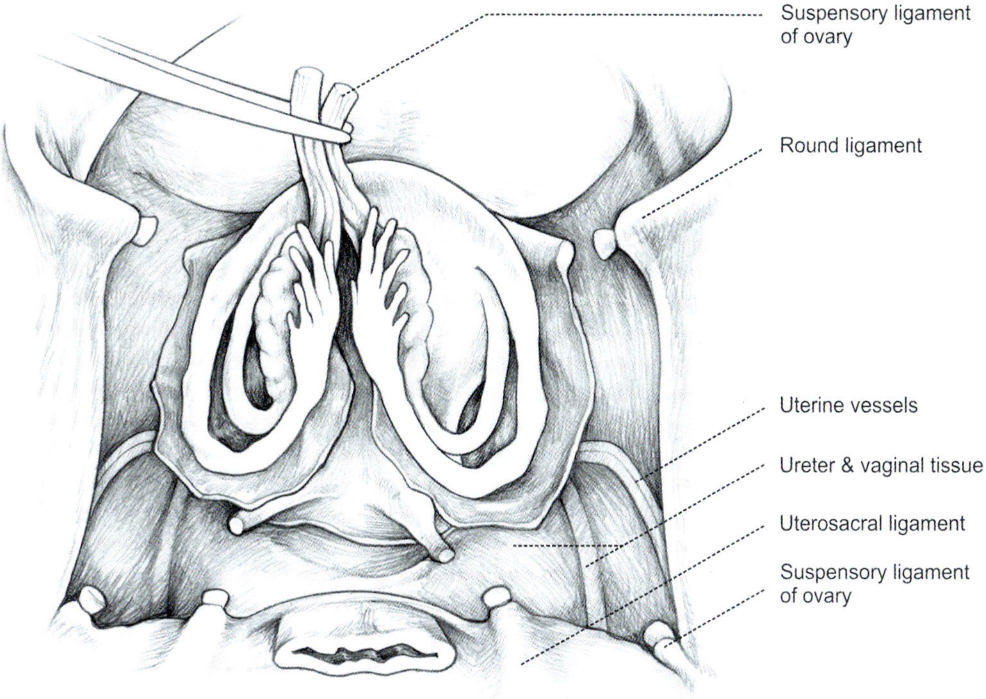

A

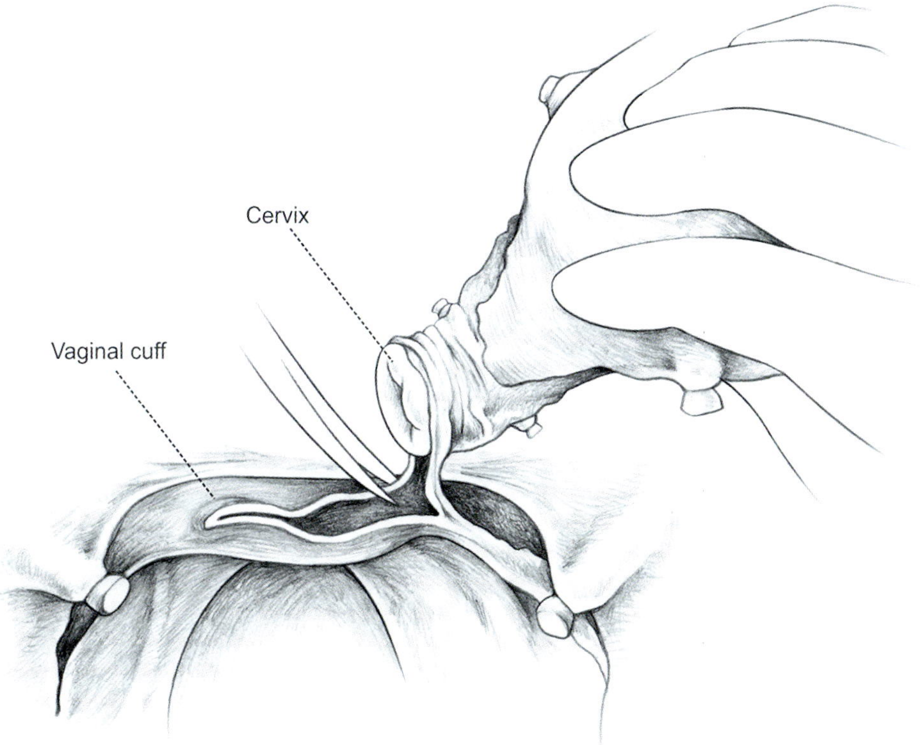

B

Figure 9.8 **A:** Divide the uterosacral ligaments. **B:** Divide the vagina at the level of the cervix.

- Divide round ligaments at the internal inguinal ring if there is any question of a malignant diagnosis.
- If there is a question of contamination or intra-abdominal sepsis at the time of surgery, consider leaving a 1-cm opening in the middle of the vaginal cuff open after closure of the peritoneum.
- Be certain to identify the ureters at the level of the suspensory ligament of the ovary as they cross over the pelvic brim and medial to the ligament.
- Be aware of the proximity of the ureter as it courses beneath the uterine blood supply—"water under the bridge" (see Fig. 9.7).
- Upward traction on the uterus throughout the procedure will facilitate dissection of tissue planes to safely complete the hysterectomy.
- Fully separate the bladder from the vagina below the level of the cervix prior to removal of the uterus to avoid injury to the bladder dome.
- Suspend the vaginal cuff by incorporating the uterosacral ligaments into the cuff angles to prevent future vaginal prolapse in the clinical setting.

URINARY DIVERSION

- See Chapter 8.

TESTICLES (ORCHIECTOMY)

INCISION

Begin the skin incision at the pubic tubercle and extend laterally on a line 45° from vertical for 9 cm (Fig. 9.9).

OPERATIVE PROCEDURE

Continue your incision down to the external oblique fascia and find the external inguinal ring. Incise the external oblique fascia in line with the fibers with a scalpel (see Fig. 4.1). Use forceps and Metzenbaum scissors in the same way you begin an open inguinal hernia taking care to preserve the ilioinguinal nerve. Incision and initial dissection mirror beginning steps for open inguinal hernia. Slit the external oblique in line with its fibers to a point above the internal ring. Isolate the spermatic cord (see Fig. 4.3) and place a tight 0.5-in. Penrose drain around the cord.

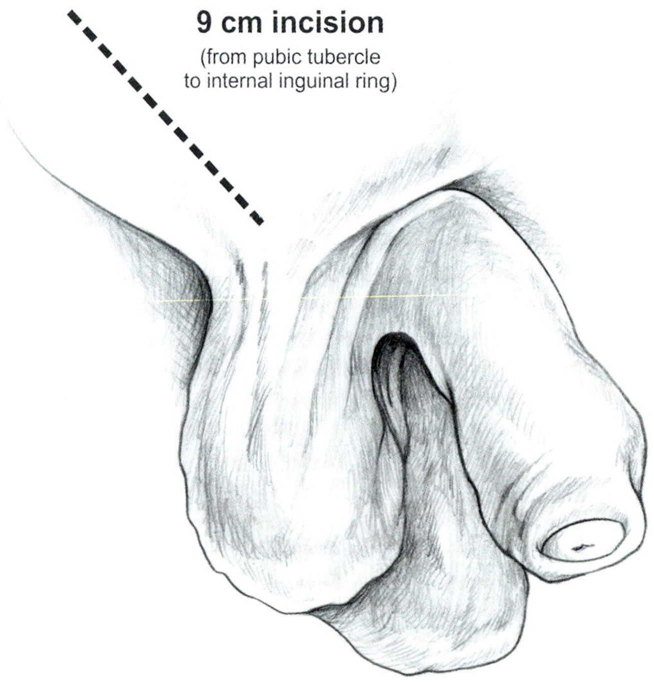

Figure 9.9 Incision for orchiectomy.

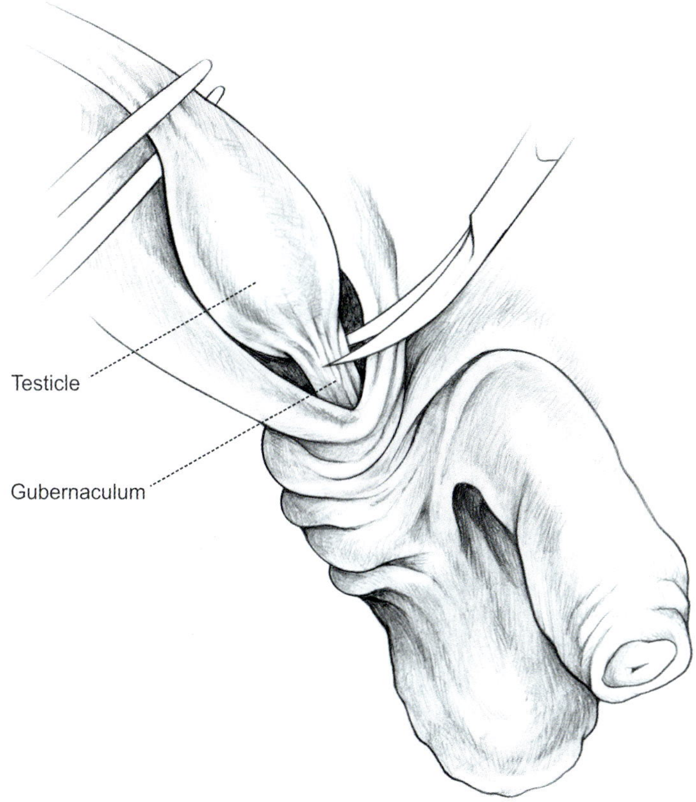

Figure 9.10 Division of the gubernaculum.

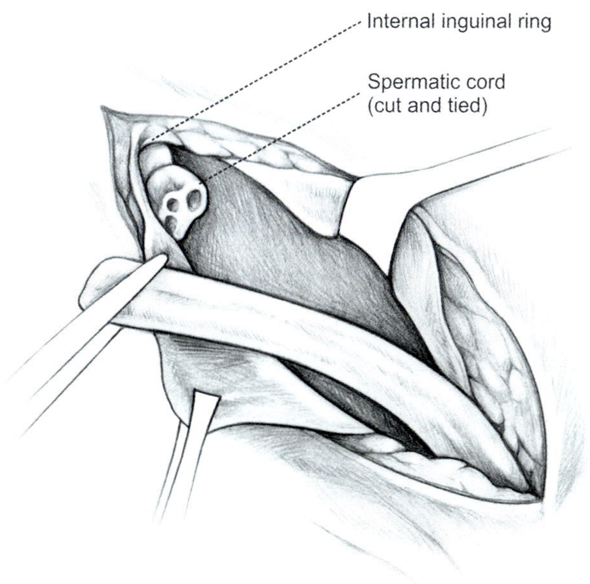

Figure 9.11 Proximal ligation and division of the spermatic cord.

Using traction on the cord, continue to isolate the cord and elevate the testicle out of the scrotum into the surgical field. Blunt and sharp dissection along with scrotal pressure will help to deliver the testicle into the wound (so be certain to prep the scrotum and perineum into the surgical field in the OR). Divide the gubernaculum (Fig. 9.10). Examine the testicle to confirm the diagnosis of a testicular tumor before ligating and dividing the cord (Fig. 9.11). The patient with a testicular cancer may eventually require a retroperitoneal node dissection. Leave some length on the nonabsorbable ligature ends so the node dissection can be determined to have removed all of the cord structures.

CLINICAL HIGHLIGHTS

- Most common presentation of testicular cancer is a painless testicular mass.
- Further surgical treatment for testicular cancer is determined by preoperative imaging, by tumor markers, and by tumor pathology.

- Blood supply to the testicle comes from three vessels forming collateral circulation:
  - Gonadal (testicular) artery
  - Ductus deferens artery
  - Cremasteric vessels

- If the testicular artery is occluded, the testicle will not develop gangrene but will likely atrophy.

- Testicular venous drainage: right to inferior vena cava (IVC) and left to left renal vein (see Figs. 8.4 and 8.5).
  - Unilateral varicocele on the right may indicate right-sided retroperitoneal tumor.
  - Left-sided varicoceles are very common and may be associated with infertility or chronic pain. If the left varicocele is new, it may indicate a left renal tumor.

10

Aorta, Iliac Arteries, and Blood Supply to the Lower Extremity (Vascular Exposure)

Outline

Find a few rubber bands and cut them to open the loop into a single strand. Remember that in the operating room, the rubber bands would instead be atraumatic plastic vessel loops to gently control blood vessels. Tag the newly created vessel loops with a mosquito clamp.

All illustrations in this chapter are depicted of the RIGHT lower extremity.

ASCENDING AORTA, AORTIC ARCH, AND DESCENDING THORACIC AORTA

INCISION

Taking advantage of previously dissected areas, flip the chest plate superiorly, remove the lungs and heart, and expose the ascending aorta (see Fig. 3.2).

OPERATIVE PROCEDURE/OBSERVATION

Place an index finger into the ascending aorta and identify the brachiocephalic (innominate) artery. Clear it off until it bifurcates into the right carotid and right subclavian arteries. Note the large size of the brachiocephalic (innominate) artery in relation to the left common carotid and left subclavian arteries, which are the next branches off the aortic arch. Palpate within the aorta to identify atherosclerotic plaque at the takeoff of the left subclavian artery. Note that the left subclavian artery arises from the posterior aspect of the aortic arch. Opposite the left subclavian, identify the ligamentum arteriosum. Identify the left recurrent laryngeal nerve distal to the ligamentum arteriosum (see Figs. 3.4 and 3.10).

CLINICAL HIGHLIGHTS

- The brachiocephalic (innominate) artery has a significantly larger diameter than the left subclavian artery. This causes the right subclavian artery to be shorter than the left. Remember that transmitted pressure is dependent on length of the tube and diameter of the tube. Based on a larger diameter of the brachiocephalic (innominate) artery and a shorter right subclavian artery, the right brachial artery pressure will more accurately reflect pressure in the aorta. This will also avoid falsely lower blood pressures in the left subclavian due to more common atherosclerotic stenosis at the takeoff of the left subclavian. Thus, all blood pressures on adults should be taken from the right arm to assure accuracy.

- Trachea and brachiocephalic (innominate) artery are adjacent to each other. Tracheal erosion from excessive pressure in an endotracheal tube can cause a tracheoinnominate fistula (see Fig. 3.10).

- So-called bovine arch anatomy refers to a common origin for the innominate and *left* common carotid arteries in 27% of people.

- Remember that the right recurrent laryngeal nerve loops under the right subclavian artery. Right subclavian aneurysmal disease, trauma or nerve invasion from malignant disease may induce a paresis or paralysis of the right vocal cord.

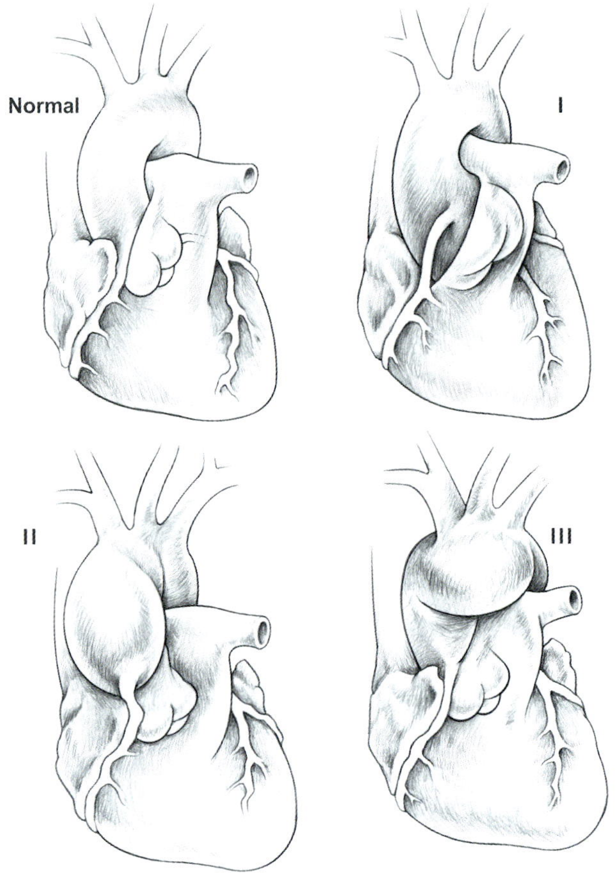

Figure 10.1 Ascending aortic aneurysms: type I, II, or III.

- The left recurrent laryngeal nerve loops under the arch of the aorta distal to the ligamentum arteriosum. Aneurysmal disease of the arch, trauma, or nerve invasion from malignant disease may cause paralysis of the left vocal cord.

- Ascending and aortic arch aneurysms or trauma are classified as type I, II, or III as shown (Fig. 10.1).

DESCENDING THORACIC AORTA

INCISION

Prepare the specimen as noted in number 1 above.

OPERATIVE PROCEDURE/DISSECTION

Dissection of the descending thoracic aorta centers on identification and careful dissection of the intercostal branches. Identify the proximal level at T10 (thoracic vertebra number 10). At this level or one level above or below, the intercostal artery will typically give rise to the anterior spinal artery of Adamkiewicz. Careful dissection of these intercostal tributaries in this area may lead the surgeon to a vessel that tracks along with the nerve rootlet through the foramen to give rise to the anterior spinal artery (Figs. 10.2 and 10.3). Thoracoabdominal aneurysms can take the form of any of four types (Fig. 10.4). The inferior pulmonary ligament forms the division between the proximal and distal descending aorta. Continue exposure of the aorta as it passes through the aortic hiatus of the diaphragm by incising the diaphragm radially toward the aorta and then divide the diaphragmatic crus. The esophagus is anterior to the thoracic aorta. Segmental blood supply to the esophagus may be identified leading directly from the aorta to the esophagus (see Fig. 10.2).

CLINICAL HIGHLIGHTS

- Blood supply to the spinal cord is typically predominantly supplied at the level of T10. That can vary anywhere proximally on the descending aorta to all the way down to the right internal iliac artery. Typically, the artery of Adamkiewicz is supplied at T10 or within two intercostal levels. However, it may infrequently be supplied by lumbar arteries from the abdominal aorta or even below that at the level of the iliac arteries.

- Be aware that the greatest risk for paralysis postoperatively occurs from thoracic aortic repair. Less likely is abdominal aortic repair, followed by least likely iliac arterial surgery.

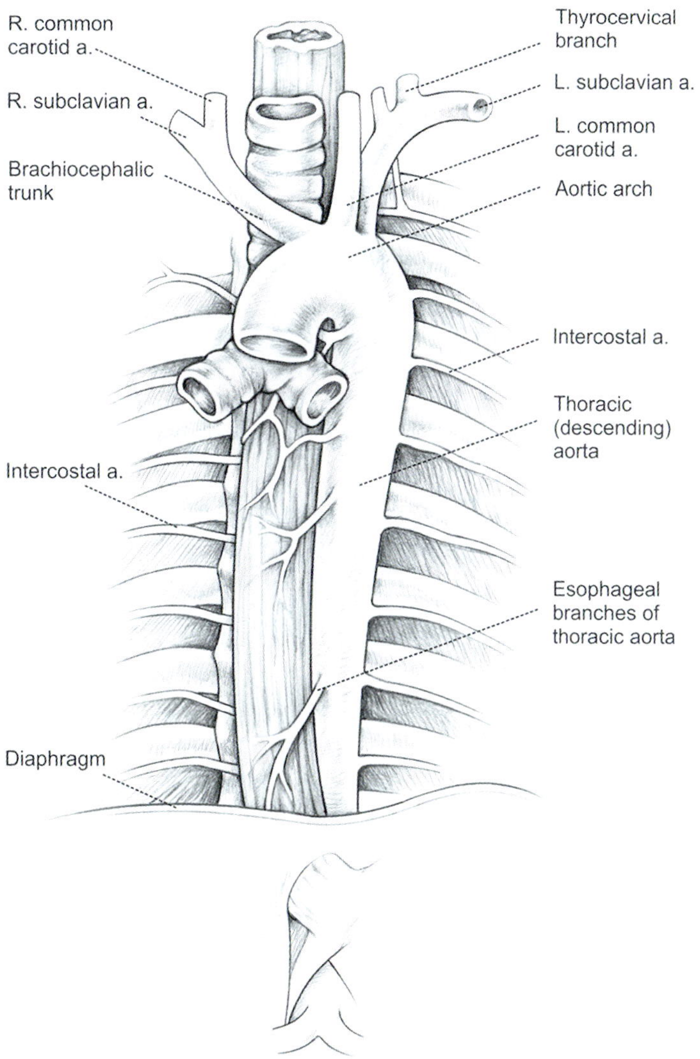

Figure 10.2 Aortic segmental branching pattern to esophagus and intercostal arteries.

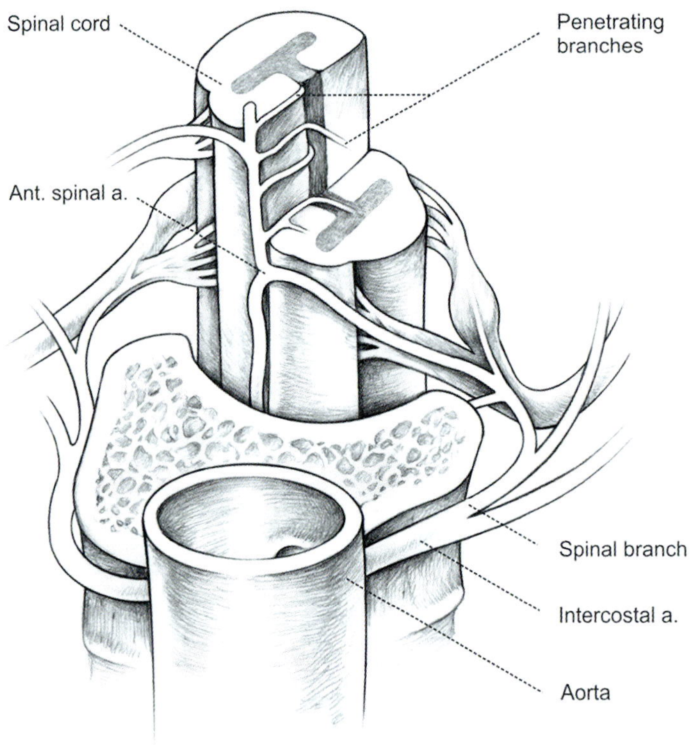

Figure 10.3 Anterior spinal artery of Adamkiewicz.

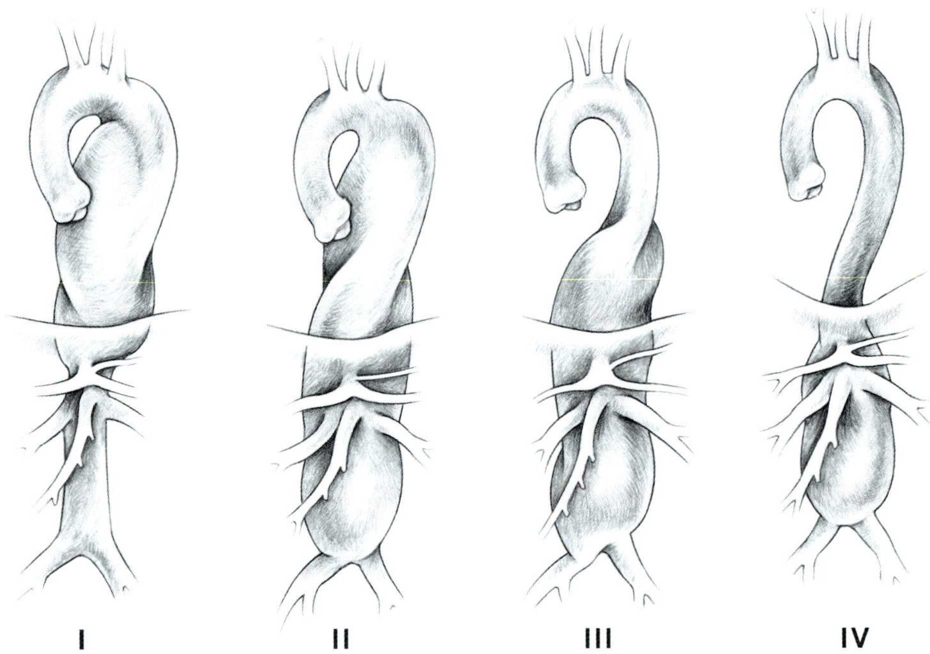

Figure 10.4 Thoracoabdominal aortic aneurysms: type I, II, III, or IV.

- The anterior spinal artery of Adamkiewicz is the predominant blood supply to the spinal cord as shown (see Fig. 10.3). Note the penetration of the spinal cord by the anterior spinal artery for blood supply.

- Blood supply to the spinal cord via the anterior spinal artery is also segmental. It begins with contributions from the vertebral arteries proximally, intercostal arteries, and lumbar arteries distally.

ABDOMINAL AORTA

INCISION

Prepare the specimen as in section number 1 of this chapter. Also, flip the abdominal wall inferiorly and remove portions of previously dissected organs to gain clear visualization of the abdominal aorta (see Figs. 3.2 and 5.1).

OPERATIVE PROCEDURE/DISSECTION

Beginning at the level of the diaphragm, clean off the abdominal aorta to identify the segmental branches in this order: celiac axis, superior mesenteric artery (SMA), right and left renal arteries, left and right gonadal arteries, and the inferior mesenteric artery (IMA). Starting at the top, note that you may find ascending phrenic vessels coming directly off the aorta. The first major branch will be the celiac axis giving off the splenic, left gastric, and common hepatic arteries. The celiac ganglia and plexus are found in this area. A great deal of dense soft tissue coverage may be encountered while dissecting in this area including the median arcuate ligament and celiac plexus nerve tissue (see Fig. 8.3).

Distal to the celiac axis, identify the SMA. If not previously performed, identify the tributaries to the small bowel and colon. Very small branches to the right and left adrenal glands may be encountered. The next major arterial supply is to the right and left kidneys. Note that the right kidney is supplied by the right renal artery, which passes posterior to the inferior vena cava (IVC) (see Fig. 8.3). The left renal vein, which is a major branch from the IVC passes anterior to the aorta below the left renal artery (see Fig. 8.1). Below this level, the gonadal arteries, first the left and then the right, takeoff directly from the aorta.

The IMA is the last major branch from the abdominal aorta (see Fig. 5.7). Also note segmental arterial supply to the ureters from the renal arteries, right and left gonadal, direct branches from the aorta and iliac arteries, and ascending branches from the superior vesical artery (Fig. 8.4). In a fresh specimen, any of these major branches off the aorta may be identified and followed distally to their target organs. However, in our setting, the surgeon has likely already removed most of these organs.

CLINICAL HIGHLIGHTS

- Remember that chronic atherosclerotic intestinal ischemia must have significant disease in two of the three major arteries that supply the gut: celiac axis, SMA, and IMA.

- SMA ischemia can be either chronic atherosclerotic or acute embolic. On cross-table lateral angiogram, atherosclerotic disease will appear to be an arrowhead pointed distally, and embolic disease will be a meniscus pointed proximally.

- Remember the right renal artery passes posterior to the IVC. The left renal vein is the typical proximal extent of abdominal aortic aneurysms (AAA). The left renal vein is inferior to the left renal artery.

- The IMA is typically ligated and usually is not reperfused during abdominal aortic aneurysm repair. 7% of patients will have ischemia of the sigmoid and left colon due to inadequate collateral circulation and will require reimplantation of the IMA.

COMMON, EXTERNAL, AND INTERNAL ILIAC ARTERIES

INCISION

Expose the pelvic arteries as noted in number 3 above.

OPERATIVE PROCEDURE/DISSECTION

Open the peritoneum over the external iliac artery from the origin of the common iliac artery to the inguinal ligament. Note the takeoff of the deep circumflex iliac artery and the inferior epigastric artery (Fig. 10.5). Note that relatively few branches come off the external iliac artery and that it lies anterior and lateral to the external iliac vein. In a like fashion, identify and remove the overlying tissue from the internal iliac artery and its branches. Identify the following branches: superior vesical, obturator, internal pudendal, and superior and inferior gluteal arterial branches (see Fig. 10.6). Female

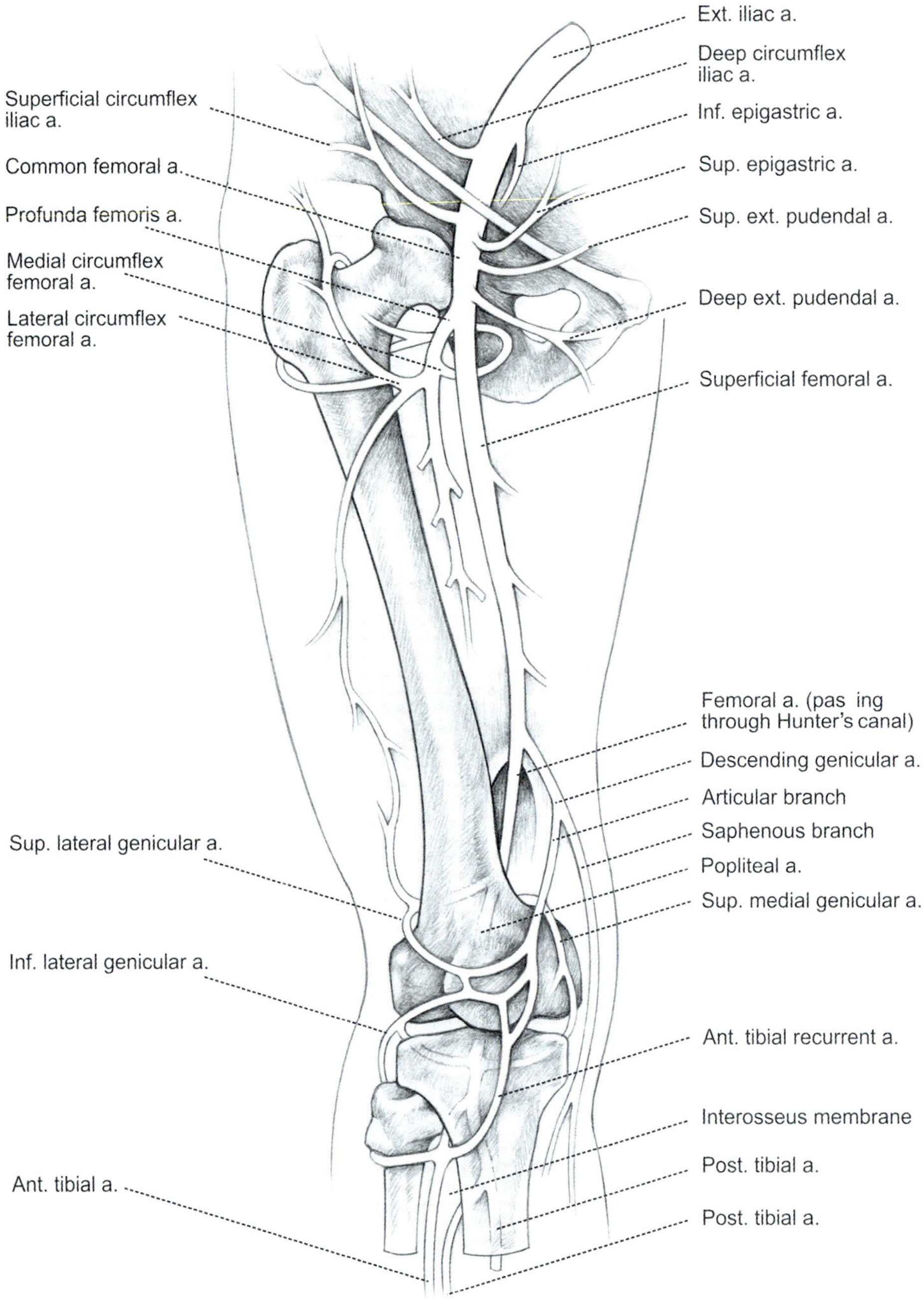

Figure 10.5 Illustration showing femoral artery and collateral circulation of hip, thigh, and knee.

specimens will have uterine arterial branches. Male specimens will have the vas (ductus) deferens traversing the pelvis from an inferior posterior to superior anterior position. This diagonal run of the vas deferens may be identified in the following diagram (see Fig. 4.10).

CLINICAL HIGHLIGHTS

- Common iliac arterial disease is most commonly treated using endovascular techniques.
- Pelvic perfusion requires only one intact internal iliac artery. Trauma or exsanguinating pelvic hemorrhage may be treated by unilateral internal iliac artery ligation. However, note that rarely the anterior spinal artery of Adamkiewicz can have its origin from an internal iliac artery.
- Severe unilateral iliac disease may be treated by a femoral-femoral bypass graft. This graft avoids entrance into the abdominal cavity.

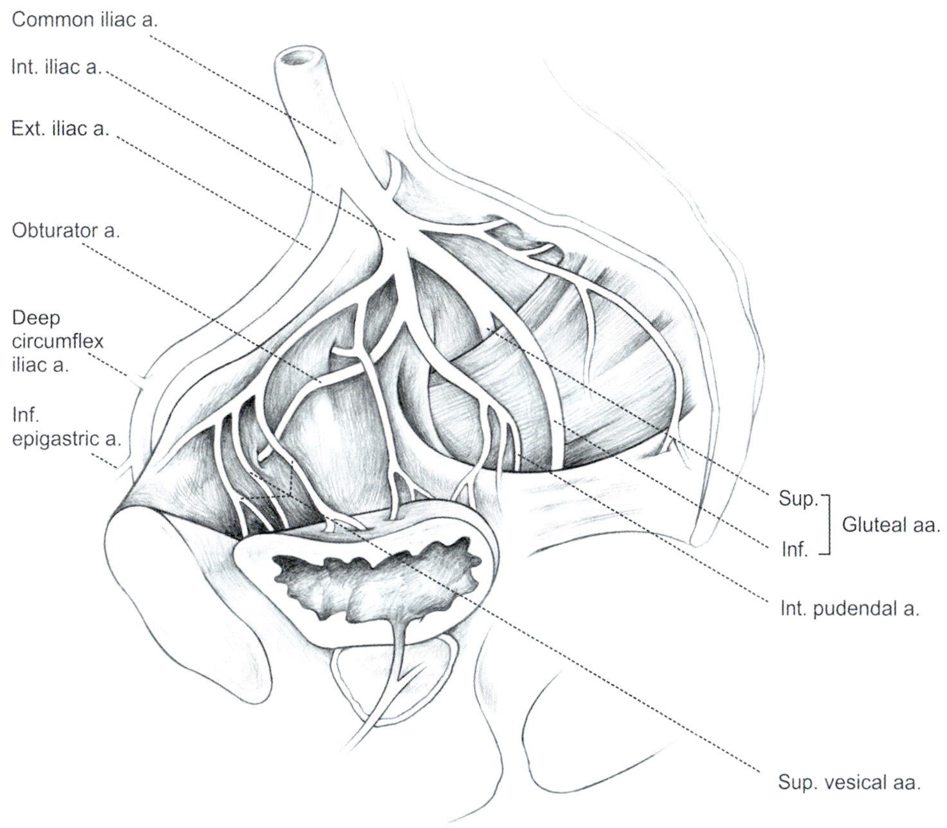

Figure 10.6 Side view of branches of internal iliac artery.

- When performing aorto-bi-iliac bypass graft, be sure to reperfuse at least one internal iliac and both external iliac arteries.
- Remember that the ureters typically cross over the iliac arteries near the bifurcation of internal and external iliac arteries.
- Further information on the obturator artery and canal can be found in Hernia, Chapter 4, Section 4.

COMMON, SUPERFICIAL, AND DEEP FEMORAL ARTERIES

INCISION

A 10-cm vertical incision placed midway between the symphysis pubis and the anterior superior iliac spine (ASIS) is to be made. The dissection is carried deeply to the femoral vessels (Fig. 10.7). Vascular surgeons in the OR may make this incision obliquely.

OPERATIVE PROCEDURE/DISSECTION

Remember, the NAVEL acronym (Fig. 10.8). The femoral nerve lies lateral to the femoral artery followed by the femoral vein in medial progression. Clinically, important named branches will arise from the femoral artery either laterally or posteriorly. From the common femoral artery, one may optionally identify the superficial circumflex iliac artery arising laterally and the superficial epigastric artery arising medially. Many other muscular branches such as the superficial external pudendal artery and the deep external pudendal artery may be seen arising from the medial common femoral artery. These branches may become clinically significant as collaterals with progressive peripheral vascular disease (see Fig. 10.5).

The common femoral artery and superficial femoral artery should be exposed from their anterior surface (Fig. 10.8). Continue dissection inferiorly until the artery descends behind the sartorius muscle. Encircle the superficial femoral artery distally and dissect it proximally. Take care to identify the deep femoral artery (profunda), which will arise laterally

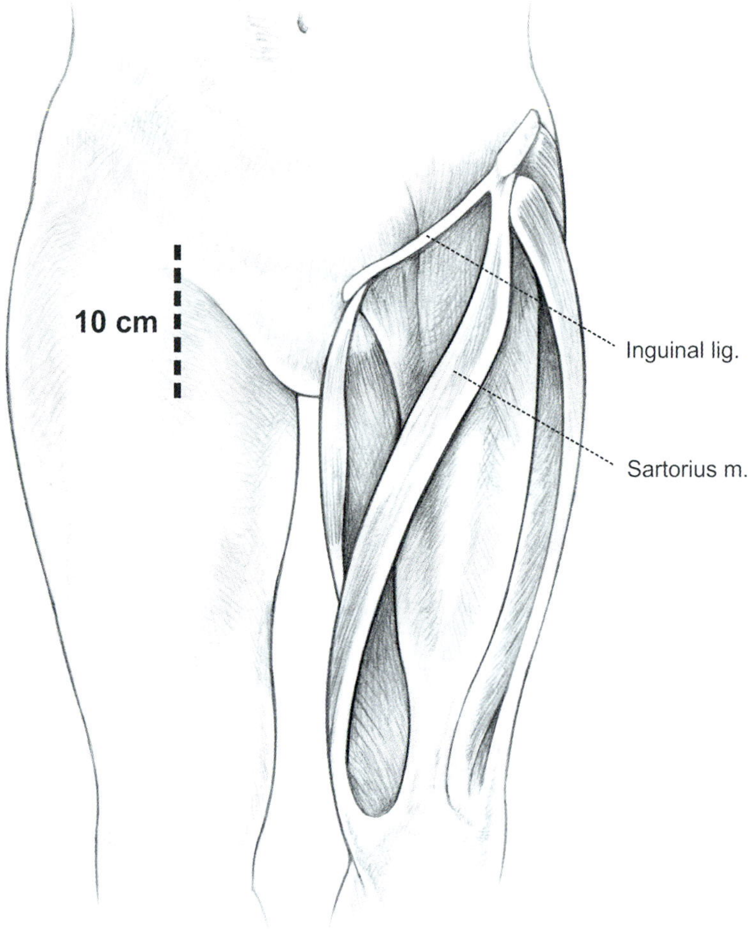

Figure 10.7 Incision for exposure of femoral vessels; insertion and course of sartorius muscle.

(Figs. 10.5 and 10.9). Continue proximal circumferential dissection of the common femoral artery to the ilioinguinal ligament. The ilioinguinal ligament may be divided vertically 1-2 cm to gain exposure to the distal external iliac artery. Proceed with exposure of the distal external iliac artery. Remember in the operating room to leave a portion of the ilioinguinal ligament intact if possible. For purposes of cadaver dissection, this ligament may be entirely divided. A femoral limb of an aortobifemoral graft may pass through this empty space area adjacent to the common femoral artery (Fig. 10.9).

Next, identify the deep femoral (profunda femoris) artery and dissect it distally as it passes adjacent and into the pectineus and adductor longus muscles. The proximal profunda femoris artery (PFA) can be approached anteromedially (proximal PFA) or anterolaterally (distal PFA) to the sartorius muscle. Deep to the superficial femoral vessels between the vastus medialis and the abductor longus muscles, expose the profunda femoris vessels. Watch for branches of the femoral nerve. Optional: Identify the medial circumflex femoral artery and the lateral circumflex femoral artery. You may identify the ascending, transverse, and descending branches of the lateral femoral circumflex artery.

Sartorial muscle flaps are very useful in the setting of postbypass groin infection. It is a vascularized soft tissue flap used to cover infected grafts. The sartorius muscle originates on the ASIS. Bluntly isolate sartorius at its origin and transect it off ASIS. Mobilize it sufficiently to cover the common femoral artery and suture the medial edge of the sartorius to the inguinal ligament to complete the flap (Fig. 10.10).

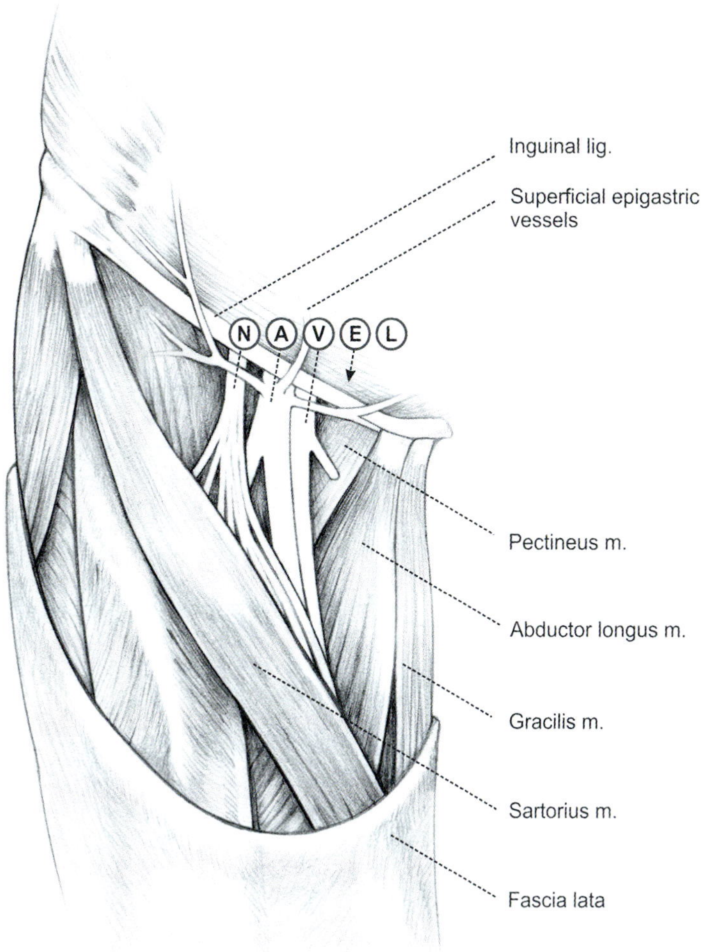

Figure 10.8 NAVEL lateral to medial at the groin.

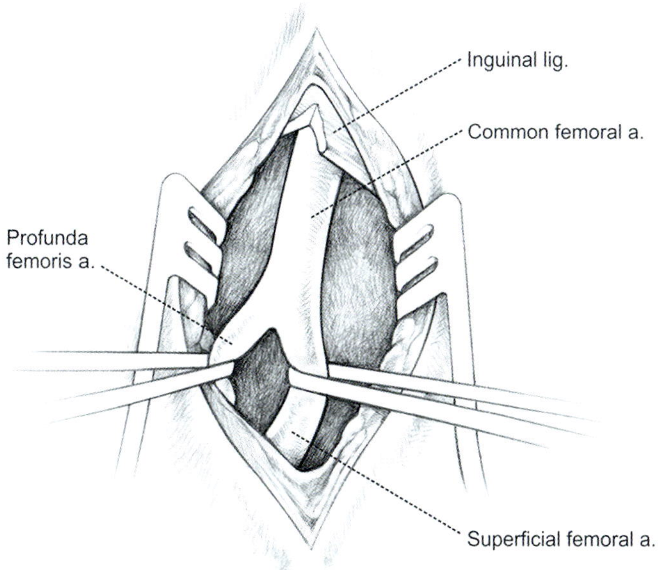

Figure 10.9 Common, superficial, and deep (profunda) femoral arteries with vessel loops.

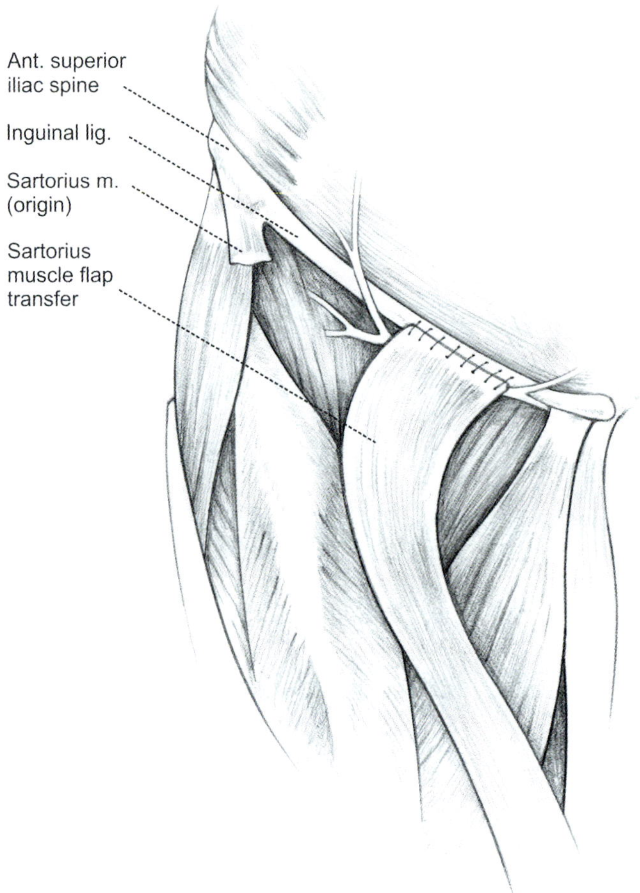

Figure 10.10 Sartorius muscle flap transposed from ASIS to ilioinguinal ligament.

CLINICAL HIGHLIGHTS

- Remember the acronym NAVEL (see Fig. 10.8).
- This acronym stands for Nerve, Artery, Vein, Empty space, Lacunar ligament. This acronym identifies the structures of the femoral canal beginning laterally and progressing medially.
- The femoral Vein is always medial to the femoral artery.
- The Empty space medial to the femoral vein is where femoral hernias can occur.
- The sartorius muscle may be mobilized and retracted medially to give good soft tissue coverage when faced with groin infection after arterial bypass.
- Remember when mobilizing the sartorius muscle that the lateral femoral cutaneous nerve is located on the proximal portion of this muscle 1.5 cm medial to the ASIS.
- Remember, to gain arterial exposure for limb salvage, any muscle may be dissected and mobilized or transected to accomplish this task.
- Remember, the blood supply of the sartorius comes from the femoral vessels and enters the muscle posteriorly at the junction of the upper 1/3 and lower 2/3.
- The PFA can be used for inflow for infrainguinal bypass if the groin is hostile from previous procedures or a limited length of autogenous vein is available.
- Optional: Look at the femoral nerve and notice that it fans out into small fibers (see Fig. 10.8).

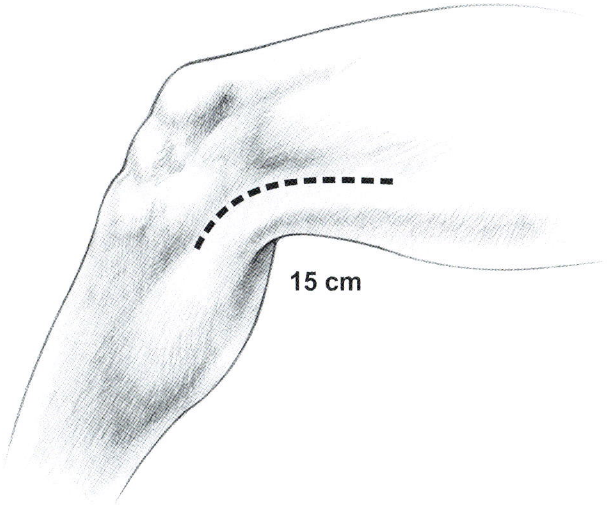

Figure 10.11 Incision for exposure of above-knee popliteal artery.

ABOVE-KNEE POPLITEAL ARTERY

INCISION

The surgeon and an assistant will need to work together to gain this exposure. The legs must be abducted from each other and the dissection leg rotated externally. A 3-foot 2 × 4 board over the contralateral knee or plastic wedge under the contralateral hip may be used to aid rotation of the leg to gain exposure. Now a 15-cm incision may be made in the medial thigh above the knee (Fig. 10.11).

OPERATIVE PROCEDURE/DISSECTION

The dissection is then carried deep to the fascia, which is incised vertically to mimic the skin incision. Incise along the anterior border of the sartorius muscle. The sartorius and semimembranosus muscles are retracted posteriorly. Retract the vastus medialis muscle and adductor magnus tendon anteriorly (Fig. 10.12). For additional proximal exposure, divide the adductor magnus tendon. Notice the adductor hiatus (Hunter's canal) through which the superficial femoral artery traverses to become the popliteal artery. Encircle the popliteal artery and identify descending genicular branches (see Fig. 10.5). These collateral branches may dilate and become significant collateral flow in the presence of peripheral vascular disease. Extend this dissection distally. To gain access to the popliteal artery distally, it will be necessary to mobilize or transect the medial head of the gastrocnemius muscle, but do not divide the muscle for this portion of the dissection. Note that the vein lies deep to the popliteal artery.

CLINICAL HIGHLIGHTS

- The adductor hiatus (Hunter's canal) is a common anatomical landmark for peripheral vascular atherosclerosis.

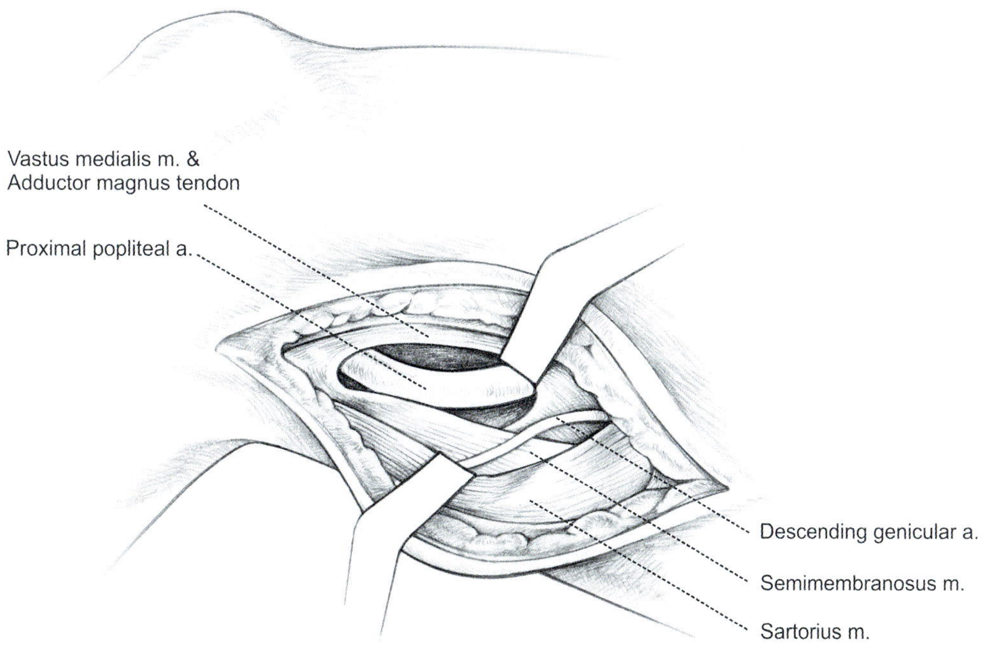

Figure 10.12 Adductor hiatus.

- Branches of the popliteal vein may interfere with popliteal artery dissection. These branch veins may be divided to improve popliteal artery exposure.

- Muscle repair of the medial head of gastrocnemius or any other transected muscle must avoid compromise of any arterial bypass or arterial repair. For all lower extremity bypasses, be sure to have the leg fully extended when measuring length of the bypass conduit.

- Remember that when dealing with limb salvage, exposure is essential. Overlying muscular structures must be retracted or divided to gain arterial exposure.

BELOW-KNEE POPLITEAL ARTERY

INCISION

The surgeon and an assistant may attempt to flex the knee. This may be difficult. If external rotation of the lower extremity is also difficult, a short 2 × 4 board over the contralateral knee or plastic wedge under the contralateral hip may be used as noted in the previous section. A 10-cm medial knee and upper calf incision is performed in the middle of the medial surface. Watch for the greater saphenous vein (Fig. 10.13).

OPERATIVE PROCEDURE/DISSECTION

Incise the tibial (crural) fascia 1 cm. posterior to the tibia. Retract the gastrocnemius muscle posteriorly (Fig. 10.14). The medial head of gastrocnemius inserts on the distal medial femoral condyle. To gain better exposure of the distal popliteal artery in the cadaver anatomy lab, divide the medial head of the gastrocnemius muscle if it has not already been done. For more distal exposure, divide the tibial attachments of the soleus muscle. Dissection is then carried down to the popliteal artery, which is encircled and cleared off medially and distally (Fig. 10.15). For more proximal exposure of the popliteal artery, cut the semimembranosus, semitendinosus, sartorius, and gracilis tendons. The popliteal vein must be retracted if it is in the way. Any branch veins may be divided to gain access to the artery. Note also that collateral geniculate arteries may be also identified. The distal portion of the popliteal artery will descend through the tendinous arch of the soleus muscle (see Fig. 10.15). Divide the soleus muscle to identify the takeoff of the anterior tibial artery, which is located anterolaterally to the distal popliteal artery.

CLINICAL HIGHLIGHTS

- The exposure described above is commonly utilized for femoral-popliteal bypass procedures.

- It is important to be very familiar with this exposure for limb salvage for peripheral vascular disease.

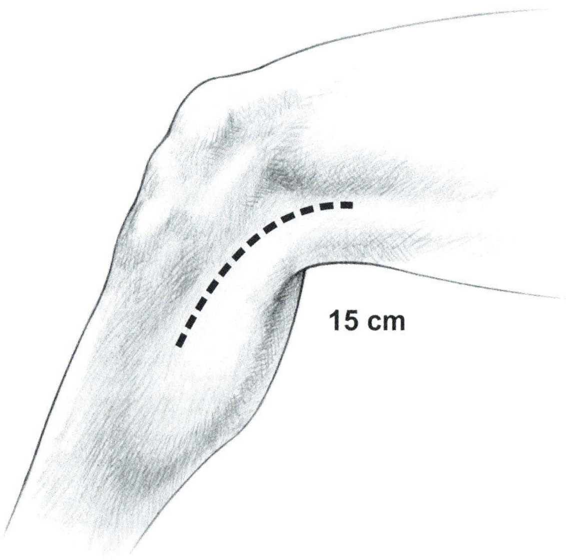

Figure 10.13 Incision for exposure of below-knee popliteal artery.

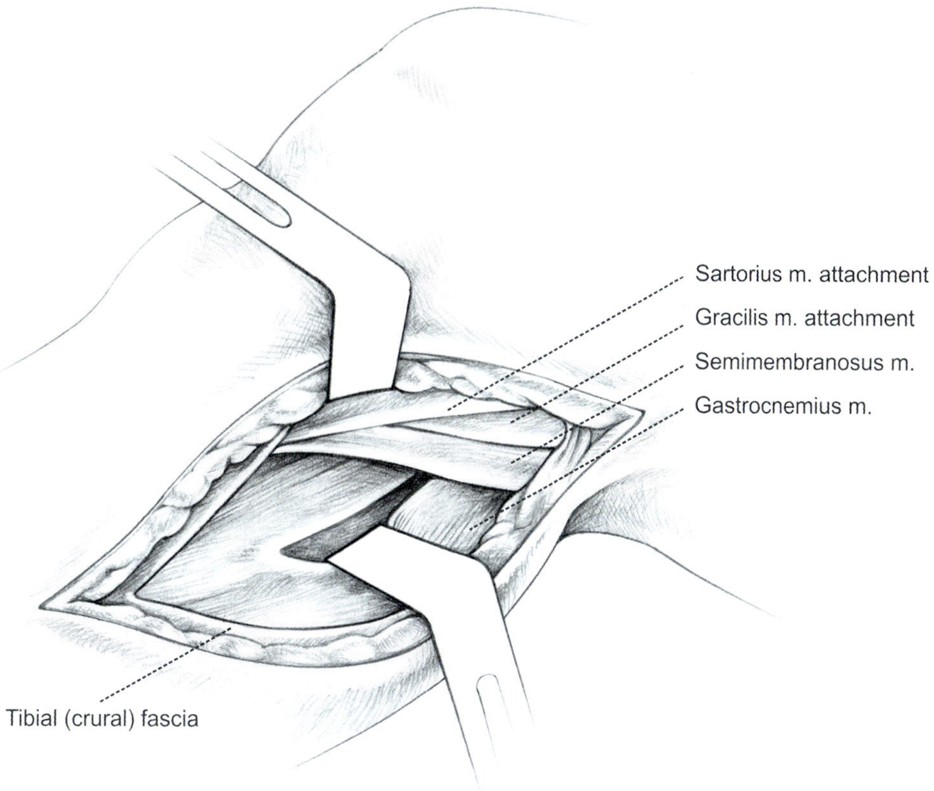

Figure 10.14 Gastrocnemius muscle retraction to expose popliteal artery.

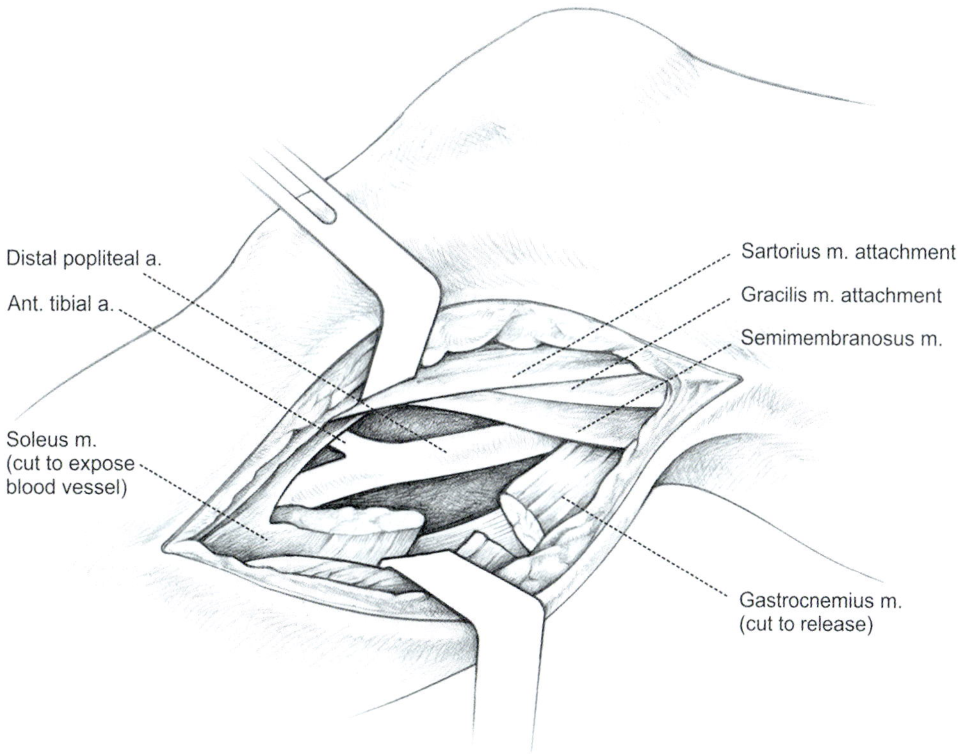

Figure 10.15 Division of medial head of gastrocnemius and soleus to expose distal popliteal artery and trifurcation of tibial arteries.

- Popliteal artery trauma or aneurysms may be approached by the posterior approach to the popliteal artery. The posterior approach is usually a Z incision using a lateral vertical above-knee and medial vertical below-knee limbs connected by a transverse incision through the popliteal fascia. The transverse incision follows Langer lines in this area.

PROXIMAL ANTERIOR TIBIAL ARTERY

INCISION

Straighten the leg for dissection and rotate it medially. A 2 × 4 board or plastic wedge under the ipsilateral hip may aid exposure in this portion of the dissection. A 12-cm incision is placed midway between tibia and fibula in a vertical fashion. Dissection is carried down to the deep fascia, which is incised vertically (Fig. 10.16A).

OPERATIVE PROCEDURE/DISSECTION

Identify the intermuscular septum between the extensor digitorum longus laterally and the tibialis anterior medially (Fig. 10.16B). Using a self-retaining clamp such as a Weitlaner, continue the dissection down to the interosseous septum where the anterior tibial artery and vein will be identified. Expose a 4-cm section of the anterior tibial artery and loop the artery with a vessel loop made from a rubber band. This is how exposure and control of the artery may be maintained in the operating room when using atraumatic vessel loops (Fig. 10.16C).

CLINICAL HIGHLIGHTS

- To gain access for a vascular conduit descending from the proximal thigh, the interosseous membrane must be breached. This breach in the membrane must be fully adequate so compression of the vascular bypass does not occur.
- Remember that the deep peroneal nerve runs alongside the anterior tibial artery.
- Examine the peroneal nerve: common peroneal nerve comes around the fibula 2 fingerbreadths below the fibular head; it innervates the anterior compartment (tibialis anterior and extensor digitorum

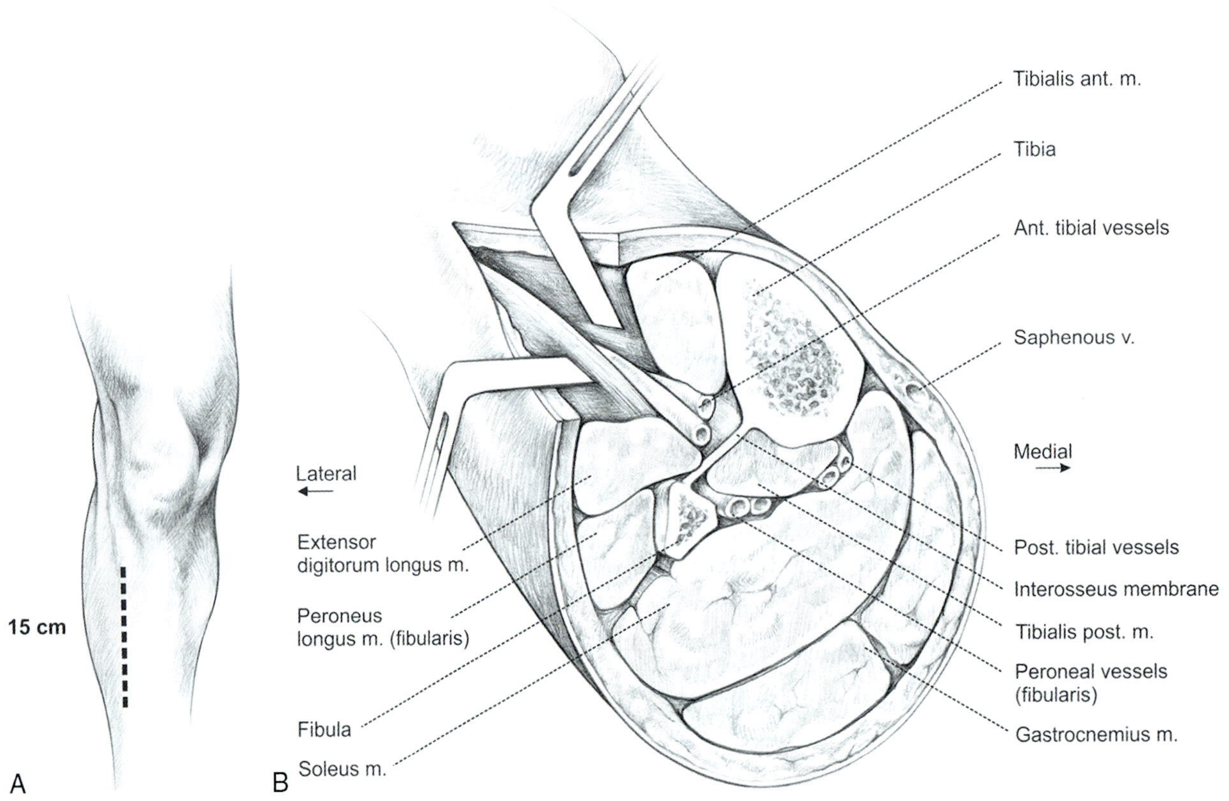

Figure 10.16 A: Incision for exposure of proximal anterior tibial artery. **B:** Cross-section view of exposure of proximal anterior tibial artery.

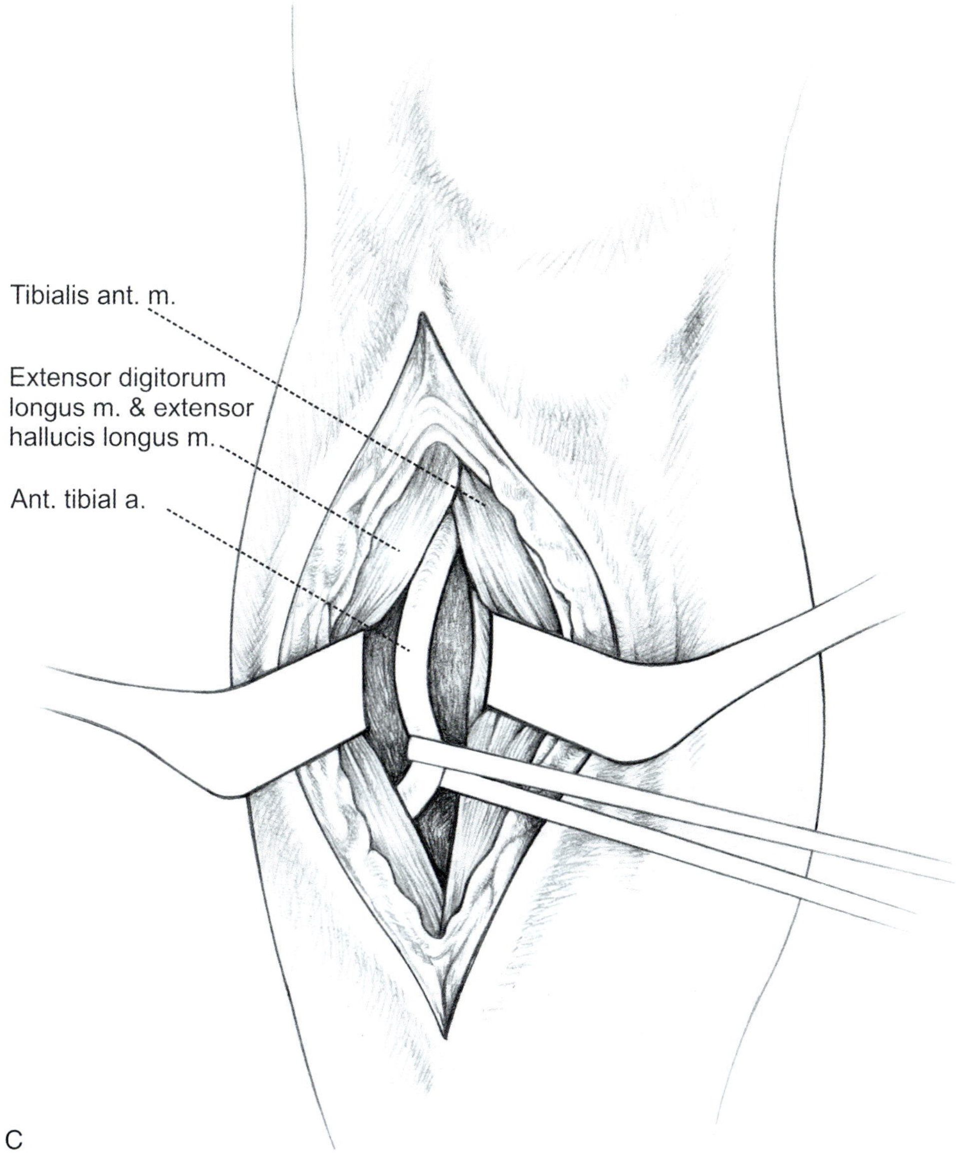

Figure 10.16 (Continued) **C:** Vessel loop controlling proximal anterior tibial artery.

longus); it continues to the first web space for sensory innervation.

- A compartment syndrome may exhibit foot drop or numbness of the first web space of the foot.
- Be sure to pad the fibular head for surgery in the supine position to avoid postoperative foot drop.

DISTAL ANTERIOR TIBIAL ARTERY

INCISION

An 8-cm incision lateral to the tibia at the level of the proximal foot extending superiorly is performed with the leg straight and the foot plantar flexed by an assistant (Fig. 10.17).

OPERATIVE PROCEDURE/DISSECTION

The deep fascial layers are divided vertically and the anterior tibial artery is identified. Remember that the deep peroneal nerve runs alongside the anterior tibial artery. The artery is just lateral to the tibialis anterior tendon, which inserts on the medial foot. Anywhere that a soft patent area of the artery is identified may be used for bypass grafting. This incision will gain exposure to any part of the distal anterior tibial artery (Fig. 10.18).

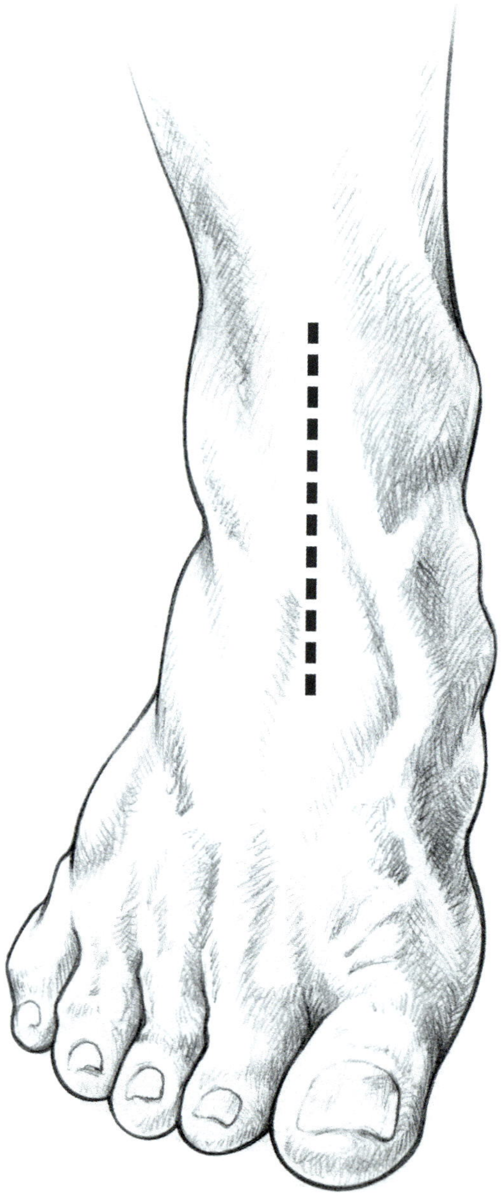

Figure 10.17 Incision for exposure of distal anterior tibial artery.

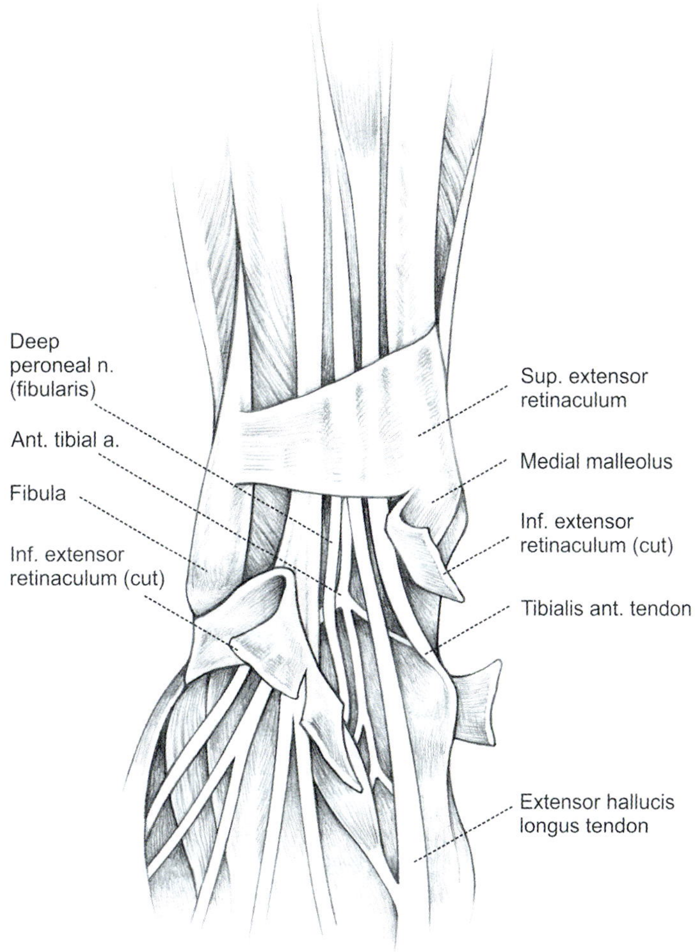

Figure 10.18 Exposure of distal anterior tibial artery and adjacent deep peroneal nerve.

CLINICAL HIGHLIGHTS

- Parchment-thin skin and underlying tissues are commonly found in ischemic conditions of the leg. Very gentle manipulation of these tissues is essential in the operating room to avoid postoperative wound necrosis.

PROXIMAL POSTERIOR TIBIAL ARTERY

INCISION

The lower extremity must be rotated laterally. This may be aided by a 2 × 4 board and a plastic wedge under the contralateral hip (used previously). A 10-cm vertical incision in the skin is performed in the middle of the medial leg below the knee (Fig. 10.19A, proximal).

OPERATIVE PROCEDURE/DISSECTION

Incise the deep fascia vertically and identify the tibia. Follow the tibia posteriorly and retract the gastrocnemius below it. Sweep the soleus and gastrocnemius posteriorly (Fig. 10.19B). Identify a 4-cm section of artery and control it with a vessel loop. Blunt dissection can aid the identification of the vessels by sweeping along the fascial planes to identify the proximal posterior tibial artery.

CLINICAL HIGHLIGHTS

- Distal bypass anastomosis at this level may be rather deep in the muscle mass of the leg. One may consider a somewhat more distal anastomotic level depending on the pattern of atherosclerosis for that patient.

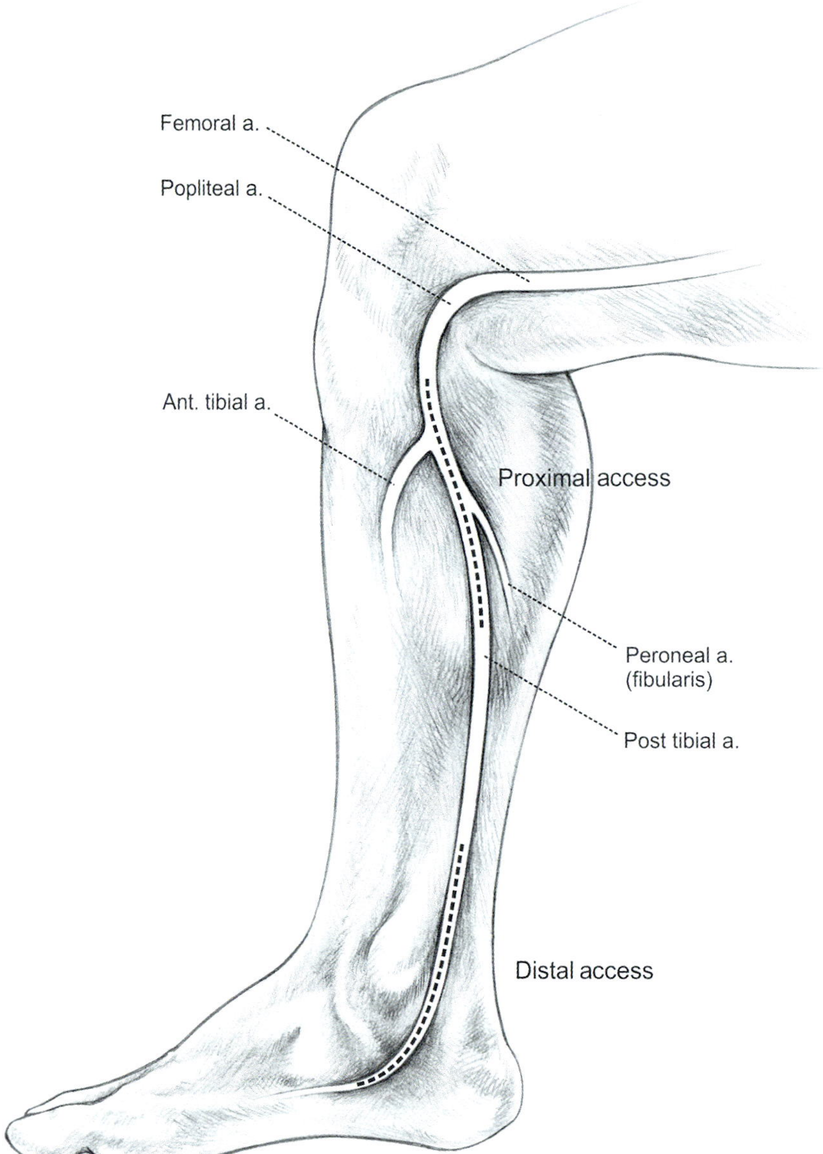

Figure 10.19 **A:** Incisions for exposure of proximal and distal posterior tibial artery.

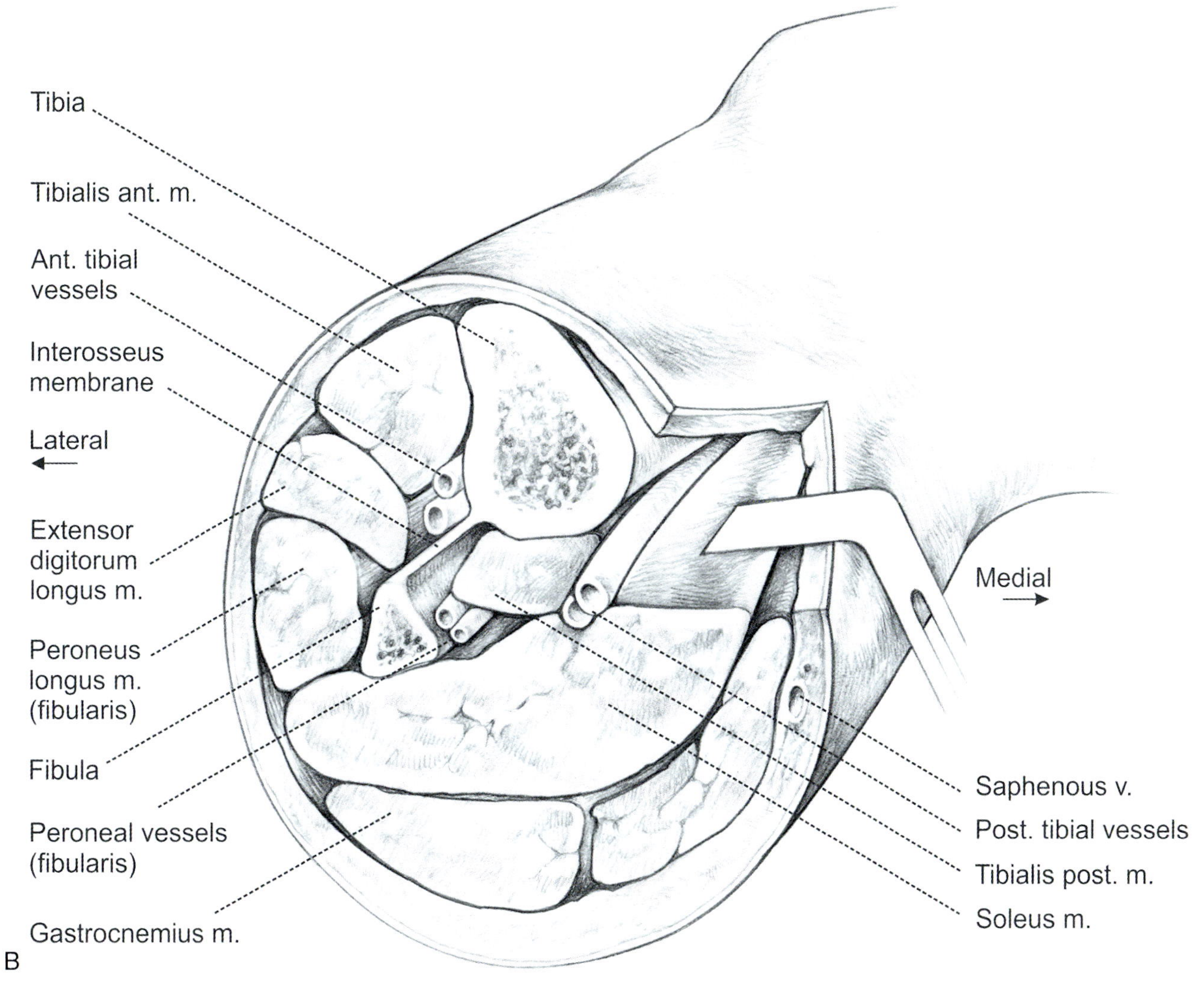

Figure 10.19 (Continued) **B:** Cross-section view of exposure of proximal posterior tibial artery.

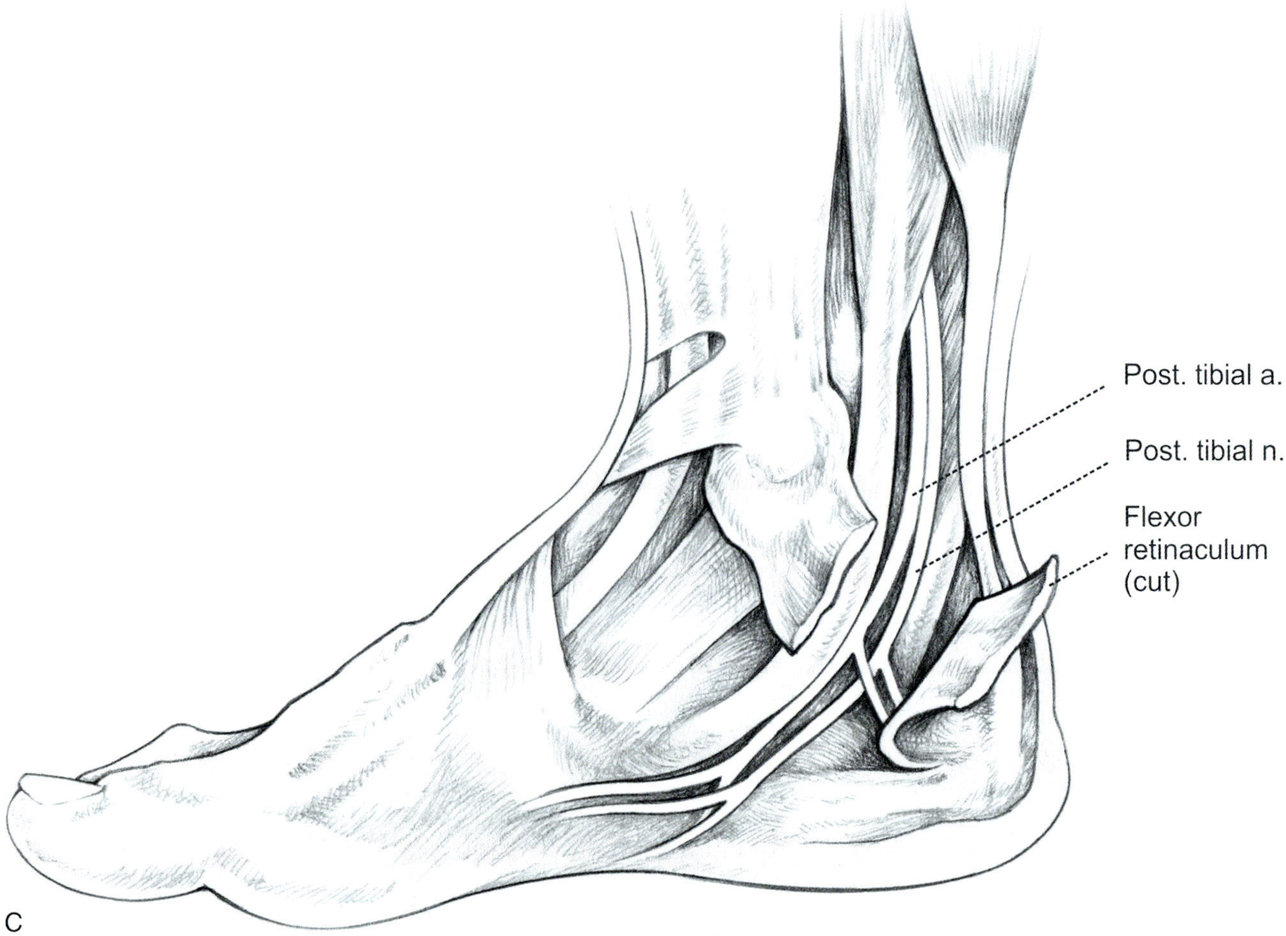

Figure 10.19 (Continued) **C:** Exposure of distal posterior tibial artery and adjacent tibial nerve.

DISTAL POSTERIOR TIBIAL ARTERY

INCISION

The straightened leg must be rotated externally. A 2 × 4 or plastic wedge under the contralateral hip as used previously may be considered. A 6-cm vertical incision posterior to the medial malleolus is performed (see Fig. 10.19A, distal).

OPERATIVE PROCEDURE/DISSECTION

Dissection is carried down through the deep fascia (flexor retinaculum) vertically. The posterior tibial artery is easily identified at this level and cleared for 4 cm (Fig. 10.19C). Use vessel loops to isolate and control the artery. Proximal exposure of the posterior tibial artery may be gained by proximal extension of this incision. The medial soleus muscle may be incised to aid in this exposure.

CLINICAL HIGHLIGHTS

- Distal bypass grafting at this level is aided by the superficial nature of the artery and easy exposure.

PERONEAL ARTERY

INCISION

The leg must be straightened and medially rotated. It will be necessary to place a cloth or a plastic wedge under the ipsilateral hip and have an assistant medially rotate the foot. A 12-cm vertical incision is placed over the middle 1/3 of the lateral fibula (Fig. 10.20).

OPERATIVE PROCEDURE/DISSECTION

Dissection is carried down to the deep fascia, which is incised vertically. Sweep extensor digitorum longus anteriorly and peroneus longus and brevis posteriorly. Identify the fibula with palpation. An 8-cm section of fibula must be cleared of its muscular and fascial attachments sharply and bluntly. Using a rib cutter or a Gigli saw, remove the 8-cm segment of the triangular fibular bone. Be very careful to not disturb the underlying peroneal artery and veins (Fig. 10.21). Expose a 4-cm section of artery and vessel loop it.

CLINICAL HIGHLIGHTS

- The surgeon can now see the deep and protected anatomical location of the peroneal artery. This should be kept in mind when considering bypass grafting to the artery at this level.
- Dense, muscular and fascial attachments are present to the fibula. The deepest portion of these attachments must be carefully dissected. The best technique is to identify the fibula and stay directly on bone with a periosteal elevator to release all fascial and muscular attachments. In this fashion, it is possible to avoid damage to the peroneal vessels.
- Remember that the deep peroneal nerve will be found within two fingerbreadths of the fibular head at the knee. Staying below this level will insure preservation of the peroneal nerve.
- Anterior and posterior communicating branches of the distal peroneal artery constitute runoff of this artery and are usually found near the level of the lateral malleolus.

SAPHENOUS VEIN

INCISION

The previous arterial dissection at all levels on the medial leg will aid identification of the saphenous vein but will not be positioned directly over the saphenous vein.

Identify the greater saphenous vein at the anterior portion of the medial malleolus. Perform a 5-cm incision to identify the greater saphenous vein at the ankle (Fig. 10.22). Follow the saphenous vein all the way to the fossa ovale, which is the junction of the greater saphenous vein with the femoral vein. Multiple and variable tributaries converge into the greater saphenous vein at the fossa ovale. Many variations of the greater saphenous vein will be encountered including dual greater saphenous veins around the level of the knee. However, the classic greater saphenous system is identified in the medial leg and thigh (Fig. 10.23).

Multiple sites of direct perforating veins may be identified perforating through the fascia of the leg and thigh (Fig. 10.24). These branches are ligated in a formal saphenous vein stripping procedure and when a nonreversed (in situ) saphenous vein bypass is performed.

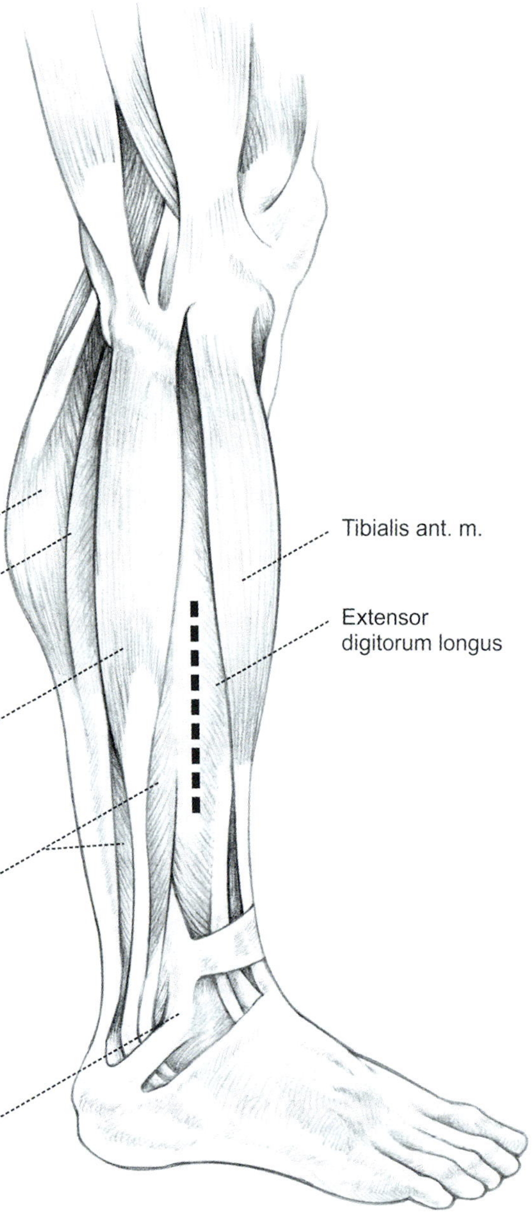

Figure 10.20 Incision for exposure of peroneal artery.

CLINICAL HIGHLIGHTS

- Remember to mark the patient preoperatively for the course of the saphenous vein as well as major perforators.
- Remember to preoperatively have vascular lab analysis and duplex examination of the deep femoral and deep tibial veins. Patency of these veins is essential to establish prior to any superficial vein stripping procedure.
- Pump action of the deep veins is supported by the one-way venous valves to promote venous blood return from the lower extremity. Valvular incompetence may lead to venous varicosities (see Fig. 10.24).
- A saphenous vein cutdown will give IV access and is used in the pediatric patient population or older population when other access options are unusable (see Fig. 10.22).

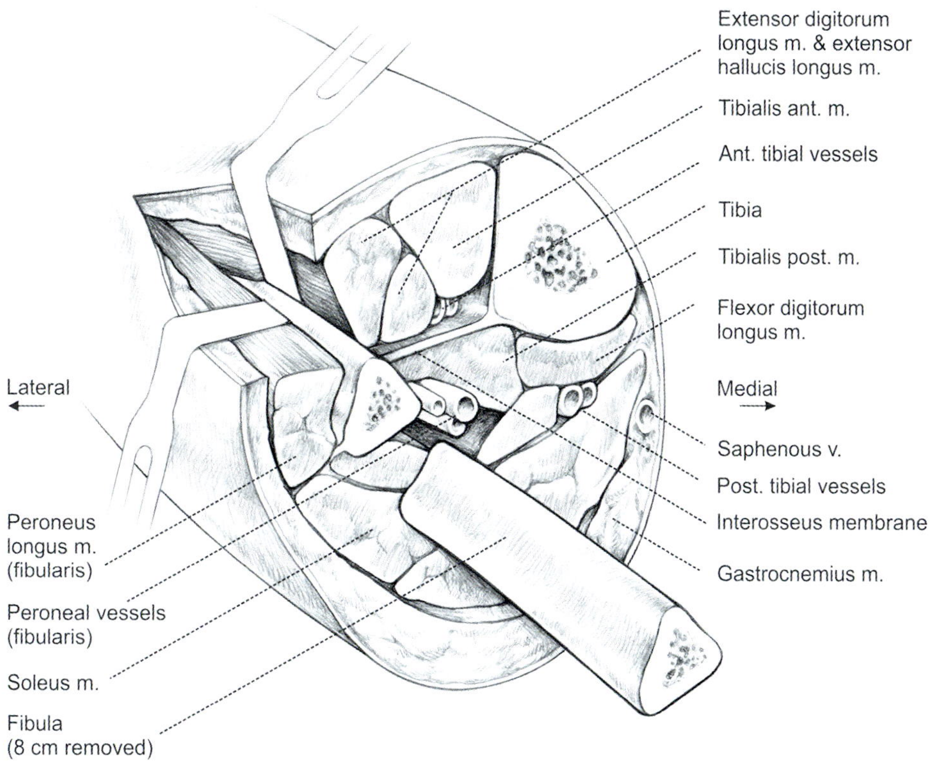

Figure 10.21 Cross-section view of exposure of peroneal artery and removal of a section of the fibula.

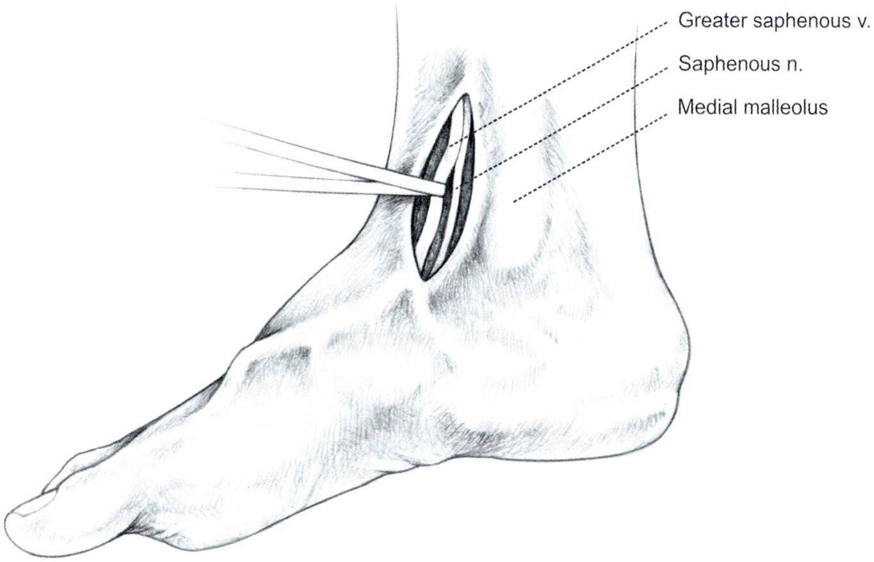

Figure 10.22 Incision at the ankle for exposure of greater saphenous vein and venous access cutdown.

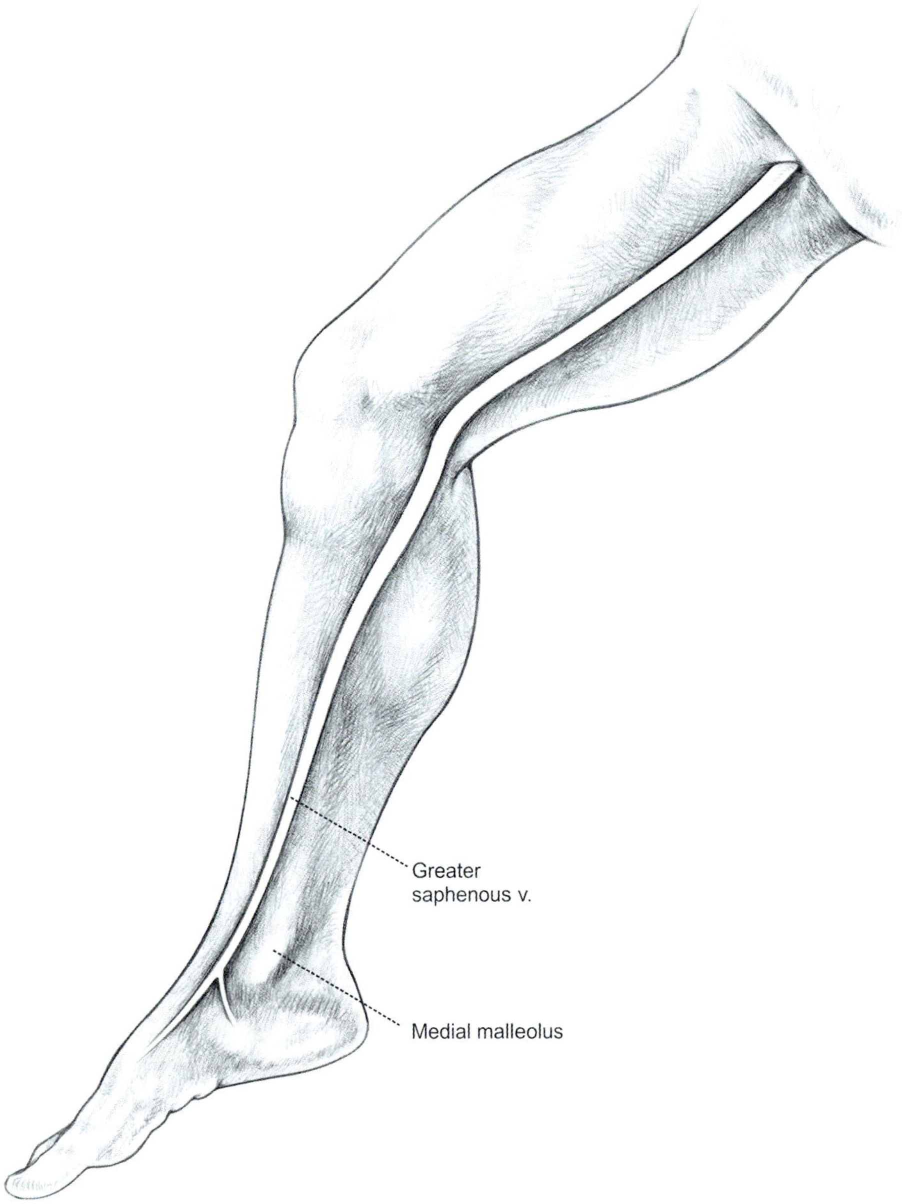

Figure 10.23 Medial location of the greater saphenous vein from the ankle to the groin.

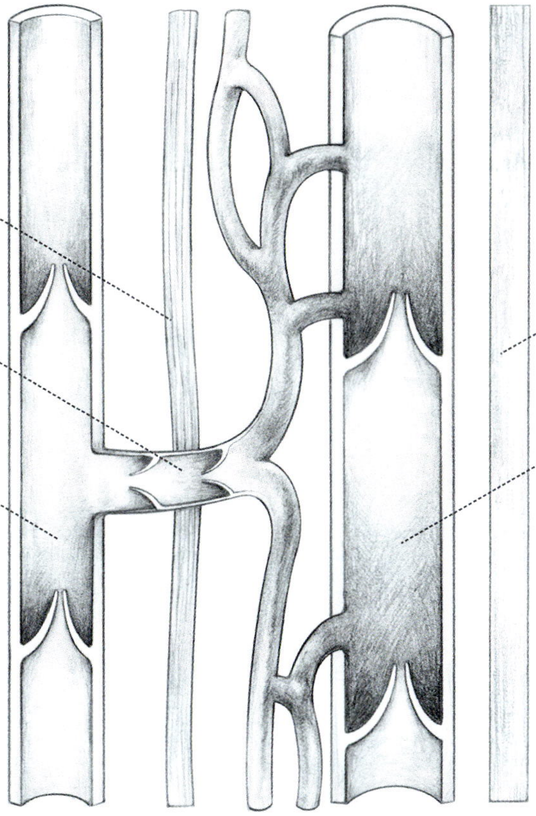

Figure 10.24 Deep and superficial venous architecture with valvular pattern.

FASCIOTOMY

INCISION

This procedure may be done independently or at the beginning of the lower extremity section. A medial skin incision is performed 1 cm posterior to the tibia (see Fig. 10.19A, proximal incision).

OPERATIVE PROCEDURE

From this medial incision, fasciotomy of the superficial posterior muscle group is performed beginning with a small nick with a knife. Extend the fasciotomy proximally and distally using a Metzenbaum scissors in line with the fibers of the fascia. Retract the gastrocnemius and split fibers of the soleus to identify the fascia of the deep posterior muscle group (see Fig. 10.19B). Slit this fascia with a knife and scissors as previously performed.

An anterolateral skin incision is performed as illustrated (see Fig. 10.16A). Fasciotomies here will be performed proximally and distally with the same knife/scissors technique. The first fasciotomy may be of the lateral compartment. To access the lateral compartment, elevate the skin flap laterally. Incise between peroneus longus and brevis muscles (see Fig. 10.20). When incising the fascia, be careful to stay tight on the fascial plane to avoid nerve injury. Then incise the anterior compartment between tibialis anterior and extensor digitorum longus muscles (see Fig. 10.16B).

CLINICAL HIGHLIGHTS

- In the OR, expect dramatic muscle swelling if compartment syndrome is present.
- The common peroneal nerve is located 1-2 finger breadths below the fibular head. This subfascial nerve needs to be preserved to prevent foot drop caused by injury to the deep branch. Injury will also cause loss of sensation in the first web space of the foot. Keep this in mind when positioning a patient in the OR. The fibular head area must be padded and protected.

Works Consulted

1. Abernathy C, Harken AH. *Surgical Secrets*. 2nd ed. Philadelphia, PA: Mosby; 2010.
2. Abrahams PH, McMinn RMH, Hutchings RT, Marks SC. *McMinn's Color Atlas of Human Anatomy*. 4th ed. London, England: Mosby; 2008.
3. Abrahams PH. *McMinn's interactive clinical anatomy* [CD-ROM]. London, England: Mosby; 2001.
4. Agur AMR, Dalley AF. *Grant's Atlas of Anatomy*. 11th ed. Philadelphia, PA: Williams & Wilkins; 2017.
5. Anson BJ. *Surgical Anatomy*. Vol. 1. Philadelphia, PA: Saunders; 1971.
6. Beattie EJ, Bloom ND, Harvey JC. *Thoracic Surgical Oncology*. Edinburgh, England: Churchill Livingstone; 1992.
7. Berger DH, Feig BW, Fuhrman GM. *The M.D. Anderson Surgical Oncology Handbook*. Boston, MA: Little, Brown; 2015.
8. Best CH, Taylor NB. *The Human Body: Its Anatomy & Physiology*. 4th ed. London, England: Chapman & Hall; 1970.
9. Blackbourne LH. *Surgical Recall*. Baltimore, MD: William & Wilkins; 2015.
10. Bunnell S, Boyes JH. *Bunnell's Surgery of the Hand*. 5th ed. Philadelphia, PA: J.B. Lippincott; 1970.
11. Cady B, Steele GD. *General Surgical Oncology*. Philadelphia, PA: W.B. Saunders; 1992.
12. Child CG. *The Liver & Portal Hypertension*. Philadelphia, PA: Saunders; 1968.
13. Clements CD. *Gray's Anatomy*. 13th ed. Philadelphia, PA: Lea & Feibiger; 1984.
14. Colborn GL, Skandalakis JE. *Clinical Gross Anatomy: A Guide for Dissection, Study, & Review*. Pearl River, NY: Parthenon Group; 1993.
15. Colcock BP, Braasch JW. *Surgery of the Small Intestine in the Adult*. Philadelphia, PA: W.B. Saunders; 1968.
16. Colombo R, Naspro R. Ileal conduit as the standard for urinary diversion after radical cystectomy for bladder cancer. *Eur Urol Suppl*. 2010;9(10):736-744.
17. Crippa S, Salvia R, Warshaw AL, et al. Mucinous cystic neoplasm of the pancreas is not an aggressive entity: lessons from 163 resected patients. *Ann Surg*. 2008;247(4):571-579.
18. Culp DA, Flocks RH. *Surgical Urology*. 5th ed. Chicago, IL: Year Book Medical; 1985.
19. Daly JM, Cady B, Low DW. *Atlas of Surgical Oncology*. St. Louis, MO: Mosby; 1993.
20. Davis JH, Sheldon GF, Drucker WR. *Surgery: A Problem-Solving Approach*. Vol. 2. 2nd ed. St. Louis, MO: Mosby; 1995.
21. Department of Surgery, Michigan State University. *Surgical Anatomy for Medical Students: Study Guide*. Los Angeles, CA: Educational Clearinghouse Association for Surgical Education; 1992.
22. DeVita VT, Hellman S, Rosenberg SA. *Cancer: Principles & Practice of Oncology*. Vols. 1-2. 3rd ed. Philadelphia, PA: Lippincott; 2015.
23. Du Plessis DJ, McGregor AL. *A Synopsis of Surgical Anatomy*. 11th ed. Bristol, England: J. Wright; 1975.
24. Dudley H, Rob C. *Operative Surgery*. Vol. 2. 4th ed. Oxford, England: Butterworth-Heinemann; 1992.
25. Edis AJ, Ayala LA, Egdahl RH. *Manual of Endocrine Surgery*. 2nd ed. Berlin, Germany: Springer-Verlag; 1975.
26. Eiseman B, ed. *Prognosis of Surgical Disease*. Eastbourne, England: Saunders; 1980.
27. Eiseman B, Wotkyns R. *Surgical Decision Making*. Philadelphia, PA: Saunders; 2004.
28. Feil W. *Atlas of Surgical Stapling*. Heidelberg, Germany: Barth; 2000.
29. Flocks RH, Culp DA, VerVais P. *Surgical Urology*. Chicago, IL: Year Book Medical; 1967.
30. Grant JCB. *An Atlas of Anatomy*. 6th ed. Baltimore, MD: Williams & Wilkins; 1972.
31. Gray H, Pick TP, Howden R. *Anatomy, Descriptive & Surgical*. 2nd ed. New York, NY: Gramercy Books. Distributed by Random House; 2012.
32. Gray SW, Skandalakis JE, McClusky DA, McClusky JE. *Atlas of Surgical Anatomy for General Surgeons*. Baltimore, MD: Williams & Wilkins; 1985.
33. Green DP, Rockwood CA. *Fractures*. Vol. 2. Philadelphia, PA: Lippincott; 1975.
34. Greene FL, Ponsky JL. *Endoscopic Surgery*. Philadelphia, PA: W.B. Saunders; 1994.
35. Chafee EE, Greisheimer EM. *Basic Physiology and Anatomy*. 2nd ed. Philadelphia, PA: J.B. Lippincott; 1969.
36. Healey JE, Hodge J. *Surgical Anatomy*. 2nd ed. Philadelphia, PA: Mosby; 1990.
37. Hollinshead WH. *Anatomy for Surgeons*. Vol. 1. Hagerstown, MD: Harper & Row; 1954.

38. Jamieson GG. *Anatomy of General Surgical Operations*. Edinburgh, England: Churchill Livingstone; 2006.
39. Johnson J, MacVaugh H, Waldhausen JA. *Surgery of the Chest*. Chicago, IL: Year Book Medical; 1970.
40. Johnson J, Kirby CK. *Surgery of the Chest*. Chicago, IL: Year Book Medical; 1964.
41. Julian OC. *Cardiovascular Surgery*. Chicago, IL: Year Book Medical; 1971.
42. Kretschmer KP. *The Intestinal Stoma: Indications, Operative Methods, Care, Rehabilitation*. Philadelphia, PA: W.B. Saunders; 1978.
43. Machado NO, Al Qadhi H, Al Wahibi K. Intraductal papillary mucinous neoplasm of pancreas. *N Am J Med Sci*. 2015;7(5):160-175.
44. Maingot R. *Abdominal Operations*. Vol. 1. 7th ed. New York, NY: Appleton-Century-Crofts; 1980.
45. Malt RA, ed. *Cholecystectomy and Cholangiography, Parotidectomy*. Boston, MA: Little, Brown; 1976. Surgical Techniques Illustrated 1(1).
46. Malt RA, ed. *Anterior Resection of the Rectum, Sphincter-Saving Rectal Resection*. Boston, MA: Little, Brown; 1977. Surgical Techniques Illustrated 2(2).
47. Malt RA, ed. *Surgery for Chronic Pancreatitis*. Boston, MA: Little, Brown; 1977. Surgical Techniques Illustrated 2(4).
48. Malt RA, ed. *Right Lower Lobectomy, Left Lower Lobectomy*. Boston, MA: Little, Brown; 1980. Surgical Techniques Illustrated 4(2).
49. Malt RA, ed. *Portosystemic Shunts*. Boston, MA: Little, Brown; 1980. Surgical Techniques Illustrated 4(3).
50. Masquelet AC, Dorn L. *An Atlas of Surgical Anatomy*. New York, NY: Taylor & Francis; 2005.
51. Mayo CW. *Surgery of the Small & Large Intestine*. Chicago, IL: Year Book Medical; 1962.
52. McMinn RMH. *Last's Anatomy: Regional & Applied*. 9th ed. Edinburgh, England: Churchill Livingstone; 2011.
53. McQuillan T, Manolas S, Hayman J, Kune G. Surgical significance of the bile duct of Luschka. *Br J Surg*. 1989;76(7): 696-698.
54. Misra SP, Dwivedi M. Pancreaticobiliary ductal union. *Gut*. 1990;31(10):1144-1149.
55. Moore KL, Dalley AF. *Clinically Oriented Anatomy*. 4th ed. Philadelphia, PA: Lippincott Williams & Wilkins; 2018.
56. Mubarak SJ, Owen CA. Double-incision fasciotomy of the leg for decompression in compartment syndromes. *J Bone Joint Surg Am*. 1977;59(2):184-187.
57. Netter FH, Brass A, Jones HR; Ciba Pharmaceutical Company. *The Ciba Collection of Medical Illustrations*. Vols. 1-7. West Caldwell, NJ: Commissioned & published by Ciba; 1969.
58. Netter FH, Dalley AF. *Atlas of Human Anatomy*. 2nd ed. Teterboro, NJ: Icon Learning Systems; 2001.
59. Netter FH. *Interactive Atlas of Clinical Anatomy* [CD-ROM]. Summit, NJ: DxR Development Group, Inc.; 1997.
60. Nyhus LM, Baker RJ, Fischer JE. *Mastery of Surgery*. Vols. 1–2. Boston, MA: Little, Brown; 1990.
61. O'Leary PF. *The Physiological Basis of Surgery*. 2nd ed. Baltimore, MD: Williams & Wilkins; 2012.
62. Paradis NA. *Cardiac Arrest: Science & Practice of Resuscitation Medicine*. Baltimore, MD: Williams & Wilkins; 2007.
63. Pearson FG. *Esophageal Surgery*. Philadelphia, PA: Churchill Livingstone; 2002.
64. Pearson FG. *Thoracic Surgery*. New York, NY: Churchill Livingstone; 2002.
65. Pemberton B, Colborn GL, Skandalakis JE. *Workbook of Surgical Anatomy*. New York, NY: McGraw-Hill; 1991.
66. Puestow CB. *Surgery of the Biliary Tract, Pancreas & Spleen*. Chicago, IL: Year Book Medical; 1964.
67. Roche Laboratories. *Surgery Illustrated*. Nutley, NJ: Roche Laboratories; 1985.
68. Rowe MI. *Essentials of Pediatric Surgery*. St. Louis, MO: Mosby-Year Book; 1995.
69. Salinas NL, Brennan JA. Penetrating and blunt neck trauma. In: Holt GR, ed. *Resident Manual of Trauma to the Face, Head, and Neck*. Alexandria, VA: American Academy of Otolaryngology—Head and Neck Surgery Foundation; 2012:164-176.
70. Schnelldorfer T, Sarr MG, Adams DB. What is the Duct of Luschka?: A systematic review. *J Gastrointest Surg*. 2012;16(3):656-662.
71. Schwartz S. *Modern Technics in Surgery. Abdominal Surgery. Installment 1*. Mount Kisco, NY: Futura; 1980.
72. Schwartz SI, Shires TG, Spencer FC, Storer EH. *Principles of Surgery*. 3rd ed. New York, NY: McGraw-Hill; 1979.
73. Schwartz SI, Shires GT. *Principles of Surgery*. Vols. 1-2. 6th ed. New York, NY: McGraw-Hill; 2004.
74. Scott-Conner CEH, Dawson DL. *Operative Anatomy*. Philadelphia, PA: J.B. Lippincott; 2009.
75. Scott-Conner CEH; Society of American Gastrointestinal Endoscopic Surgeons. *The SAGES Manual: Fundamentals of Laparoscopy, Thoracoscopy, & GI Endoscopy*. New York, NY: Springer; 2006.
76. Scott-Conner CEH. *The SAGES Manual Fundamentals of Laparoscopy & GI Endoscopy*. New York, NY: Springer; 2006.
77. Shakelford RT. *Surgery of the Alimentary Tract*. Vols. 1–3. 2nd ed. Philadelphia, PA: Saunders; 1981.
78. Shields TW, Samson PC. *General Thoracic Surgery*. 2nd ed. Philadelphia, PA: Lea & Febiger; 2011.
79. Silen W, Cope Z. *Cope's Early Diagnosis of the Acute Abdomen*. 18th ed. New York, NY: Oxford University Press; 2010.
80. Skandalakis JE. *Surgical Anatomy: The Embryologic & Anatomic Basis of Modern Surgery*. London, England: McGraw-Hill; 2002.
81. Skandalakis JE. *Surgical Anatomy and Embryology*. Philadelphia, PA: Sanders; 1993. The Surgical Clinics of North America 73(4).
82. Skandalakis LJ. *Modern Hernia Repair: The Embryological & Anatomical Basis of Surgery*. New York, NY: Parthenon; 1995.
83. Skandalakis JE, Gray SW, Rowe JS Jr. Surgical anatomy of the submandibular triangle. *Am Surg*. 1979;45(9):590-596.
84. Skandalakis JE, Gray SW, Rowe JS Jr. *Anatomical Complications in General Surgery*. New York, NY: McGraw-Hill; 1983.
85. Skandalakis JE, Skandalakis LJ, Skandalakis PN. *Surgical Anatomy & Technique: A Pocket Manual*. 2nd ed. New York, NY: Springer; 2014.
86. Stecco C, Macchi V, Lancerotto L, Tiengo C, Porzionato A, De Caro R. Comparison of transverse carpal ligament and flexor retinaculum terminology for the wrist. *J Hand Surg Am*. 2010;35(5):746-753.
87. Tam P, Wei W. Myocutaneous sartorius flap for major congenital abdominal wall defects. *Pediatr Surg Int*. 1989;4(2):143-146.
88. Terblanche J. *Hepatobiliary Malignancy: Its Multidisciplinary Management*. London, England: Edward Arnold; 1994.
89. Thorek P. *Anatomy in Surgery*. 2nd ed. London, England: J.B. Lippincott; 1963.
90. Tzovaras G, Rowlands BJ. Transduodenal sphincteroplasty and transampullary septectomy for sphincter of Oddi dysfunction. *Ann R Coll Surg Engl*. 2002;84(1):14-19.

91. Urschel HC, Cooper JD. *Atlas of Thoracic Surgery*. New York, NY: Churchill Livingstone; 2006.

92. Weigelt JA, Lewis FR. *Surgical Critical Care*. Philadelphia, PA: Saunders; 1996.

93. Welch CE. *Surgery of the Stomach & Duodenum*. Chicago, IL: Year Book Medical; 1973.

94. Welch CE. *Surgery of the Stomach & Duodenum*. Chicago, IL: Year Book Medical; 1959.

95. Wise RA, Baker HW. *Surgery of the Head & Neck*. Chicago, IL: Year Book Medical; 1976.

96. Wood WC, Skandalakis JE. *Anatomic Basis of Tumor Surgery*. St. Louis, MO: Quality Medical; 2010.

97. Zollinger RM, Zollinger RM Jr, Cutler EC. *Atlas of Surgical Operations*. New York, NY: Macmillan; 1967.

98. Zollinger RM, Zollinger RM Jr. *Zollinger's Atlas of Surgical Operations*. 8th ed. New York, NY: McGraw-Hill; 2016.

Index

Note: Page numbers in *italics* indicate figures.

B

G

H

I

J

K

L

M

N

R

S